Recent Results in Cancer Research 117

Recent Results in Cancer Research

Volume 107
R. D. Issels, W. Wilmanns (Eds.):
Application of Hyperthermia in the Treatment of Cancer
1988. 118 figures, 56 tables. XII, 277. ISBN 3-540-18486-4

Volume 108
H.-J. Senn, A. Glaus, L. Schmid (Eds.):
Supportive Care in Cancer Patients
1988. 62 figures, 97 tables. XII, 342. ISBN 3-540-17150-9

Volume 109
W. Hinkelbein, G. Bruggmoser, R. Engelhardt (Eds.):
Preclinical Hyperthermia
1988. 182 figures, 40 tables. XI, 261. ISBN 3-540-18487-2

Volume 110
P. Schlag, P. Hohenberger, U. Metzger (Eds.):
Combined Modality Therapy of Gastrointestinal Tract Cancer
1988. 105 figures, 122 tables. XVII, 301. ISBN 3-540-18610-7

Volume 111
H. Scheurlen, R. Kay, M. Baum (Eds.):
Cancer Clinical Trials: A Critical Appraisal
1988. 37 figures, 53 tables. XI, 272. ISBN 3-540-19098-8

Volume 112
L. Schmid, H.-J. Senn (Eds.): AIDS-Related Neoplasias
1988. 23 figures, 35 tables. IX, 97. ISBN 3-540-19227-1

Volume 113
U. Eppenberger, A. Goldhirsch (Eds.):
Endocrine Therapy and Growth Regulation of Breast Cancer
1989. 26 figures, 17 tables. IX, 92. ISBN 3-540-50456-7

Volume 114
P. Boyle, C. S. Muir, E. Grundmann (Eds.): Cancer Mapping
1989. 89 figures, 64 tables. XI, 277. ISBN 3-540-50490-7

Volume 115
H.-J. Senn, A. Goldhirsch, R. D. Gelber, B. Osterwalder (Eds.):
Adjuvant Therapy of Primary Breast Cancer
1989. 65 figures, 94 tables. XVI, 296. ISBN 3-540-18810-X

Volume 116
K. W. Brunner, H. Fleisch, H.-J. Senn (Eds.):
Bisphosphonates and Tumor Osteolysis
1989. 22 figures, 6 tables. IX, 78. ISBN 3-540-50560-1

V. Diehl M. Pfreundschuh
M. Loeffler (Eds.)

New Aspects in the Diagnosis and Treatment of Hodgkin's Disease

With 71 Figures and 91 Tables

Springer-Verlag Berlin Heidelberg New York
London Paris Tokyo Hong Kong

Professor Dr. med. Volker Diehl
Professor Dr. med. Michael Pfreundschuh
Dr. med. Dipl. Phys. Markus Loeffler

I. Medizinische Universitätsklinik Köln
Joseph-Stelzmann-Straße 9, 5000 Köln 41, FRG

ISBN-13:978-3-642-83783-8 e-ISBN-13:978-3-642-83781-4
DOI: 10.1007/978-3-642-83781-4

Library of Congress Cataloging-in-Publication Data
New aspects in the diagnosis and treatment of Hodgkin's disease/V. Diehl, M. Pfreundschuh, M. Loeffler (eds.). p. cm.- (Recent results in cancer research; 117)
Based on the First International Symposium on Hodgkin's Lymphoma in Cologne, Oct., 2-3, 1987. Includes Index.
ISBN-13:978-3-642-83783-8 (U.S.: alk. paper)
1. Hodgkin's disease-Congresses. I. Diehl, V. (Volker), 1938-. II. Pfreundschuh, M. (Michael) III. Loeffler, M. (Markus), 1954-. IV. International Symposium on Hodgkin's Lymphoma (1st: 1987: Cologne, Germany) V. Series.
[DNLM: 1. Hodgkin's Disease-diagnosis-congresses. 2. Hodgkin's Disease-therapy-congresses. W1 RE106P v. 117/WH 500 1987] RC261.R35 vol. 117 [RC644] 616.99'4 s-dc20 [616.99'446] DNLM/DLC 89-11302

2125/3140-543210 - Printed on acid-free paper.

Preface

In the field of Hodgkin's lymphoma, many new data have been collected during the last decade both on the cell of origin of this disease and on more effective therapies to cure the majority of patients even in the advanced stages. Therefore, it seems to be justified to compile these new data in a special volume of *Recent Results in Cancer Research*. This volume summarizes the contributions presented at the First International Symposium on Hodgkin's Disease that took place in Cologne (FRG) on October 2-3, 1988.

There is little doubt that the Hodgkin and Reed-Sternberg (H and RS) cells and their variants represent the malignant population in Hodgkin's lymphoma; however, there is still a fierce debate as to the possible cell of origin of H and RS cells. Many of the problems confounding earlier research into this question were related to the difficulty or virtual impossibility, of obtaining purified populations of H and RS cells. Most of the recent progress stems from the establishment of permanent cell lines of H and RS cells in culture. Though permanent cell lines may degenerate from their ancestors and may not be representative of the original cells, the data accumulated demonstrate that the cells in vitro reflect very well the biological and immunological aspects of H and RS cells in vivo. The marker spectrum of Hodgkin-derived cell lines in vitro and that of H and RS cells in vivo, as well as gene rearrangement studies of the cell lines, suggest that H and RS cells in vivo are lymphoid in origin. Even though gene rearrangement studies of biopsy material are not yet conclusive, immunological marker studies with monoclonal antibodies suggest that H and RS cells are activated lymphoid cells of either T- or B-cell type. Taking into account the data presented and summarized in this volume, we cannot totally exclude the derivation of H and RS cells from other cells (e. g., macrophages), especially in the light of some functional studies (e. g., antigen presentation) and the reactivity with monoclonal antibodies that recognize antigens related to the monocytoid/myeloid lineage. However, all contributors agree upon the conclusion that at the current state of our knowledge we must regard H and RS cells as lymphoid.

Just as revolutionary as the new data concerning the cell of origin in Hodgkin's lymphoma have been the changes in the diagnostic and therapeutic approaches to this disease. The safest ground for diagnosing Hodgkin's disease is still conventional morphology, but immunohistology is of help in confirming the diagnosis, especially in cases which cannot be classified using morphological criteria. Diagnostic laparotomy with splenectomy is no longer generally recommended. The new diagnostic strategies presented here - immunoscintigraphy and bone marrow scintigraphy with magnetic resonance imaging - may possibly help to replace this invasive method completely in the near future.

The improvement of therapy in Hodgkin's lymphoma during the last two decades is mostly due to the development of effective combination chemotherapy protocols. More than 20 years after its inauguration, MOPP is still the golden standard by which all new chemotherapy protocols have to be measured. The gains accomplished by new drug combinations, if they exist at all, are minor if they are projected onto the total population of patients with Hodgkin's disease. With remission rates between 70% and 90% and cure rates less than 50%, the major challenge for clinical research of Hodgkin's lymphoma in the 1990s is the definition of prognostic subgroups by clinical or laboratory parameters in order to select patients who benefit from more aggressive therapy and those in whom the intensity, and thus the toxicity of treatment can be reduced.

Oncologists dealing with adult patients are impressed by the results obtained in pediatric patients, which are presented here. Of course, the approaches used for children cannot simply be transferred to adult patients, but much can be learned from the tailoring of the diagnostic and therapeutic strategy to the individual patient's needs. The recent results of clinical trials demonstrate the value of risk factor analysis. It becomes evident that the staging system according to Ann Arbor may no longer serve as a strategy for making therapeutic decisions, and a more refined system of prognostic subgrouping is badly needed. The problem is that different risk factors were reported for different treatment strategies. To define intrinsic or therapy-independent risk factors, we need an intergroup analysis of risk factors. Such an analysis can only be accomplished by accumulating the data of several large international study groups. It is one of the most important results of the *First International Symposium on Hodgkin's Lymphoma* that the participating representatives of the most important institutions and cooperative groups have agreed to contribute their data for an international workshop on risk factor analysis which is now being carried out.

Failure of primary treatment need no longer be fatal. The therapeutic advances of conventional salvage therapy and the innovative approach of bone marrow transplantation are discussed in a special section. Finally, the psychological problems that arise in treating our patients and the intermediate and long-term sequelae of treatment deserve further attention. These sequelae will answer the question whether our therapeutic efforts, which succeed in curing most of the patients, are really worthwhile. This can only be the case if we not only free the patients from their disease but also provide them with a psychological and somatic quality of life that justifies the efforts and the costs involved in conquering the disease.

The attendance and the enthusiastic comments made us believe that the First International Symposium on Hodgkin's Lymphoma in Cologne succeeded in summarizing the most important goals that we have achieved during the last two decades and in defining the questions that remain to be answered in the future to complete our knowledge for the sake of the patients.

I do not want to finish this preface without first thanking the Ministry of Research and Technology of the Federal Government (BMFT), The German Research Society (DFG), the Thyssen Foundation, the Boehringer-Ingelheim Foundation, the German Association of Medical Oncology (AIO), and the City of Cologne for their financial and administrative support. My personal thanks go to our Hodgkin Study Group team, especially to Mrs. Nisters-Backes and Olga Pavlovic and to our students, who guaranteed the flawless organization of a meeting that we hope many of the participants will remember for a long time to come.

Cologne, March 1989

For the editors
Volker Diehl

[illegible] and longer fish life. The third [illegible] salvage therapy and new [illegible] that now contain [illegible] and in a [illegible] biological problems that arise in [illegible] and the need to educate and long-term perspective of the [illegible]. These aspects will not [illegible] after therapeutic efforts, which succeed in [illegible] worthwhile. The continuity of the [illegible] the current medical care that [illegible] individual and somatic quality of life that [illegible] of a general approach in the [illegible].

[illegible] and discussions brought about [illegible] Symposium on Hodgkin's Disease [illegible] summarizing the most important [illegible] during the last two decades and to defining [illegible] to be answered in the future to improve [illegible] the patients.

[illegible] without first thanking the [illegible] Research and Technology and Federal [illegible] Research Society [illegible] Foundation, the German [illegible] and the City of Cologne [illegible] personally thanks [illegible] Mrs. [illegible] who made the [illegible] hope [illegible] for a long time to come.

[illegible] For the editors

[illegible]

Contents

*List of Contributors**

* The address of the principal authors is given on the first page of each contribution.

[1] Page on which contribution begins.

Pathology and Cell Biology of Hodgkin's Disease

Pathology of Hodgkin's Disease: Anything New?

D. H. Wright

Department of Pathology, Southampton General Hospital, University of Southampton, Tremona Road, Southampton SO6 4XY, Great Britain

Following many years of controversy and confusion, the Rye classification (Lukes et al. 1966 a) appeared to provide a sound basis for the identification and categorisation of the subtypes of Hodgkin's disease (HD). This classification was widely adopted by pathologists and clinicians throughout the world and provided an area of agreement in contrast to the chaos and controversy that accompanied the subsequent attempt to restructure the classification of non-Hodgkin's lymphomas (NHLs). The Rye classification established a framework on which substantial advances were made in the study and treatment of Hodgkin's disease. To some extent, we are indebted to Samuel Wilkes for introducing the term 'Hodgkin's disease' since, without this unifying eponym, attempts to achieve an international classification might have sunk into the quagmire of semantics (Rose 1981). It is, however, anachronistic that in the late twentieth century our whole categorisation of lymphoreticular neoplasms into HD and NHLs is based on a small series of morbid anatomical descriptions of lymphadenopathy made in the early nineteenth century. [A pathologist, Herbert Fox (1926), looked at histological material from the six patients whom Hodgkin had described in 1832. He diagnosed three cases as HD, one as tuberculosis, one as syphilis and one as NHL.] While Hodgkin provided an eponym (as has Burkitt in more recent years) that allowed us to diagnose, categorise and investigate a disease before we understood its histogenesis, the inherent weakness of this position is beginning to cause problems.

It is not within the remit of this presentation to discuss immunohistochemistry, cell culture and gene rearrangement studies, which are dealt with by other participants in the symposium. I shall keep within the bounds of conventional gross pathology and light microscopy but, in doing so, shall benefit from the illumination provided by studies using these new techniques. It is, perhaps, a paradox that some of the techniques that have done so much to unravel the problems of NHL appear, at least initially, to have increased the confusion surrounding the histogenesis of HD. In this situation conventional morphological techniques, aided and supported by newer methods, provide the only secure base from which the study of HD can go forward.

The title of this presentation asks 'Anything New'. Since most studies of Hodgkin's disease identify and define their cases on the criteria laid down by Lukes et

Recent Results in Cancer Research. Vol. 117

Table 1. Schematic representation of variation in morphological features in histological types of Hodgkin's disease. (Lukes et al. 1966b)

Histological groups	Lymphocytes	Histio-cytes	Eosino-phils	Plasma cells	Fibrillar reticulum	Collagen	Reed-Sternberg cells
Lymphocyte/histiocyte predominant							
Nodular	+ + + + + + + +	+	0	0	0	0	+
Diffuse	+ + + + + +	+ + +	0	0	0	0	+
Nodular sclerosis	+ to + + + +	+ to + +	+	+	+	+ to + + + +	+
Mixed	+	+ + +	+ +	+	+ +	0	+ +
Diffuse fibrosis	0	+	+	+	+ + + + +	0	+ +
Reticular	+	0	+ +	+	+	0	+ + + +

al. (1966b), and subsequently incorporated in the Rye classification, I shall use this paper as the conceptual base from which to explore subsequent developments. Lukes et al. (1966b) considered HD to be a single disease process with varying histological manifestations and proposed that the numerous histological types represent differences in the hosts' attemps to 'prevent induction of malignant neoplasia'. They identified six subtypes of HD (reduced to four groups in the Rye classification) and tabulated the constituent cells that characterise each type (Table 1). Variant forms of Reed-Sternberg (RS) cells were described in lymphocyte/histiocyte predominant HD and in nodular sclerosis, but central to the diagnosis of all types of HD was the identification of characteristic RS cells with 'large inclusion-like nucleoli, thick nuclear membranes, with perinuclear halos, and abundant eosinophilic, to amphoteric cytoplasm' (Fig. 1). Later studies showed that RS cells or, more correctly, RS-like cells could be found in a number of reactive and neoplastic diseases other than HD. It then became a dictum that a diagnosis of HD should only be made when RS cells are seen in an appropriate cellular setting for one of the subtypes of HD.

It has long been suspected that nodular lymphocyte/histiocyte predominant HD (NLPHD) might be a different disease from other types of HD. Lukes et al. (1966b) noted the long survival of patients with NLPHD even without therapy. The disease is characterised by polylobated cells (polylobated RS cells, popcorn cells, Fig. 2a, b) that may be present in large numbers, whereas classic RS cells are 'usually rare and difficult to find'. Many authors, however, stress the need to search numerous sections, if necessary, to identify classic RS cells before making a diagnosis of HD.

Further evidence that NLPHD is a different disease from other types of HD was provided by the report of Miettinnen et al. (1983) that 5 of 51 patients with this disease developed large cell NHL, 4–11 years after the onset of NLPHD, and that only one of these patients had received radiotherapy. Two further patients developed other types of HD. The progression of a proportion of patients with NLPHD to large cell NHL is now well recognised (Trudel et al. 1987). More recent studies, using immunohistochemistry, have shown that the polylobated RS

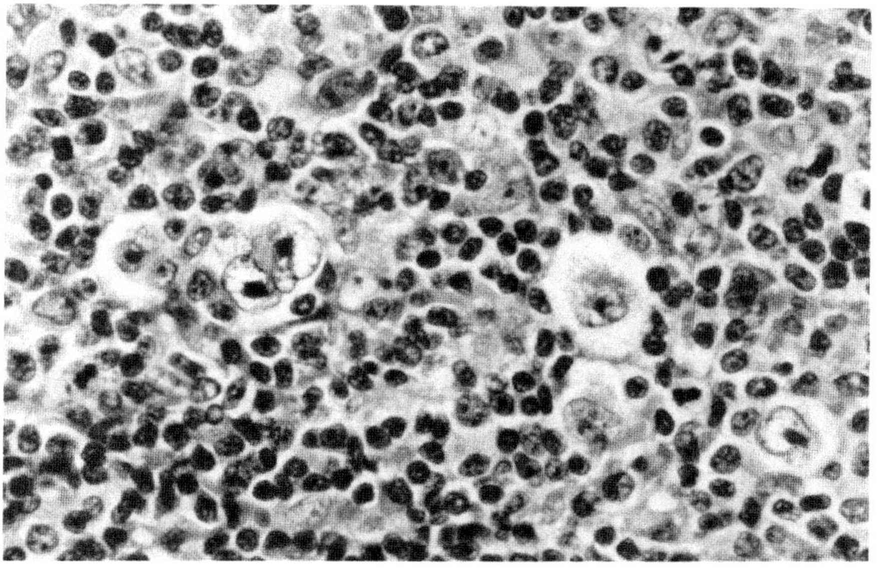

Fig. 1. Mixed cellularity HD showing a classic RS cell and two mononuclear Hodgkin's cells

cells of NLPHD have a different phenotype from classic RS cells. In particular, they express cytoplasmic J-chain, indicating a B-cell origin, although, surprisingly, the cells in any one case do not show light chain restriction (Timens et al. 1986), suggesting that they represent a polyclonal, non-neoplastic proliferation. These studies make it clear that cells with the immunophenotype of classic RS cells do not occur in NLPHD and that the apparent identification of these cells, in NLPHD, was based on delusion. It is the writer's view that pathologists anxious to identify classic RS cells in order to fulfill the diagnostic dictum that RS cells must be identified before making a diagnosis of HD will eventually find a polylobated RS variant sufficiently like a classic RS cell in order to satisfy their conscience (Fig. 2b). The polylobated variant RS cell is characteristic and when seen in the appropriate cellular setting is sufficient in itself for the diagnosis of NLPHD.

In non-specific lymphadenitis, germinal centres expanded and disrupted by lymphocytes, designated as progressively transformed germinal centres (PTGCs), may be seen. These nodules bear a close resemblance to the nodules of NLPHD, except that polylobated RS cells are not seen. Poppema et al. (1979) noted the occurrence of PTGCs and NLPHD in the same patient and suggested that PTGCs might be the origin of NLPHD. They noted that, unlike other types of HD, NLPHD arises in the B-cell areas of lymph nodes. Subsequent studies have noted the association of PTGCs with NLPHD and occasionally with other subtypes of HD (Burns et al. 1984; Osborne and Butler 1984). From a practical point of view, it is important that PTGCs should be differentiated from NLPHD since the latter may progress to NHL or HD and requires local therapy, whereas the former is often a benign self-limiting disease.

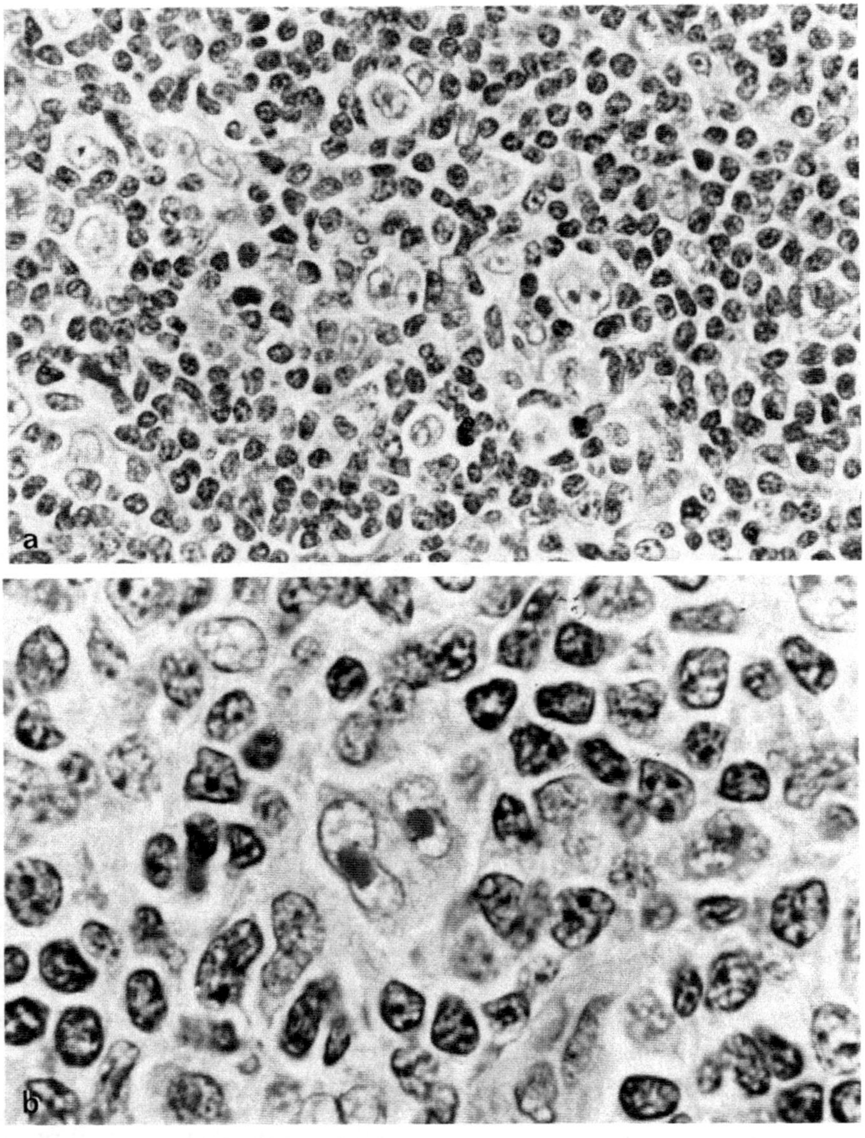

Fig. 2. a Lymphocyte/histiocyte predominant Hodgkin's disease showing several polylobated RS cells. **b** Higher power of tumour illustrated in **a** showing a cell resembling a classic RS cell. Such cells are difficult to find and have a different immunophenotype from classic RS cells despite their morphological similarity

In the Rye classification NLPHD was placed in the same category as diffuse lymphocyte/histiocyte predominant HD (DLPHD). Lukes et al. (1966b) noted that histiocytes were often more abundant in the diffuse disease, although the polylobated RS cells were morphologically the same in both. More recent studies have shown that the polylobated cells are immunophenotypically different in the two diseases (Stein et al. 1986). Trudel et al. (1987) also noted differences in the presentation of the two diseases. NLPHD usually involves a single anatomical site, whereas DLPHD often presents at a high stage. These differences may be due, at least in part, to the confusion between DLPHD and other types of HD as well as NHL. Thus, in a study of 659 cases of HD, Colby et al. (1981) noted that 'the differentiation of LP [lymphocyte predominance] from MC [mixed cellularity] rests solely in some cases on the subjective difficulty in finding RS cells in these variants'. DLPHD might also be confused with the lymphoepithelioid variant of T-cell lymphoma (Lennert's lymphoma) in which rare 'RS cells' might be found (Suchi et al. 1987). Further studies are required to determine whether DLPHD, excluding other types of HD and NHL, is a different disease from NLPHD, particularly since nodular and diffuse LPHD can occur together in the same lymph node (Lukes et al. 1966b).

Nodular sclerosing Hodgkin's disease (NSHD) is identified both by bands of fibrous tissue, that divide the tumour into islands, and the presence of characteristic lacunar RS cells. Lukes et al. (1966b) noted that the process may be observed in a cellular phase in which the formation of collagen bands, with circumscription of cellular nodules, is limited to one portion of the specimen. There has, however, been controversy in the literature as to whether biopsies showing lacunar cells but a complete absence of sclerosis should be categorised as cellular phase NS or MC. Kadin et al. (1971) felt that the use of the term 'cellular phase NS' was justified by the frequent finding of typical sclerotic lesions elsewhere in the patient. Strum and Rappaport (1971) observed progression in serial biopsies from five of seven patients with cellular phase NS to typical sclerosing NS. These observations support the concept that cellular phase NS is a variant of classic NSHD. From a practical point of view, however, Colby et al. (1981), in a follow-up study of 659 cases of HD, found that cellular phase NS had clinical features and a survival rate more akin to MC.

Age, stage and treatment now appear to be more important in determining prognosis than histology (Colby et al. 1981). This is undoubtedly due to the greater effectiveness of therapy but may, in part, be due to the way in which the histological subtypes are categorised. In a study of cases submitted to the British National Lymphoma Investigation (BNLI) NSHD, the largest subtype in most series, could be divided into prognostically significant subgroups on the basis of the composition of the cellular areas (Bennett et al. 1981). More recently, the same group has divided NSHD into two grades with grade 2 showing areas of lymphocyte depletion or numerous pleomorphic Hodgkin's cells. They have shown significantly worse actuarial survivals for grade 2 than grade 1 (Haybittle et al. 1985; Bennett et al. 1985). Since grade 2 NS comprised 22% of their cases (cf. lymphocyte depleted 1.5%) this could be a clinically important subcategorisation. These results are at variance of the findings of Colby et al. (1981), who found that within the NS subgroup it was only sclerosis that correlated with survival. This difference could be due to differences in the criteria used to identify NS cases and in the clinical and treatment differences between the Stanford cases (Colby et al. 1981) derived from a single hospital and the BNLI patients (Bennett et al. 1985), who are derived from a large multicentre study.

Recently, Strickler et al. (1986) have identified a syncytial variant of NSHD. This presumably corresponds to grade 2 NS. In drawing attention to this variant the authors were more concerned with its diagnostic, rather than its prognostic, significance, since syncytial NSHD may mimic non-Hodgkin's lymphomas and metastatic neoplasia.

Lacunar RS cells are characteristic of NSHD (Fig. 3a, b). They often have very complex multilobated nuclei with eosinophilic nucleoli of variable size. Their cytoplasm is abundant and stains poorly in both cytological and histological preparations. It is the artefactual shrinkage of this cytoplasm that forms the lacunae around these cells. Lukes et al. (1966b) noted that diagnostic RS cells are usually infrequent and difficult to find in NSHD but are still necessary for the diagnosis. This has been the accepted dogma since that time. How many pathologists, however, on seeing a biopsy with characteristic sclerosis and lacunar RS cells would not diagnose NSHD? What else could it be? Those of timid disposition might like to search until they find a classic RS cell, or at least a cell that bears some resemblance to a classic RS cell. None of the immunohistochemical studies reported have identified separable lacunar and classic RS cells in NSHD. It is the writer's opinion that carefully identified lacunar cells, in the appropriate setting, are, in themselves, diagnostic of NSHD.

While it is clear that there is still some uncertainty in the definition and delineation of NSHD this does not affect the majority of cases and this is the least frequently misdiagnosed subgroup of HD (Miller et al. 1982). In contrast, a number of recent publications have identified a high rate of misdiagnosis of lymphocyte-depleted Hodgkin's disease (LDHD) (Bennett et al. 1985; Kant et al. 1986; Myskow et al. 1987). Misdiagnosis occurs between other types of HD and with NHL. The precise border between LDHD and MCHD and some cases of grade 2 NSHD is not always easily defined (Fig. 4). Distinction between LDHD and NHL, particularly of the T-cell phenotype, can be extremely difficult, if not impossible. T-cell lymphomas are often infiltrated by large numbers of reactive cells including eosinophils and, in some cases, contain multinucleated RS or RS-like cells (Suchi et al. 1987). At our present state of knowledge it would appear that immunohistochemistry and gene rearrangement studies are unable to solve these problem cases and highlight our need to understand the histogenesis of the RS cell. Some assistance in the separation of LDHD from NHL might be provided by the clinical features. Kant et al. (1986) noted that in nine of ten cases of NHL misdiagnosed as LPHD the presentation was unusual for HD such as bulky abdominal disease, epitrochlear lymphadenopathy or hypercalcaemia. It would appear that the poor prognosis attributed to LDHD is due, at least in part, to the inclusion of cases of NHL in this category (Kant et al. 1986; Myskow et al. 1987).

The Rye classification of HD amalgamated both the diffuse fibrosis and reticular subtypes of HD recognised by Lukes et al. (1966b) into the LD category. Neiman et al. (1973) reported 13 cases of LDHD (10 diffuse fibrosis and 3 reticular) and noted a difference between these two forms of the disease. Diffuse fibrosis, in

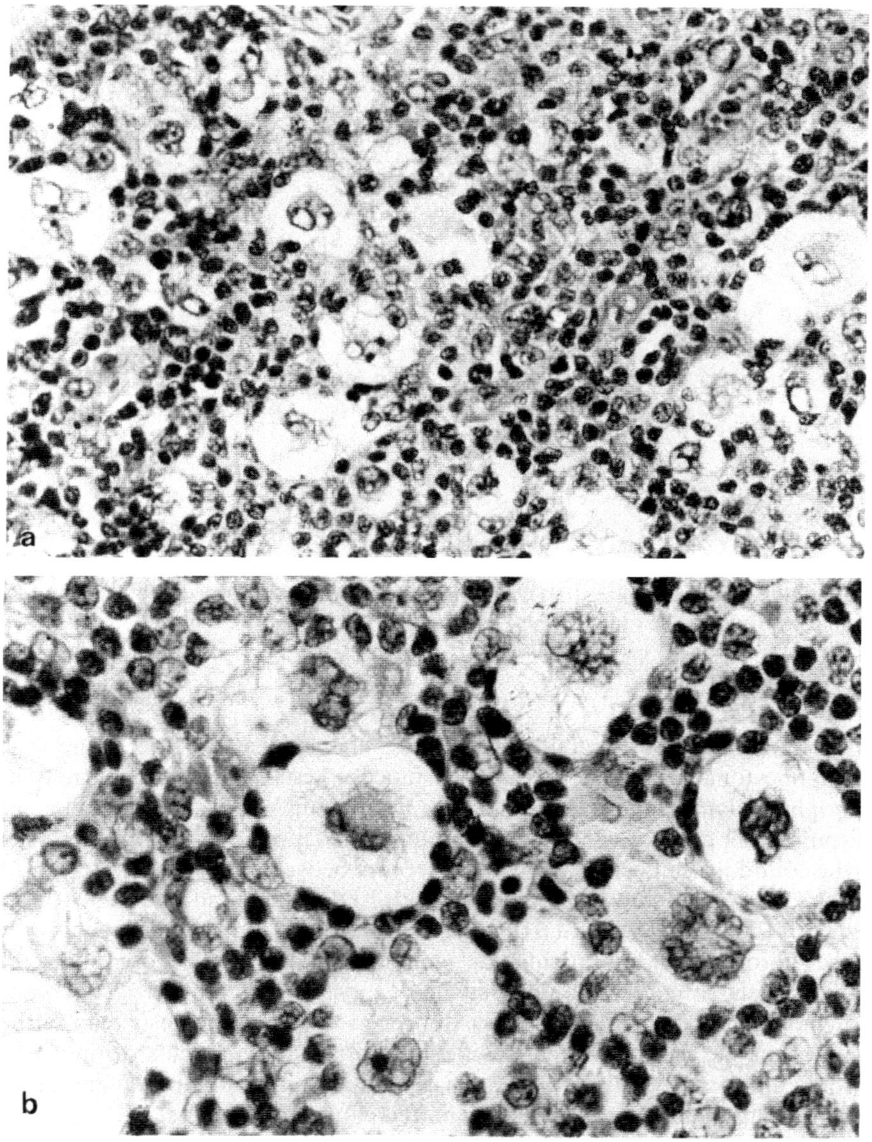

Fig. 3. **a** Nodular sclerosing HD showing numerous lacunar RS cells. These cells have abundant, watery cytoplasm that shrinks during histological processing leaving a clear space (lacuna) around the cell. **b** Higher power of the tumour shown in **a** to illustrate the complex multilobated nuclei of the lacunar RS cells, most of which have small nucleoli

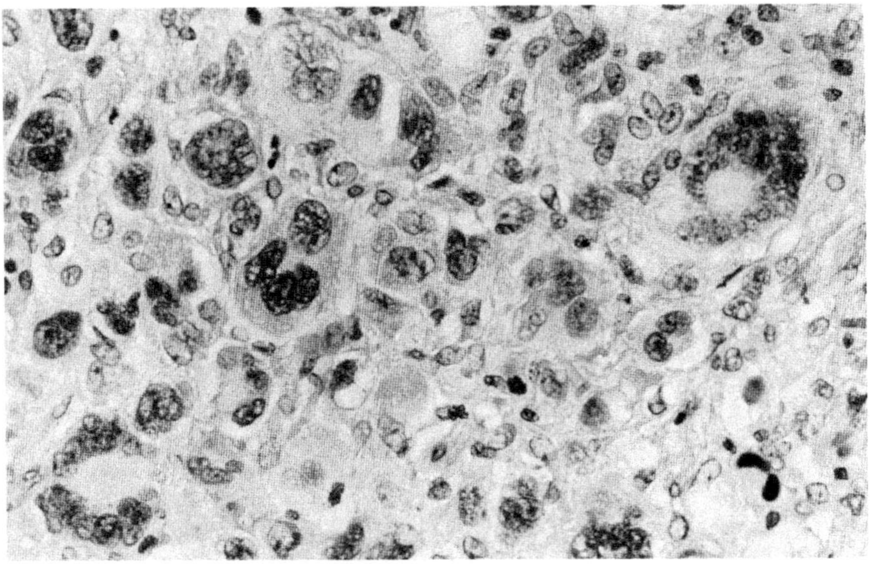

Fig. 4. Lymphocyte-depleted HD, reticular type showing numerous multinucleated RS cells some with wreath-like nuclei. Few lymphocytes are seen. The majority of the intervening cells are histiocytes

their experience, was a rapidly fatal disease characterised by fever, pancytopaenia, lymphocytopaenia and abnormal liver function tests. The patients usually had no peripheral lymphadenopathy, the diagnosis being established by bone marrow aspiration or autopsy. Conflicting findings were reported by Bearman et al. (1978), who found no clinical or survival differences between the reticular and diffuse fibrosis group in a study of 39 patients with LDHD. In a more recent review, however, of 25 patients with LDHD, Greer et al. (1986) found that patients with diffuse fibrosis had less peripheral lymphadenopathy and more bone marrow involvement compared with the reticular subtype. Amongst patients who received chemotherapy, median survival was longer in the diffuse fibrosis than in the reticular subtype. The separation of diffuse fibrosis and the reticular forms of LDHD would appear to be justified on the basis of their histopathological and gross anatomical differences and also their reported clinical differences.

Summary and Conclusions

This presentation deals with the gross and microscopic pathology of HD. Recent advances in immunohistochemistry, gene rearrangement studies and cell culture are not discussed, except where they shed light on the pathology. Clinical and pathological experience, over the past 2 decades, suggests that HD should be divided into six subtypes, as originally proposed by Lukes et al. (1966b), rather than the four subtypes included in the Rye classification.

Nodular lymphocyte/histiocyte predominant HD forms a clinicopathological entity separate from the other subtypes. It most frequently presents at a single nodal site and, even without therapy, progresses only slowly over a period of many years. A proportion of the patients (in the region of 10%) develop large cell NHL and a smaller number develop other types of Hodgkin's disease. This progression is not due to therapy since it most frequently occurs in untreated patients. Characteristic polylobated RS cell variants are seen in NLPHD. These differ from classic RS cells in that they have a B-cell phenotype, they do not show light chain restriction and, therefore, they do not appear to be a clonal proliferation. Although current dogma states that classic RS cells must be identified before a diagnosis of HD, including NLPHD, is made, it is the author's contention, supported by immunohistochemistry, that this type of RS cell does not occur in NLPHD. Polylobated RS cell variants in the appropriate cellular setting are, in themselves, diagnostic of NLPHD. They also serve to differentiate NLPHD from progressive transformation of germinal centres, an unusual proliferative expansion that may occur in association with HD but which, in itself, appears to be an entirely benign, reactive process.

Diffuse lymphocyte/histiocyte predominant HD (DLPHD) differs from NLPHD in its diffuse growth pattern and the frequent presence of larger numbers of histiocytes. Polylobated RS cells are characteristic of both diseases. In some biopsies nodular and diffuse areas are seen in the same lymph node. Despite these similarities, the two diseases differ clinically (NLPHD is usually stage 1, DLPHD is frequently of a higher stage) and in their immunohistochemistry (fewer RS cell variants in DLPHD contain J-chain than in NLPHD). The reasons for these differences are not apparent and require further investigation. It may be due, at least in part, to the inclusion of cases of MCHD and NHL in the DLPHD category.

Nodular sclerosis is the commonest category of HD in most reported series. Lacunar RS cells characterise this type of HD, although current dogma insists that classic RS cells must be seen before the diagnosis of HD is made. It is the author's contention that lacunar RS cell variants, in the appropriate setting, are sufficient, in themselves, for the diagnosis of NSHD.

Cellular phase NSHD may show a temporal or a sequential association with classic NSHD, suggesting that both are manifestations of the same disease process. Despite this, cellular phase NS has been reported to have clinical features and a response to therapy more akin to MCHD. NSHD may be divided into grade 1 and grade 2, the latter showing areas of lymphocyte depletion or numerous pleomorphic Hodgkin's cells. Grade 2, which constitutes a substantial proportion of all cases of HD, has a significantly worse prognosis than grade 1.

A high rate of misdiagnosis has been reported in studies of LDHD. Histological reviews have shown the inclusion of other subtypes of HD and of many cases of NHL in this category. This reflects the ill-defined boundary between some cases of MCHD or NSHD grade 2 and LDHD. It may be extremely difficult, if not impossible, to separate LDHD from some variants of T-cell lymphoma. Such cases highlight the need to identify the precise nature of the RS cell and to determine the pathogenesis of HD. LDHD consists of two subtypes recognised by Lukes et al. (1966b) as diffuse fibrosis and reticular HD. These are morphologically distinct, and clinical studies suggest that they have significant differences in their clinical presentation and survival.

In recent years studies of immunohistochemistry, molecular biology and cell culture, using Hodgkin's tissue, have produced conflicting data that have confused attempts to determine the histogenesis of RS cells and the pathogenesis of HD. In attempting to unravel this problem some of the old observations, concerning the pathology of HD, should be borne in mind. Hodgkin's disease typically involves the axial lymph nodes, thymus and spleen, initially occupying the T-cell areas of these organs. It progresses in a predictable fashion between contiguous lymph node groups, and blood-borne spread to other sites appears to occur from the spleen. Involvement of the gastrointestinal tract, or other mucosal sites, and the skin are uncommon and when they occur are usually late manifestations. These anatomical and behavioural characteristics may well reflect the physiology of the benign analogue of the RS and Hodgkin's cell and could provide clues to the histogenesis of these cells.

For the time being the diagnosis of Hodgkin's disease remains firmly in the province of the morphologist (histopathologist) aided, in some areas, by immunohistochemistry.

References

Bearman RM, Pangalis GA, Rappaport H (1978) Hodgkin's disease, lymphocyte depletion type: a clinicopathologic study of 39 patients. Cancer 41: 293-302

Bennett MH, Tu A, Vaughan Hudson B (1981) Analysis of grade 1 Hodgkin's disease (report no 6), part 2. Nodular sclerotic Hodgkin's disease. Cellular subtypes related to prognosis. Clin Radiol 32: 491-498

Bennett MH, MacLennan KA, Easterling MJ, Vaughan Hudson B, Vaughan Hudson G, Jelliff AM (1985) Analysis of histological subtypes in prognosis and survival. In: Quaglino D, Hayhoe FGJ (eds) The cytobiology of leukaemias and lymphomas. Raven, New York, pp 15-32

Burns BF, Colby TV, Dorfman RF (1984) Differential diagnostic features of nodular L&H Hodgkin's disease including progressive transformation of germinal centers. Am J Surg Pathol 8: 253-261

Colby TV, Hoppe RT, Warnke RA (1981) Hodgkin's disease: a clinicopathologic study of 659 cases. Cancer 49: 1848-1858

Fox H (1926) Remarks on the presentation of microscopic preparations made from some of the original tissue described by Thomas Hodgkin, 1832. Annals of Medical History 8: 370-374

Greer JP, Kinney MC, Cousar JB, Flexner JM, Dupont WD, Graber SE, Greco FA, Collins RD, Stein RS (1986) Lymphocyte depleted Hodgkin's disease. Clinicopathologic review of 25 patients. Am J Med 81: 208-214

Haybittle JL, Hayhoe FGJ, Easterling MJ, Jelliff AM, Bennett MH, Vaughan Hudson G, Vaughan Hudson B, MacLennan KA (1985) Review of British National Lymphoma Investigation studies of Hodgkin's disease and development of prognostic index. Lancet I: 967-972

Kadin ME, Glatstein E, Dorfman RF (1971) Clinicopathologic studies of 117 untreated patients subjected to laparotomy for the staging of Hodgkin's disease. Cancer 27: 1277-1294

Kant JA, Hubbard SM, Longo DL, Simon RM, De Vita VT, Jaffe ES (1986) The pathologic and clinical heterogeneity of lymphocyte depleted Hodgkin's disease. J Clin Oncol 4: 284-294

Lukes RJ, Craver LF, Hall TC, Rappaport H, Ruben P (1966a) Report of the Nomenclature Committee. Cancer Res 26: 1311

Lukes RJ, Butler JJ, Hicks EB (1966b) Natural history of Hodgkin's disease as related to its pathologic picture. Cancer 19: 317-344

Miettinen M, Franssila KO, Saxen E (1983) Hodgkin's disease lymphocytic predominance nodular. Increased risk for subsequent non-Hodgkin's lymphomas. Cancer 51: 2293-2300

Miller TP, Byrne GE, Jones SE (1982) Mistaken clinical and pathologic diagnosis of Hodgkin's disease. A Southwest Oncology Group study. Cancer Treat Rep 66: 645-651

Myskow MW, Robinson MC, Krajewski AS, Angus B, Proctor SJ (1987) Lymphocyte depleted Hodgkin's disease, a retrospective study. J Pathol 152: 200A

Neiman RS, Rosen PJ, Lukes RJ (1973) Lymphocyte depletion Hodgkin's disease. A clinicopathologic entity. N Engl J Med 288: 751-755

Osborne BM, Butler JJ (1984) Clinical implications of progressive transformation of germinal centers. Am J Surg Pathol 8: 725-733

Poppema S, Kaiserling E, Lennert K (1979) Hodgkin's disease with lymphocytic predominance, nodular type (nodular paragranuloma) and progressively transformed germinal centres - a cytohistological study. Histopathology 3: 295-308

Rose M (1981) Curator of the dead. Thomas Hodgkin (1798-1866). Peter Owen, London

Stein H, Hansmann ML, Lennert K, Brandtzaeg P, Gatter KC, Mason DY (1986) Reed-Sternberg and Hodgkin's cells in lymphocyte predominant Hodgkin's disease of nodular subtype contain J-chain. Am J Clin Pathol 86: 292-297

Strickler JG, Michie SA, Warnke RA, Dorfman RF (1986) The 'syncytial variant' of nodular sclerosing Hodgkin's disease. Am J Surg Pathol 10: 470-477

Strum SB, Rappaport H (1971) Interrelations of the histologic types of Hodgkin's disease. Arch Pathol 91: 127-139

Suchi T, Lennert K, Tu L-Y, Kikuchi M, Sato E, Stansfeld AG, Feller AC (1987) Histopathology and immunohistochemistry of peripheral T-cell lymphomas: a proposal for their classification. J Clin Pathol 40: 995-1015

Timens W, Visser L, Poppema S (1986) Nodular lymphocyte predominance type of Hodgkin's disease is a germinal center lymphoma. Lab Invest 54: 457-461

Trudel MA, Krikorian JG, Neiman RS (1987) Lymphocyte predominance Hodgkin's disease: a clinicopathologic reassessment. Cancer 59: 99-106

Immunology of Hodgkin and Reed-Sternberg Cells

H. Stein, R. Schwarting, F. Dallenbach, and D. Dienemann

Institut für Pathologie, Klinikum Steglitz, Freie Universität Berlin, Hindenburgdamm 30, 1000 Berlin, FRG

Introduction

Hodgkin and Reed-Sternberg (H and RS) cells, the tumor cells in Hodgkin's disease, have been related to nearly every cell in the hematolymphoid system (a selection of the theories propounded is listed in Table 1). This prompted Clive Taylor (1983) to make the following statement:

Table 1. Presumed cellular origin of RS cells

Cell type	References
Sinus endothelial cells	Reed 1902[a]
Lymphoblasts	Mallory 1914[a]
Monocyte-like cells	McJunkin 1928[a]
Megakaryocytes	Medlar 1931[a]
Myeloblasts	Lewis 1941[a]
Histiocytes	Bessis 1948[a]; Rappaport 1966; Mori and Lennert 1969; Kaplan and Gartner 1977; Kadin et al. 1978; Isaacson 1979; etc.
T cells	Order and Hellmann 1972; Biniaminow and Ramot 1974
B cells	Leech 1973; Garvin et al. 1974; Boecker et al. 1975
Follicular dendritic cells	Curran and Jones 1978
Interdigitating cells	Poppema 1979 (personal communication); Hansmann and Kaiserling 1981; Kadin 1982; Hsu et al. 1985
Myeloid precursor cells	Stein et al. 1982b
Myelomonocytic precursor cells	Diehl et al. 1982
New lymphoid cell type	Stein et al. 1982a
Dendritic cells of the Steinmann type	Fisher et al. 1983
Antigen-presenting cells	Fisher et al. 1985
Activated lymphoid cells of either B- or T-cell type	Stein et al. 1984, 1985a

[a] Cited in Taylor (1974).

Recent Results in Cancer Research. Vol. 117

"If our 100-year quest for the cell of origin of Hodgkin's disease has taught us anything, it is the importance of maintaining an open mind, preferably with a cheerful capacity for changing it according to the dictates of fashion. Every cell has had its day; as prospective candidates, all cells are equal, but (as we shall see) some are more equal than others."

History of Antigenic Markers in Hodgkin's Disease

The history of H and RS cell-associated markers begins in 1982. In this year, an antigen was recognized in H and RS cells that was absent from all cells in the hematolymphoid system except granulocytes and their later precursors (Stein et al. 1982a). In the first workshop on human leucocyte differentiation antigens detected by monoclonal antibodies (mAbs), this antigen was found to be identical with x-hapten and was designated CD15 (Bernard et al. 1984). Two and three years later, this antigen was redetected in H and RS cells, respectively, by several authors using the commercial antibody Leu-M1. In 1982, our group (Schwab et al. 1982) described the H and RS cell-associated antigen Ki-1, and one year later the H and RS cell-associated antigens Ki-24 and Ki-27 (Stein et al. 1983). The Ki-1 (CD30) antigen was found to be most constantly expressed in H and SR cells, and we therefore decided to first study this antigen in greater detail. The original Ki-1 antibody produced relatively weak staining reactivity. This prompted us to modify the APAAP technique developed in David Mason's lab in Oxford (Cordell et al. 1984) in order to increase its sensitivity. This aim was achieved by repeating the incubation with the bridging antibody and the APAAP complex once or twice, and by using a modified new fuchsin development instead of the fast red reaction. With this method, it could be shown that the Ki-1 antigen is present in nearly all H and RS cells in almost every case studied. However, despite the development of the highly sensitive APAAP technique, the staining with the Ki-1 antibody remained difficult. Furthermore, the Ki-1 antibody was not effective on paraffin sections. This, and the need to have an antibody against a second epitope for establishing an ELISA for the demonstration of fluid-phase Ki-1 antigen, induced us to produce five additional antibodies with Ki-1 reactivity. These antibodies led to the acceptance of the Ki-1 antigen as a new T- and B-cell activation marker, and it was designated "CD30" (Beverley 1987).

Ber-H2: A New Monoclonal Antibody Directed at a Formol-Resistant Epitope on the Ki-1 (CD30) Antigen

Among the new anti-Ki-1 antibodies, we found one (Ber-H2) which stained H and RS cells as strongly as the CD8 antibody stains suppressor and cytotoxic T cells (Schwarting et al. 1987). This antibody could be shown as reactive with an epitope distinct from that recognized by the Ki-1 antibody. The greatest advantage of the new Ber-H2 antibody is that the epitope to which this antibody binds is formalin-resistant. Figures 1–3 illustrate the staining pattern achieved with the Ber-H2 antibody in paraffin sections. The staining pattern ranges from surface membrane-like

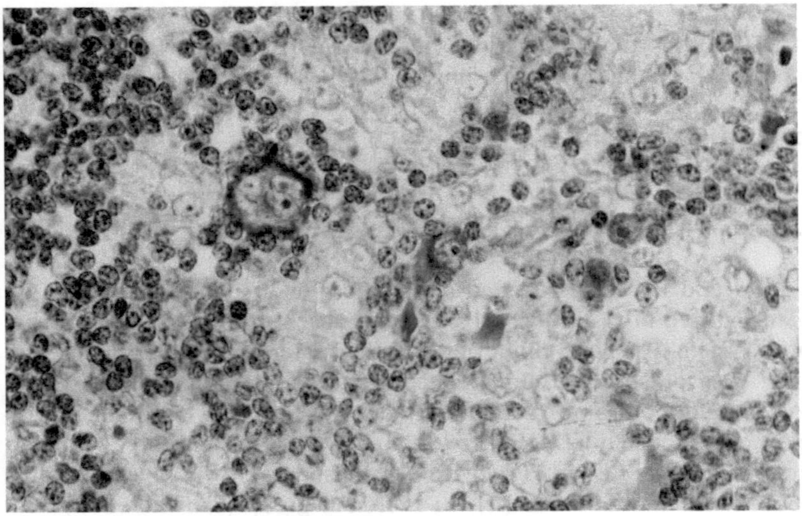

Fig. 1. Paraffin sections of Hodgkin's disease, mixed cellularity, epithelioid cell-rich, immunostained with the mAb Ber-H2. The H and RS cells show an intense labeling of their surface membrane. APAAP, ×450

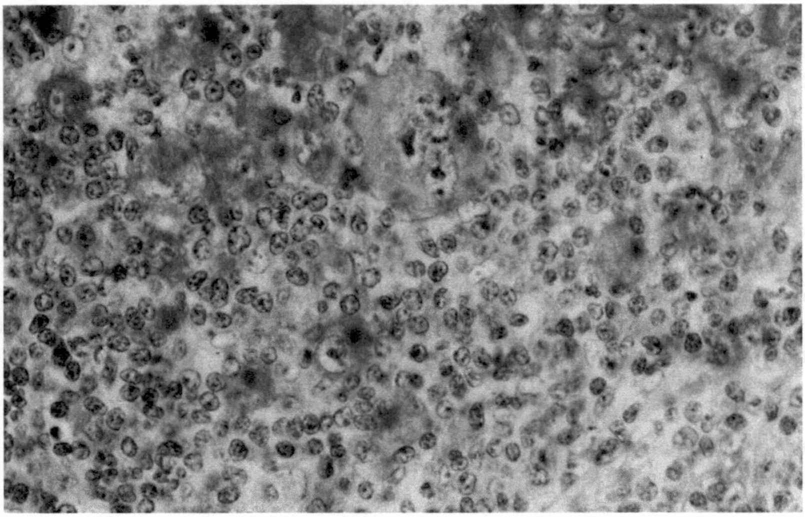

Fig. 2. Paraffin section of Hodgkin's disease, nodular sclerosis, immunostained with the mAb Ber-H2. The tumor cells show a moderately strong staining of the surface membrane and a diffuse, weak cytoplasmic labeling with a dot-like accentuation in the Golgi region. APAAP, ×450

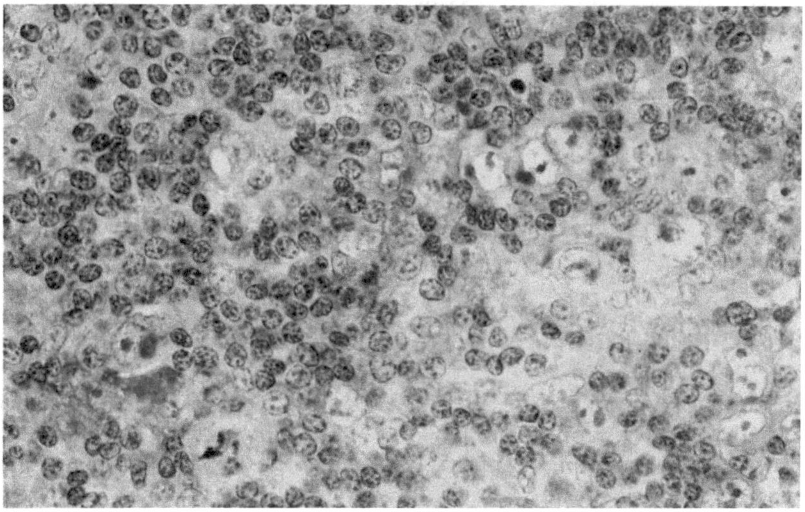

Fig. 3. Paraffin section of nodular lymphocyte-predominant Hodgkin's disease (nLPHD), immunostained with Ber-H2. The tumor cells display a dot-like labeling only in the Golgi region. APAAP, ×450

Table 2. Reactivity of H and RS cells of various histological types of Hodgkin's disease with the monoclonal antibody Ki-1 in frozen sections

Histological type investigated	No. of cases	No. of cases with Ki-1-positive H and SR cells
Lymphocyte-predominant type		
Nodular subtype	12	10
Other than nodular subtype	2	2
Nodular sclerosis	52	52
Mixed cellularity	31	31
Epithelioid cell-rich type	5	5
Lymphocyte-depleted type	8	8
Total	110	108

to dot-like with or without a diffuse cytoplasmic staining. The immunostaining of frozen sections of a larger series of Hodgkin's disease cases revealed that the Ber-H2 antibody was positive in the majority of or in all tumor cells in 108 of 110 cases (Table 2). The two negative cases fell into the lymphocyte predominance category. In paraffin sections, the antibody works with nearly the same reliability in the nodular sclerosis and mixed cellularity categories. However, in the category of lymphocyte predominance, only one-third of the cases showed a staining of the tumor cells with Ber-H2 (Table 3).

Table 3. Detection of the Ki-1 (CD30) antigen in paraffin sections of the three main histological types of Hodgkin's disease

	No. of cases	No. of cases with Ber-H2-positive H and RS cells
LP	25	8 (32%)
NS	40	39 (98%)
MC	58	56 (97%)
Total	123	103

Ber-H2-Reactive Cells in Normal Lymphoid Tissue

In normal lymphoid tissue, the Ber-H2 antibody stains large blastoid cells preferentially localized around and between B-cell follicles and, to a lesser extent, at the inner edge of germinal centers. The latter cells usually show a weaker staining than those outside of the B-cell follicles. The Ber-H2 antibody was found to be unreactive with T-cell and B-cell precursors, as well as with resting and circulating B- and T-lymphocytes of the blood and peripheral lymphoid tissue, but positive with a varying proportion of lectin-stimulated T- and B-blasts (Schwarting et al. 1987). Remarkably, the largest blasts show the strongest labeling with Ber-H2. This finding has been confirmed in the Third Workshop on Leukocyte Differentiation Antigens Recognized by Monoclonal Antibodies (Beverley 1987). The reactivity of the Ber-H2 antibody with normal cells is summarized in Table 4.

Ber-H2 Reactivity with Activated Monocytes and Tissue Macrophages

There is no agreement about the reactivity of antibodies with Ki-1 specificity with macrophages (Pfreundschuh et al. 1987). We therefore investigated the reactivity of monocytes with the mAb Ki-1 and Ber-H2 before and after stimulation with phorbol esters, conditioned medium, gamma interferon, and lipopolysaccharides (LPS). In all these experiments, the stimulated and non-stimulated monocytes did not react with the mAb Ki-1 or Ber-H2, provided that the immunostaining reaction was performed under conditions in which binding of the antibodies via Fc-gamma receptors was avoided (Table 5). However, these studies revealed a very small proportion of strongly Ber-H2-positive large cells with hairy-like cytoplasmic processes and lobated nuclei. The nature of these cells has not yet been identified. It should be mentioned in this context that Andreesen et al. reported in this meeting (see Andreesen et al. this volume) the appearance of Ki-1 antibody reactivity on monocytes after culturing them on plastic surfaces in the presence of gamma interferon and LPS for longer than 5 days. Our own experiments were stopped on day 3. The experiment by Andreesen et al. further revealed that the stainability of the activated monocytes could only be seen if gamma interferon was included in the substances used for stimulation. Since it is well known that especially gamma interferon induces the presence of high-affinity Fc-gamma receptors (Guyre et al. 1983), it remains to be seen whether the reactivity observed in

Table 4. Expression of Ki-1 antigen and Interleukin-2 receptor (IL-2R) in normal cells

Cells	Ki-1	IL-2R	Cells	Ki-1	IL-2R
Precursor T cells[a]	–	–	Monocytes	–	–
Precursor B cells[a]	–	–	Activated monocytes	–	+
Resting T cells	–	–	Tissue macrophages	–	+
Resting B cells	–	–	Interdigitating cells	–	–
Activated T cells	+	+	Follicular dendritic cells	–	–
Activated B cells	+[b]	+[b]			

[a] As revealed by immunolabeling fetal thymus, fetal liver, and bone marrow.
[b] Germinal center B-blasts are negative.

Table 5. Reactivity of peripheral blood cells enriched for monocytes before and after exposure to gamma interferon (IFN) and lipopolysaccharides (LPS)

Reagents applied[a]	Day 0 Untreated (%)	24 h IFN+LPS (%)	48 h IFN+LPS (%)
Buffer	0	0	0
S-HCL3 (CD11c)	80 m	80–90 s	80–98
Ber-MAC3	10 w	30 s	40–60 s
CD2	5–10	5–10	5–10
CD19	2–8	1–8	1–8
ACT-1 (CD25)	0.1–1 m	50–90 m/s	70–95 m/s
HSR-1 (CD30)	0.01–0.1	1–4 m	1–5 m
Ber-H2 (CD30)	0.01–0.1	1–4 s	1–5 s

s, strong labeling; m, moderately strong; w, weak.
[a] All labeling reactions were performed after 60-min fixation with acetone and chloroform and 30-min preincubation with rabbit-IgG.

their cultures is due to these Fc-gamma receptors or due to the antigen-binding fragment of the Ki-1 antibody. However, the negative results on unstimulated monocytes and monocytes stimulated for 3 days is consistent with tissue-staining results. In all tissues from the human body and all malignant lymphomas and other malignancies, the antibodies Ki-1 and Ber-H2 proved to be unreactive with macrophages.

Reactivity of Ber-H2 with Lymphomas Other than Hodgkin's Disease

The reactivity of the Ber-H2 antibody with malignant lymphomas other than Hodgkin's disease is shown in Table 6. It is evident from this table that the majority of non-Hodgkin lymphomas are unreactive with this antibody. Among the Ber-H2-reactive lymphomas, two groups can be distinguished: one in which only some to many, but not all, tumor cells are positive, and a second group in which all tumor cells are positive. A varying proportion of cases of mycosis fungoides and pleomorphic T-cell lymphoma, as well as all cases of angioimmunoblastic T-cell lym-

Table 6. Ki-1 expression in malignant lymphomas

All tumor cells Ki-1 negative	No. of cases studied	Some to many tumor cells positive[a]	No. of cases studied	All or almost all tumor cells positive	No. of cases studied
T lympho-blastic/ALL	35	MF, pleomorphic	8	Hodgkin's disease	110[b]
B lympho-blastic/ALL	15	Pleomorphic T (small and large)	35	Anaplastic large cell lymphoma (formerly malignant histiocytosis)	62
T-CLL	3	AILD-type	10	Lymphomatoid papulosis	25
B-CLL	42	Lennert's lymphoma	10		
Centrocytic (small cleaved)	25				
Hairy cell leukemia	57				
MF, lympho-cytic	17				
Centroblastic-centrocytic (follicular center cell)	52				
Centroblastic (large non-cleaved[c])	33				

ALL, acute lymphoblastic leukemia; CLL, chronic lymphatic leukemia; AILD, angio-immunoblastic lymphadenopathy with dysproteinemia; MF, mycosis fungoides.
[a] Ki-1 expression restricted to large cells.
[b] Two cases of HDLP were negative.
[c] Few exceptions with weak Ki-1 expression.

phoma, Lennert's lymphoma, and lymphomatoid papulosis, belong to the first group. In this group, the reactivity of Ber-H2 is usually confined to the large H- or RS-like cells. In the second group, the Ber-H2 antibody (like the Ki-1 antibody) stains all tumor cells. The tumor cells in this second group are all large and bizarre (anaplastic). These tumors were diagnosed on histological grounds as malignant histiocytosis, anaplastic carcinoma, interdigitating reticulum cell sarcoma, or some other type of neoplasia (Stein et al. 1985b; Delsol et al. 1988). The study of other antigenic markers in these Ki-1 antigen-positive anaplastic large-cell lymphomas (Ki-1$^+$ALCL) indicates that nearly all, if not all, of these tumors are related to lymphoid cells rather than to other cell types (Stein et al. 1985b, 1986; Delsol et al. 1988).

Table 7. Growth fraction in Hodgkin's disease. (Gerdes et al. 1987)

Type of Hodgkin's disease	No. of cases	$Ki\text{-}1^+$ and $Ki\text{-}67^+$ cells	
		Median	Range
LP	3	78	76–80
NS	27	83	53–95
MC	5	80	73–98

Table 8. Phenotypes of H and RS cells of Hodgkin's disease

	Ki-1	CD3 CD2	CD4	CD8	CD19 CD22	IL-2R CD25	
I	+	–	–	–	–	+/–	Null
II	+	+	+	–	–	+	T4
III	+	+	–	+	–	+	T8
IV	+	–	–	–	+	–/+	B
V	+	+	+	–	+	+/–	Mixed

Origin and Heterogeneity of H and RS Cells

The Ki-1 antigen distribution, as revealed with the Ber-H2 antibody, further supports the hypothesis that the Ki-1-positive H and RS cells represent neoplastic-activated lymphoid cells of either T-cell or B-cell derivation. This hypothesis implies that most of the H and RS cells proliferate and express IL-2 receptors and T- and B-cell antigens, as do physiologically activated (Ki-1-positive) T-blasts and B-blasts. As shown in Table 7, the vast majority of H and RS cells are in fact Ki-67 positive and are thus proliferating (Gerdes et al. 1987). In earlier labeling experiments, we (Stein et al. 1982b) and others were not able to demonstrate T-cell antigens, except the CD4 antigen, on H and RS cells. However, by increasing the resolution power and the sensitivity of our immunodetection system (i.e., APAAP), we (Stein et al. 1985b, 1986; Falini et al. 1987) were able to show that H and RS cells express T-cell antigens in some cases, and B-cell antigens in a minority of cases. It was striking to find two cases of Hodgkin's disease with CD8-positive and CD4-negative tumor cells. Because of these findings, it appears that Hodgkin's disease is heterogeneous. Five phenotypes can be distinguished (see Table 8).

Correlation between Immunophenotype of H and RS Cells and Histological Type of Hodgkin's Disease

Attempts to correlate these phenotypes with morphological categories were not successful, with the exception of the lymphocyte predominance category. Poppema (1980) was the first to describe the presence of the B-cell-associated J chain in H and RS cells in a single case of lymphocyte-predominant Hodgkin's disease. By extending the J-chain labeling to a larger number of cases, we were able to demon-

Table 9. Reactivity of H and RS cells in Hodgkin's lymphomas

Type of Hodgkin's lymphoma	CD 30 (Ki-1) Ber-H2	J chain	CD15 3C4	LN-1	L26	UCHL1 MT1
LP n+d	6/18	22/43	6/43	23/23	9/10	0/18
NS	39/40	0/25	41/41	2/16[a]	6/15[a]	0/16
MC	56/58	0/20	44/53	10/58[a]	2/12[a]	0/35

All specimens were routinely fixed and embedded in paraffin.
[a] Only few H/RS cells were positive.

strate that the tumor cells of nodular lymphocyte-predominant Hodgkin's disease express J chain in the majority of cases (Stein et al. 1986). Sherrod et al. (1986) recently reported that the anti-B-cell antibodies LN-1 and LN-2, effective in paraffin sections, are reactive with the tumor cells in all cases of lymphocyte-predominant Hodgkin's disease. In our own studies, these results were confirmed in principle (see Table 9). With the availability of the anti-B-cell antibody L26 (Ishii et al. 1984), effective on paraffin sections, we demonstrated that the tumor cells in eight of ten cases of lymphocyte-predominant Hodgkin's disease could also be stained with this antibody. The antibody UCHL1, directed at a formol-resistant epitope of a T-cell-associated antigen, was consistently negative with the tumor cells. The above findings indicate B-cell derivation of the tumor cells in lymphocyte-predominant Hodgkin's disease. The link of this subtype of Hodgkin's disease with the B-cell system is further supported by the finding that the tumor cells are usually located in enlarged B-cell follicle-like structures, which have been designated "progressively transformed germinal centers" by Poppema et al. (1979a, b). The association of these transformed B-cell follicles with lymphocyte-predominant Hodgkin's disease was first emphasized by Poppema et al. (1979a, b). Immunohistological studies using antibodies reactive with follicular dendritic cells further confirmed this relationship. The enlarged nodules in lymphocyte-predominant Hodgkin's disease proved to be composed of a dense meshwork of follicular dendritic cells (Stein et al. 1982a, b; Abdulaziz et al. 1984).

Distinction of Germinal Center-Related and Inter-/Perifollicular Forms of Hodgkin's Disease

In comparing the distribution of the tumor cells in the different categories of Hodgkin's disease, it can be shown that the tumor cells are localized within the enlarged B-cell follicles in nodular lymphocyte-predominant Hodgkin's disease, whereas the Ki-1-positive cells tend to be distributed around and between the B-cell follicles in the other Hodgkin's disease types. In normal lymphoid tissue, the majority of Ki-1-positive cells are also preferentially localized around and between B-cell follicles, with only few Ki-1-positive cells being present at the rim of the germinal centers. Immunolabeling reactions of adjacent sections of normal lymphoid tissue, using anti-B- and -T-cell antibodies effective on formalin-fixed tissue, revealed that the peri- and interfollicular normal Ki-1-positive cells are heterogeneous. Some express B-cell antigens and others T-cell antigens, while the weakly Ki-1-positive cells at the outer rim of germinal centers seem to exhibit B-cell antigens only. These findings are consistent with the B-cell phenotype of the tumor cells in lymphocyte-predominant Hodgkin's disease, nodular subtype, and with the heterogeneous phenotype of the other types of Hodgkin's disease.

Cytokine, Secretion: Characteristic of Hodgkin's Disease

As pointed out earlier (Stein et al. 1985b), an attractive hypothesis to explain why these B-cell- and T-cell-derived types of Hodgkin's disease differ from non-Hodgkin lymphomas of T-cell and B-cell type is the assumption that the tumor cells in Hodgkin's disease are capable of secreting cytokines, which are responsible for the presence of the non-malignant cells between and around H and RS cells. If this hypothesis is correct, then we can deduce that in the germinal center cell type H and RS cells present in lymphocyte-predominant Hodgkin's disease secrete a cytokine mixture different from that released by the tumor cells of the other types, because there is a huge spherical meshwork of follicular dendritic reticulum cells only in the nodular lymphocyte-predominant type (see Fig. 4). The admixed non-malignant cells in nodular sclerosis and mixed cellularity of B- and T-cell phenotype are the same. We can thus conclude that the peri- and interfollicular H and RS cells representing activated T and B cells are capable of secreting the same mixture of cytokines.

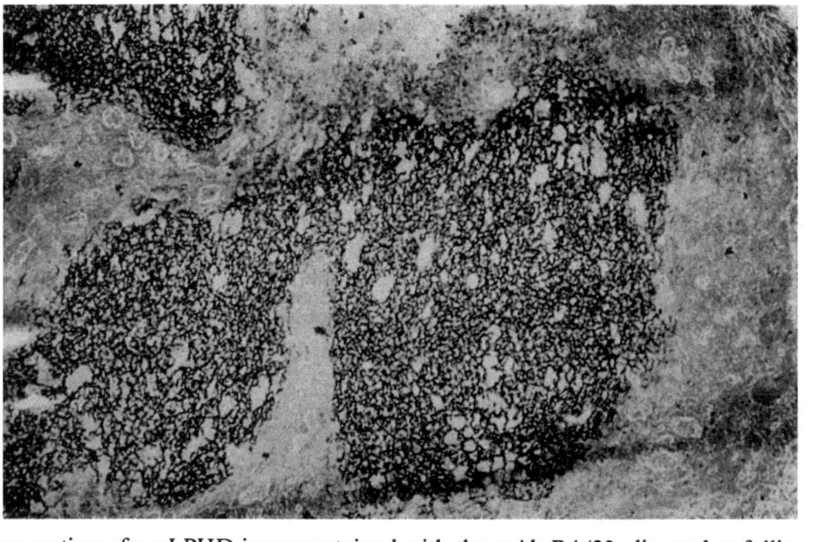

Fig. 4. Frozen section of a nLPHD immunostained with the mAb R4/23, directed at follicular dendritic cells (FDCs). Note that the nodules of nLPHD are composed of a dense meshwork of FDCs containing many H and RS cells (represented by the large holes). Three-stage PAP, ×20

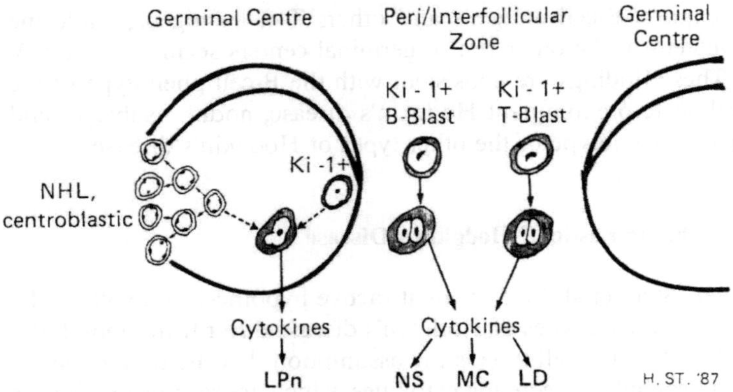

Fig. 5. Putative pathogenesis of Hodgkin's lymphomas

Hodgkin's Disease: A Group of Lymphomas Derived from Activated T or B Cells

Taken together, the data presented above support the view that H and RS cells are lymphocytic in origin and may be related to either T or B cells; phenotypically, they most closely resemble activated T and B cells in differentiation stage. The lymphocyte-predominant type of Hodgkin's disease is probably germinal center cell-derived in most, if not all, instances, whereas nodular sclerosis and mixed cellularity types seem to be heterogeneous in that their cells may be either of the B-cell or of the T-cell type (Fig. 5).

Acknowledgements. The authors wish to thank Mr B. Young for his help in the preparation of the manuscript.

References

Abdulaziz Z, Mason DY, Stein H, Gatter KC, Nash JRG (1984) An immunohistological study of the cellular constituents of Hodgkin's disease using a monoclonal antibody panel. Histopathology 8: 1-25

Bernard A, Boumsell L, Hill C (1984) Joint report on the First International Workshop on Human Leucocyte Differentiation Antigens by the investigators of the participating laboratories. In: Bernard A, Boumsell L, Dausset J, Milstein C, Schlossman SF (eds) Leucocyte typing. Springer, Berlin Heidelberg New York, p 99

Beverley PCL (1987) Activation antigens: new and previously defined clusters. In: McMichael AJ (ed) Leucocyte typing III. Oxford University Press, Oxford, pp 516-524

Biniaminow M, Ramot B (1974) Possible T lymphocyte origin of Reed-Sternberg cells. Lancet I: 368

Boecker WR, Hossfeld DK, Gallmeier WM, Schmidt CG (1975) Clonal growth of Hodgkin cells. Nature (Lond) 258: 235-236

Cordell JL, Falini B, Erber WN, Ghosh AK, Abdulaziz Z, MacDonald S, Pulford KAF, Stein H, Mason DY (1984) Immunoenzymatic labeling of monoclonal antibodies using immune complexes of alkaline phosphatase and monoclonal anti-alkaline phosphatase (APAAP) complexes. J Histochem Cytochem 32: 219-229

Curran RC, Jones EL (1978) Hodgkin's disease: an immunohistochemical and histological study. J Pathol 125: 39-51

Delsol G, Al Saati T, Gatter K, Gerdes J, Schwarting R, Caveriviere P, Rigal-Huguet F, Robert A, Stein H, Mason DY (1988) Coexpression of epithelial membrane antigen (EMA), Ki-1 and Interleukin-2 receptor by anaplastic large cell lymphomas. Diagnostic value in so-called malignant histiocytosis. Amer J Pathol 130: 59-70

Diehl V, Kirchner HH, Burrichter H, Stein H (1982) Characteristics of Hodgkin's disease-derived cell lines. Cancer Treat Rep 66: 615-632

Falini B, Stein H, Ribacchi R, Martelli MF, Grignani F, Canino S, Minelli O, Ciani C, Mori A, Fagioli M (1987) Expression of T cell antigens on Hodgkin's and Sternberg-Reed cells of Hodgkin's disease. Histopathology 11: 1229-1242

Fisher RI, Bostick-Bruton F, Sauder DN, Scala G, Diehl V (1983) Neoplastic cells obtained from Hodgkin's disease are potent stimulators of human primary mixed lymphocyte cultures. J Immunol 130: 2666-2670

Fisher RI, Diehl V, Volkman DJ (1985) Hodgkin's disease is a tumor of antigen-presenting cells. Proc Amer Assoc Cancer Res 25: 193

Garvin AJ, Spicer SS, Parmley RT, Munster AM (1974) Immunohistochemical demonstration of IgG in Reed-Sternberg and other cells in Hodgkin's disease. J Exp Med 139: 1077-1083

Gerdes J, van Baarlen J, Pileri S, Schwarting R, van Unnik JAM, Stein H (1987) Tumor growth fraction in Hodgkin's disease. Am J Pathol 129: 390-393

Guyre PM, Morganelli PM, Miller R (1983) Recombinant immune interferon increases immunoglobulin G Fc receptors on cultured human mononuclear phagocytes. J Clin Invest 72: 393-397

Hansmann ML, Kaiserling E (1981) Electron-microscopic aspects of Hodgkin's disease. J Cancer Res Clin Oncol 101: 135-148

Hsu SM, Yang K, Jaffe ES (1985) Phenotypic expression of Hodgkin and Reed-Sternberg cells in Hodgkin's disease. Am J Pathol 118: 209-217

Isaacson PG (1979) Immunochemical demonstration of J chain: a marker of B cell malignancy. J Clin Pathol 32:802-807

Ishii Y, Takami T, Yuasa H, Takei T, Kikuchi K (1984) Two distinct antigen systems in human B lymphocytes: identification of cell surface and intracellular antigens using monoclonal antibodies. Clin Exp Immunol 58: 183-192

Kadin ME, Sites DP, Levy R, Warnke R (1978) Exogenous immunoglobulin and the macrophage origin of Reed-Sternberg cells in Hodgkin's disease. N Engl J Med 299: 1208-1214

Kadin ME (1982) Possible origin of the Reed-Sternberg cell from an interdigitating reticulum cell. Cancer Treat Rep 66: 601-608

Kaplan HS, Gartner S (1977) "Sternberg-Reed" giant cells of Hodgkin's disease: cultivation in vitro, heterotransplantation, and characterization as neoplastic macrophages. Int J Cancer 19: 511-525

Leech H (1973) Immunoglobulin-positive Reed-Sternberg cells in Hodgkin's disease. Lancet II: 265-266

Mori Y, Lennert K (1969) Electron microscopic atlas of lymph node cytology and pathology. Springer, Berlin Heidelberg New York, pp 29-30

Order SE, Hellmann S (1972) Pathogenesis of Hodgkin's disease. Lancet I: 571-573

Pfreundschuh M, Tirier C, Fuchs R, Wendt F, Löffler M, Diehl V (1987) Fast alternating chemotherapy with COP/ABV/IMEP in patients with Hodgkin's disease. Third International Conference on Malignant Lymphoma, June 10-13, Lugano

Poppema S (1980) The diversity of immunohistological staining pattern of Sternberg-Reed cells. J Histochem Cytochem 28: 788-791

Poppema S, Kaiserling E, Lennert K (1979a) Hodgkin's disease with lymphocytic predominance, nodular type (nodular paragranuloma) and progressively transformed germinal centres - a cytohistological study. Histopathology 3: 295-308

Poppema S, Kaiserling E, Lennert K (1979b) Nodular paragranuloma and progressively transformed germinal centers. Virchows Arch [B] 31: 211-225

Rappaport H (1966) Tumors of the hematopoietic system. Atlas of tumor pathology, Sect 3, Fasc 8. Armed Forces Institute of Pathology, Washington D.C.

Schwab U, Stein H, Gerdes J, Lemke H, Kirchner H, Schaadt M, Diehl V (1982) Production of a monoclonal antibody specific for Hodgkin and Sternberg-Reed cells of Hodgkin's disease and a subset of normal lymphoid cells. Nature 299: 65-67

Schwarting R, Gerdes J, Stein H (1987) Ber-H2: a new monoclonal antibody of the Ki-1 family for the detection of Hodgkin's disease in formaldehyde-fixed tissue sections. In: McMichael AJ (ed) Leukocyte typing III. Oxford University Press, Oxford, pp 574-575

Sherrod AE, Felder B, Levy N, Epstein A, Marder R, Lukes RJ, Taylor CR (1986) Immunohistologic identification of phenotype antigens associated with Hodgkin and Reed-Sternberg cells. Cancer 57: 2135-2140

Stein H (1986) Useful markers for the definition of Hodgkin and Sternberg-Reed cells in Hodgkin's disease. In: Daar AS (ed) Tumor markers. Blackwell Scientific, London, pp 228-241

Stein H, Gerdes J (1986) Phänotypische und genotypische Marker bei malignen Lymphomen. Ein Beitrag zum zellulären Ursprung des Morbus Hodgkin und der malignen Histiozytose sowie Implikationen für die Klassifikation der T-Zell und B-Zell-Lymphome. Verh Dtsch Ges Pathol 70: 127-151

Stein H, Uchanska-Ziegler B, Gerdes J, Ziegler A, Wernet P (1982a) Hodgkin and Sternberg-Reed cells contain antigens specific to late cells of granulopoiesis. Int J Cancer 29: 283-290

Stein H, Gerdes J, Schwab U, Lemke H, Mason DY, Ziegler A, Schienle W, Diehl V (1982b) Identification of Hodgkin and Sternberg-Reed cells as a unique cell type derived from a newly detected small cell population. Int J Cancer 30: 445-459

Stein H, Gerdes J, Schwab U, Lemke H, Diehl V, Mason DY, Bartels H, Ziegler A (1983) Evidence for the detection of the normal counterpart of Hodgkin and Sternberg-Reed cells. Hematol Oncol 1: 21-29

Stein H, Gerdes J, Lemke H, Burrichter H et al. (1984) Hodgkin's Disease and so-called malignant histiocytosis: Neoplasms of a new cell type? Contr Oncol 19: 88-104

Stein H, Gerdes J, Lemke H, Mason DY (1985a) Hodgkin's disease: a neoplasm of activated lymphoid cells of either T cell or B cell type. In: Grignani F, Martelli MF, Mason DY (eds) Serono symposia 26: monoclonal antibodies in haematopathology. Raven, New York, pp 265-269

Stein H, Mason DY, Gerdes J, O'Conner N, Wainscoat J, Pallesen G, Gatter K, Falini B, Delsol G, Lemke H, Schwarting R (1985b) The expression of the Hodgkin's disease-associated antigen Ki-1 in reactive and neoplastic lymphoid tissue - evidence that Reed-Sternberg cells and histiocytic malignancies are derived from activated lymphoid cells. Blood 66: 848-858

Stein H, Hansmann ML, Lennert K, Brandtzaeg P, Gatter KC, Mason DY (1986) Reed-Sternberg cells and Hodgkin cells in lymphocyte-predominant Hodgkin's disease of nodular subtype contain J chain. Am J Clin Pathol 86: 292-297

Taylor CR (1974) An immunohistological study of human and murine lymphomata. Thesis, Oxford University

Taylor CR (1983) Upon the enigma of Hodgkin's disease and the Reed-Sternberg cell. In: Bennet JM (ed) Controversies in the management of lymphomas. Martinus Nijhoff, Boston, p 91

DNA Gene Rearrangement Studies in Hodgkin's Disease and Related Lymphomas: A Contribution to Their Cellular Origin

A. C. Feller and H. Griesser

Institut für Pathologie, Universität Kiel, Michaelisstraße 11, 2300 Kiel, FRG

Introduction

Since Hodgkin's disease (HD) was first described, its nature has remained enigmatic and the identity of the putative tumor cell open to debate. The neoplastic nature of this disease was finally determined by cytogenetic studies demonstrating the aneuploidy and clonality of giant cells (Seif and Spriggs 1967; Whitelaw 1969). The most generally accepted and clinically relevant concept of classification of HD is the histological differentiation of four different subtypes as proposed at the Rye Conference in 1966 (Lukes et al. 1966). Although the Hodgkin (H) and Sternberg-Reed (SR) cells are usually the minor population except in HD lymphocyte depletion (HDLD), they are now widely accepted as the neoplastic cell in this disease. However, these giant cells are by no means specific for HD. They may be observed in reactive lesions such as infectious mononucleosis and in pleomorphic T-cell lymphomas or large cell anaplastic lymphomas (LCALs) (Stein et al. 1985; Suchi et al. 1987).

Because of the phenotypic diversity of H and SR cells, almost no cell in the hematopoietic system has escaped suspicion as the normal counterpart. Electronmicroscopic findings have underlined the view of a lymphatic origin for H and SR cells (Dorfman et all 1973), whereas enzyme histochemistry and, among others, the presence of actin-like cytoplasmic fibrils have testified to their histiocytic derivation (Carr 1975; Kaplan and Gartner 1977).

More recently, immunohistochemical studies demonstrated the expression of a myeloid-associated antigen in these cells (Stein et al. 1982). Others visualized T- or B-cell markers (Diehl et al. 1983; Linch et al. 1985; Falini et al. 1987). The detection of a newly recognized antigen (Ki1/HSR1 - CD30) on H and SR cells has led to the speculation that they may derive from a small, as yet unidentified, lymphoid cell population in human lymphatic tissue (Schwab et al. 1982; Stein et al. 1983). However, this antigen could also be detected on activated histiocytic cell lines (Andreesen et al. 1984).

Recently, a number of c-DNA probes became available to study the clonal origin of T and B cells. The detection of a clonal rearrangement of the β-chain gene of the T-cell receptor (TcR) without heavy-chain gene (Ig_H) rearrangement was

Recent Results in Cancer Research. Vol. 117

thought to be T-cell specific, whereas Ig_H chain gene rearrangement without TcR β-chain rearrangement was primarily found in B cells and in B-cell lymphomas. To further elucidate the cellular origin and the neoplastic cell population in HD we used DNA rearrangement studies and compared the results in HD with those in peripheral T-cell lymphomas, large cell anaplastic lymphomas, and in cases where morphology alone could not determine whether they represent HD or peripheral T-cell lymphomas (borderline cases). Furthermore, these data were compared with the immunophenotype of the individual cases using monoclonal antibodies (mAbs).

All cases were investigated with c-DNA probes for the TcR gamma chain (Griesser et al. 1986a), TcR beta constant region, Ig_H-joining region (J_H), and Ig_k light chain constant region. DNA was digested with the three restriction enzymes *Bam*H1, *Eco*R1, and *Hin*dIII.

Results

Peripheral T-Cell Lymphoma

We investigated 34 cases of peripheral T-cell lymphoma, including pleomorphic T-cell lymphoma, T-cell lymphoma of the AILD (angioimmunoblastic lymphadenopathy with dysproteinemia) type, T-zone lymphoma, lymphoepithelioid lymphoma (Lennert's lymphoma), and T-immunoblastic lymphoma. The immunophenotype of these cases was always that of peripheral T cells with a partial antigen loss in three cases. On the DNA level all cases showed a homogeneous pattern with clonal rearrangement (R+) of TcR gamma and β-chain genes whereas Ig_H chain genes were always found in germline configuration with the exception of 6 out of 23 cases of T-cell lymphoma of the AILD type (Table 1). These six cases exhibited beside TcR gamma and β R+ additonal Ig_H R+. These patients with additional Ig_H R+ in the AILD types showed enhanced numbers of proliferating (Ki67+) CD8-positive cells compared with AILD cases with TcR gamma and β-chain R+ only. The latter group showed a clear predominance of CD4+ proliferating cells.

Table 1. Genotypic pattern of peripheral T-cell lymphomas

Peripheral T-cell lymphomas	*n*	TcR_g+ TcR_β	TcR_g+TcR_β +Ig_H
T-zone	5	5	0
AILD	23	17	6
Lymphoepitheloid (Lennert's)	4	4	0
Pleomorphic T	4	4	0
T-immunoblastic	3	3	0

TcR_β, beta chain gene of the T-cell receptor; TcR_g, gamma chain gene of the T-cell receptor; Ig_H, Ig heavy chain gene; AILD, angioimmunoblastic lymphadenopathy with dysproteinemia.

Table 2. Comparison of immunophenotypic and genotypic patterns of LCAL

Genotype	Immunophenotype				
	T	B	H	Mixed	0
T_g	1	–	1	–	1
T_g+T_β	3	–	–	1[a]	4
$T_g+T_\beta+Ig_H$	–	2	–	–	3

0, cell lineage could not be determined immunophenotypically.
[a] This case expressed B, T, and histiocytic antigens on the tumor cells.

Large Cell Anaplastic Lymphomas

Sixteen cases of primary large cell anaplastic lymphoma (LCAL) were investigated for both genotype and immunophenotype. Their common features were their histological and cytological appearance and the expression of the activation antigen CD30 (Ki1/HSR1). Furthermore, all of them expressed HLA-DR/DQ and 10 out of 15 expressed the IL2-receptor (CD25). Immunophenotypically, four cases were of T-cell origin, two cases were of B-cell origin, one was histiocytic, and one was mixed (B, T and histiocytic antigens on the tumor cells). In eight cases the tumor cells did not express lineage-specific antigens (Table 2). A similar heterogeneity was found in their genotype. Three cases had detectable clonal rearrangements of the TcR gamma-chain genes only, whereas eight cases showed TcR gamma and β R+. Five cases had additional Ig_H R+ (Table 2).

Hodgkin's Disease

Twenty-three cases were investigated and classified into the four subtypes of lymphocytic predominance (LP, $n=5$), nodular sclerosis (NS, $n=5$), mixed cellularity MC, $n=6$), and lymphocytic depletion (LD, $n=4$). The genotype was heterogeneous even within the different subtypes. The majority of cases ($n=9$) had TcR gamma R+ only. Five cases exhibited both TcR gamma and β chain R+. One case had additional Ig_H R+ and two other cases showed Ig_H R+ only. Four cases had germline configurations with all the probes used. Two of these were MC with partial infiltration, the remaining two cases belonging to the LP subtype, one of which was LP nodular paragranuloma.

Using a sensitive double immunoenzyme labeling technique (Feller and Parwaresch 1983), we tried to determine the phenotype of those cells which expressed the CD30 antigen or which were positive for the proliferation-associated antigen Ki67 (Gerdes et al. 1986), since it has been shown that H and SR cells are able to divide and are not end-stage cells (Kaplan and Gertner 1977).

In six cases T-cell antigens could be detected on the cell surface of Ki67-positive giant cells; B-cell antigens were detected in four other cases. In a further four cases lineage-specific antigen expression could not be demonstrated or a clear de-

Table 3. Gene rearrangement pattern in different subtypes of HD compared with immunophenotype of HSR cells

No.	Diagnosis	Immune phenotype of HSR cells	T_g	T_g+_β	Ig_H	Germline
1	HDLP	nd	++	–	–	
2	HDLP	nd	+	–	–	
3	HDLP	T	–	+	–	
4	HDLP	nd	–	–	–	+
5	HDLP	nd	–	–	–	+
6	HDNS	H	+	–	–	
7	HDNS	u	+	–	–	
8	HDNS	T	+	–	–	
9	HDNS	T	+	–	–	
10	HDNS	u	–	++	–	
11	HDNS	T	–	++	–	
12	HDMC	H?	+	–	–	
13	HDMC	u	+	–	–	
14	HDMC	T	+	–	–	
15	HDMC	T	–	+	–	
16	HDMC	nd	–	–	–	+
17	HDMC	nd	–	–	–	+
18	HDMC	B	–	+	+	
19	HDLD	B	+	–	–	
20	HDLD	u	+	–	–	
21	HDLD	B	+	–	–	
22	HDLD	B	–	–	++	–
23	HDLD	B	–	–	++	–

+, rearranged new band(s); ++, intensity of the new rearranged bands is nearly identical to the intensity of the germline band(s); LP, lymphocytic predominance; NS, nodular sclerosis; MC, mixed cellularity; LD, lymphocytic depletion; H, histiocytic; nd, not done; u, phenotype could not be defined.

termination of lineage was not possible due to the high admixture of histiocytes. One case showed the expression of the monocytic/histiocytic-associated antigen Ki-M8 in about 20% of cells which expressed the CD30 and/or Ki67 antigen. Six cases were not investigated using this immunostaining technique (Table 3).

Comparing the immunophenotype with the genotypic pattern revealed that three cases with T-cell phenotype had TcR gamma and β R+, whereas three other T-cell types had TcR gamma R+ only. Two B-cell types (LD) had Ig_H R+ only, one case TcR gamma R+, and another case TcR gamma and β as well as Ig_H R+. A single case with histiocytic properties presented TcR gamma R+ only.

The intensity of the newly detected bands were faint compared with the germline bands in all but five cases, so that it could be estimated that in the further cases about 5% of cells comprising the total cellular content were clonally rearranged. In five cases the intensity of the newly detected bands correlated roughly to the number of non clonally proliferating cells that was above 20%. In two of these cases the number of H and SR cells and cells expressing the activation-associated antigen CD30 was 5% or less. The third case had up to 30% CD30-positive cells. Two other cases which were of HDLD subtype with Ig_H R+ also showed strong new bands, indicating a clonal rearrangement of more than 20% of the total cellular content.

Borderline Cases of Hodgkin's Disease/T-Cell Lymphoma

Five cases could not be distinguished morphologically as representing HD or T-cell lymphomas. They showed presence of single H and SR or H and SR-like cells and a mixed cellular infiltrate with lymphocytes exhibiting nuclear pleomorphism. Four cases had detectable clonal rearrangements for both gamma- and beta-chain genes. All these cases had a predominant T-cell infiltrate which made up about 70%–90% of the total cellular infiltrate. In these cases only a few CD30-positive cells were found. One case showed gamma chain gene rearrangement only and expressed B-cell antigens on the CD30-positive giant cells.

Discussion

Concerning the cellular origin of HD, increasing evidence is given that HD, at least in part, is a special variant of non-Hodgkin's lymphomas. Immunophenotypically, it was possible to demonstrate B- or T-cell properties on the cell surface of H and SR cells (Linch et al. 1985; Falini et al. 1987; Oka et al. 1988). Morphologically, it became evident that single cases show strong association to peripheral T-cell lymphomas, thus hindering a clear distinction between these entities in individual cases. The H and SR cells thought to belong to the neoplastic cell clone were found in both peripheral T-cell lymphomas and HD (Suchi et al. 1987). Moreover, some cases which are designated as HDLD cannot be distinguished from large cell anaplastic non-Hodgkin's lymphoma due to their large number of atypical H and SR cells. That the H and SR cells belong to the tumor cell clone is documented by Sundeen et al. (1987). They demonstrated rearrangements in H- and SR-cell-enriched populations which could not be detected using the whole cellular infiltrate. The fact that some cases with a high content of H and SR cells did not show clonal rearrangement (Knowles et al. 1986) does not exclude this possibility since even large cell anaplastic lymphomas with a high content of tumor cells do not necessarily exhibit clonal rearrangement for Ig_H or TcR-β chain genes. Our series included two cases of HDLD with up to 80% H and SR cells exhibiting only TcR gamma chain R+. The fact that we detected faint new bands in the majority of cases – indicating the clonal proliferation of a small cell population – further underlines the view that H and SR cells belong to the neoplastic cell clone. Cases of HDLD, or cases with a high content of H and SR cells, almost always exhibited Ig_H chain R+ (Weiss et al. 1986; Sundeen et al. 1987; Linch et al. 1985; O'Connor et al. 1987b; Griesser et al. 1987).

We and others have reported rare cases which had detectable TcR-β chain rearrangements (Griesser et al. 1987; Sundeen et al. 1987). These cases were predominately of NS or MC. The results make it reasonable that the majority of cases had only small clonally proliferating cell populations. However, three cases in our series with a low percentage of H and SR cells showed rearranged bands of nearly the same intensity as the germline bands. This indicates that beside H and SR cells also other lymphoid cells must belong to the tumor cell clone, perhaps as precursors of the giant cells in at least some cases. This is exemplified in a case of HDLP (Table 3, case 1).

Peripheral T-cell lymphomas were immunophenotypically and genotypically homogeneous in that all had a phenotype of peripheral T cells and clonal rearrangement for TcR gamma and beta chain genes (Griesser et al. 1986). Similar results were obtained in four cases of HD. Three of these had detectable T-cell antigens on H and SR cells and thus could be designated as activated T cells. Two of these had intensive new rearranged bands similar to those in peripheral T-cell lymphomas (Table 3, cases 10, 11).

Some cases of LCAL also belong to this category. They expressed the CD30 antigen and showed a partial antigen loss, as is known from other high-grade peripheral T-cell lymphomas (Table 2). On the other hand, there are cases of HD in which the B-cell lineage of the neoplastic cells clearly could be demonstrated. They expressed B-cell-associated antigens and had clonally rearranged Ig_H chain genes without rearrangement of TcR genes. This is a characteristic constellation in B-cell lymphomas.

This genotypic pattern has been frequently described in cases of HD which exhibited clonal rearrangement. However, detailed immunophenotypic analysis was not given. Most of these cases had a high content of H and SR cells (Linch et al. 1985; Weiss et al. 1986; O'Connor et al. 1987b). This is in line with our findings where two cases with Ig_H R+ were of HDLD and expressed the B-cell antigen CD22 in the one case which could be investigated with this mAb.

Only a single case rearranged both Ig_H and TcR genes. The H and SR cells expressed a B-cell phenotype. This genotypic pattern is found in some B-cell lymphomas (up to 10% of the cases, unpublished data) and more frequently in lymphoblastic lymphomas of B-cell lineage (Pelicci et al. 1985; Neri et al. 1987; Norton et al. 1988). However, this genotype is not B cell specific as it may be found also in T-lymphoblastic lymphomas and leukemias as well as in some acute myelomonocytic neoplasias (Chen et al. 1986). We demonstrated an identical configuration in two cases of LCAL, testifying to the strong association between LCAL and HDLD. Furthermore, three other cases of LCAL in which no lineage-specific antigen expression could be found had the same genotype. Finally, some "mature" B-cell lymphomas (B-CLL and follicular center cell lymphomas) may also have rearranged TcR genes beside the Ig_H R+ (Griesser et al. 1986b; Norton et al. 1988).

The majority of our cases had rearranged their TcR gamma chain genes only. Phenotypically, H and SR cells were of T, B, or histiocytic origin. This is also true for the cases of HDLD with a high number of atypical giant cells. This accords with the results of O'Connor et al. (1987a), who did not find TcR or Ig_H genes rearranged in a considerable number of LCALs. At least some cases of HD described without TcR or Ig_H R+ may have TcR gamma R+ only, which impedes determination of their cellular origin. Even a histiocytic derivation cannot be excluded in individual cases.

Four out of five borderline cases showed the typical genotypic pattern of peripheral T-cell lymphomas and clearly differed from typical HD with faint new bands and/or gamma chain gene R+ only. These latter cases should be designated as HD.

Two cases of HDLP including one case of nodular paragranuloma did not show rearrangement. From clinical studies we know that this subtype has a very favorable prognosis and/or an indolent course, with only a few patients dying from disease-related causes (Hansmann et al. 1984; Regula et al. 1988). Comparison of the results may allow the speculation that some cases of HDLP are not malignant lymphomas but premalignant disorders.

These results make it clear that H and SR cells belong to the neoplastic cell clone although in individual cases other lymphoid cells also show clonal proliferation. Thus, HD probably is not a polyclonal disease. Some cases have phenotypic properties of peripheral T-cell lymphomas and an equivalent rearrangement pattern and thus show their strong association to peripheral T-cell lymphomas. Similar cases occur in the B-cell lineage. Other cases, irrespective of the detection of clonality by faint new bands and/or gamma chain gene rearrangement, clearly differ from the known pattern of B- and T-cell lymphomas and thus may represent "true" cases of HD. Moreover, HD is heterogeneous regarding its cellular origin and the tumor cells may be of B, T, or monocytic derivation.

References

Andreesen R, Osterholz J, Löhr GW, Bross KJ (1984) A Hodgkin cell-specific antigen is expressed on a subset of auto- and alloactivated T (helper) lymphoblasts. Blood 63: 1299-1302

Carr I (1975) The ultrastructure of the abnormal reticulum cells in Hodgkin's disease. J Pathol 115: 45

Cheng GY, Minden MD, Toyonaga B, Mak TW, McCulloch EA (1986) T cell receptor and immunoglobulin gene rearrangements in acute myeloblastic leukemia. J Exp Med 163: 414-424

Diehl V, Burrichter H, Schaadt M, Kirchner HH, Fonatsch C, Stein H, Gerdes J, Heit W, Ziegler A (1983) Hodgkin's cell lines: characteristics and possible pathogenetic implications. Hematol Oncol 1: 139-147

Dorfman RF, Rice DF, Mitchell AD, Kempson RL, Levine G (1973) Ultrastructural studies of Hodgkin's disease. Natl Cancer Inst Monogr 36: 221

Falini B, Stein H, Pileri S, Canino S, Farabbi R, Martelli MF, Grignani F, Fagioli M, Minelli O, Ciani C, Flenghi L (1987) Expression of lymphoid-associated antigens on Hodgkin's and Reed-Sternberg cells of Hodgkin's disease. An immunocytochemical study on lymph node cytospins using monoclonal antibodies. Histopathol 11: 1229-1242

Feller AC, Parwaresch MR (1983) Simultaneous enzyme-immunocytochemical detection of antigens in monocellular specimens with monoclonal antibodies. J Immunol Methods 63: 273-279

Gerdes J, Schwarting R, Stein H (1986) High proliferative activity of Reed Sternberg associated antigen Ki-1 positive cells in normal lymphoid tissue. J Clin Pathol 39: 993-997

Griesser H, Feller AC, Lennert K, Tweedale M, Messner HA, Zalcberg J, Minden MD, Mak TW (1986a) The structure of the T cell gamma chain gene in lymphoproliferative disorders and lymphoma cell lines. Blood 68: 592-594

Griesser H, Feller AC, Lennert K, Minden MD, Mak TW (1986b) Rearrangement of the β chain of the T cell antigen receptor and immunoglobulin genes in lymphoproliferative disorders. J Clin Invest 78: 1179-1184

Griesser H, Feller AC, Mak TW, Lennert K (1987) Clonal rearrangements of T-cell receptor and immunoglobulin genes and immunophenotypic antigen expression in different subclasses of Hodgkin's disease. Int J Cancer 40: 157-160

Hansmann ML, Zwingers T, Böske A, Löffler H, Lennert L (1984) Clinical features of nodular paragranuloma (Hodgkin's disease, lymphocyte predominance type, nodular). J Cancer Res Clin Oncol 108: 321-330

Kaplan HS, Gartner S (1977) "Sternberg-Reed" giant cells of Hodgkin's disease: cultivation in vitro, heterotransplantation, and characterization as neoplastic macrophages. Int J Cancer 19: 511

Knowles DM, Neri A, Pelicci PG, Burke JS, Wu A, Winberg CD, Sheibani K, Dalla-Favera R (1986) Immunoglobulin and T-cell receptor β-chain gene rearrangement analysis of Hodgkin's disease: implications for lineage determination and differential diagnosis. Proc Natl Acad Sci USA 83: 7942-7946

Linch DC, Berliner N, O'Flynn K, Kay LA, Jones HM, MacLennan K, Huehns ER, Goff K (1985) Hodgkin-cell leukaemia of B-cell origin. Lancet 78-80

Neri A, Jakobiec FA, Pelicci PG, Dalla-Favera R, Knowles DM (1987) Immunoglobulin and T-cell receptor β-chain gene rearrangement analysis of occular adnexal lymphoid neoplasms: clinical and biologic implications. Blood 70: 1519-1529

Norton JD, Pattinson J, Hoffbrand AV, Jani H, Yaxley JC, Leber BF (1988) Rearrangement and expression of T-cell antigen receptor genes in B-cell chronic lymphocytic leukemia. Blood 71: 178-185

O'Connor NTJ, Stein H, Gatter KC, Wainscoat JS, Crick J, Al Saati T, Falini B, Delsol G, Mason DY (1987a) Genotypic analysis of large cell lymphomas which express the Ki-1 antigen. Histopathol 11: 733-740

O'Connor NTJ, Crick JA, Gatter KC, Mason DY, Falini B, Stein H (1987b) Cell lineage in Hodgkin's disease. Lancet: 158

Oka K, Mori N, Kojima M (1988) Anti-Leu-3a antibody reactivity with Reed-Sternberg cells of Hodgkin's disease. Arch Pathol Lab Med 112: 139-142

Pelicci PG, Knowles DM, Favera RD (1985) Lymphoid tumors displaying rearrangements of both immunoglobulin and T cell receptor genes. J Exp Med 162: 1015-1024

Regula DP Jr, Hoppe RT, Weiss LM (1988) Nodular and diffuse types of lymphocytic predominance Hodgkin's disease. N Engl J Med 318: 214-219

Seif GSF, Spriggs AL (1967) Chromosome change in Hodgkin's disease. J Natl Cancer Inst 39: 557

Stein H, Uchanska-Ziegler B, Gerdes J, Ziegler A, Werner P (1982) Hodgkin and Sternberg-Reed cells contain antigens specific to late cells of granulopoiesis. Int J Cancer 29: 283-290

Stein H, Gerdes J, Schwab U, Lemke H, Diehl V, Mason DY, Bartels H, Ziegler A (1983) Evidence for detection of the normal counterpart of Hodgkin and Sternberg-Reed cells. Hematol Oncol 1: 21-29

Stein H, Mason DY, O'Connor N, Wainscoat J, Pallesen G, Gatter K, Falini B, Delsol G, Lemke H, Schwarting R, Lennert K (1985) The expression of the Hodgkin's disease associated antigen Ki-1 in reactive and neoplastic lymphoid tissue. Evidence that Reed-Sternberg cells and histiocytic malignancies are derived from activated lymphoid cells. Blood 66: 848-858

Suchi T, Lennert K, Tu LY, Kikuchi M, Sato E, Stansfeld AG, Feller AC (1987) Histopathology and immunohistochemistry of peripheral T-cell lymphomas: a proposal for their classification. J Clin Pathol 40: 995-1015

Sundeen J, Lipfort E, Uppenkamp M, Sussman E, Wahl L, Raffeld M, Cossman J (1987) Rearranged antigen receptor genes in Hodgkin's disease. Blood 70: 96-103

Weiss LM, Strickler JG, Hu E, Warnke RA, Sklar J (1986) Immunoglobulin gene rearrangements in Hodgkin's disease. Hum Pathol 17: 1009-1014

Whitelaw DM (1969) Chromosome complement of lymph node cells in Hodgkin's disease. Can Med Assoc J 101: 74

*Genetics of Hodgkin's Lymphoma**

C. Fonatsch, G. Gradl, and J. Rademacher

Institut für Humangenetik, Medizinische Universiät Lübeck, Ratzeburger Allee 160, 2400 Lübeck, FRG

Introduction

With the application of improved methods of cell culture and chromosome banding within the past 10 years, our understanding of the specificity of certain chromosome abnormalities in tumor cells has grown and it has been shown that specific chromosome changes are of diagnostic and prognostic significance. Enormous progress in cancer research has been accomplished by the association of cytogenetic and molecular genetic data leading to the localization of cellular oncogenes and cell differentiation genes in chromosome regions which are nonrandomly involved in marker formation.

Whereas a great number of cytogenetic data meanwhile have been established for leukemias and non-Hodgkin's lymphomas, our knowledge of chromosome abnormalities in Hodgkin's disease (HD) is still rather poor and only a few chromosome analyses using banding techniques have been published (Mitelman 1985; Fonatsch et al. 1986). This may be due to the fact that the Reed-Sternberg cells only represent a cellular minority in primary biopsies that are contaminated mainly by reactive lymphoid cells. Moreover, cytogenetic analyses are limited by the low mitotic index of Reed-Sternberg cells as well as the poor-quality banding pattern of their chromosomes. Thus, many investigations have revealed either no mitotic cells at all or only cells with a normal karyotype.

In the following, our own approach toward the cytogenetic and molecular genetic background of HD is presented and also of secondary neoplasias, like acute nonlymphocytic leukemia (ANLL), which frequently occur in HD patients after therapy.

The aims of our study were as follows.

Investigations in HD-Derived Lymphoma Lines. The karyotypes of HD-derived cell lines, established by Diehl's group, were analyzed in order to delineate specific chromosome regions which are nonrandomly involved in chromosome aberra-

* The work was supported in part by research grants from the Deutsche Forschungsgemeinschaft (Fo 91/11-1).

tions. In addition, we tried to correlate such characteristic chromosome rearrangements with gene rearrangements by the use of Southern blot and in situ chromosomal hybridization.

Investigations in Lymphoblastoid Cell Lines (LCLs) from HD Patients. A study in cooperation with Kirchner from Hannover was performed on Epstein-Barr virus-transformed B-lymphocytes from HD patients after in vitro treatment with cytostatics used in HD therapy. In this experimental system the sensitivity of certain chromosomes and chromosome regions to mutagenic/carcinogenic agents was tested and the question was raised whether in vitro induced chromosome aberrations in LCLs correspond to aberrations occurring in vivo in secondary neoplasias after HD therapy.

Materials and Methods

Cell culture techniques and chromosome preparations were done as previously described (Fonatsch et al. 1980, 1986). Molecular genetic studies were performed using the methods described by Fonatsch et al. (1987).

Results and Discussion

Investigations in HD-Derived Lymphoma Lines

Figures 1-3 show karyotypes of three HD-derived lymphoma lines. In case L 540 (Fig. 3), primary material in the form of bone marrow and blood was also available for cytogenetic investigation. Since the karyotypes of both primary material and

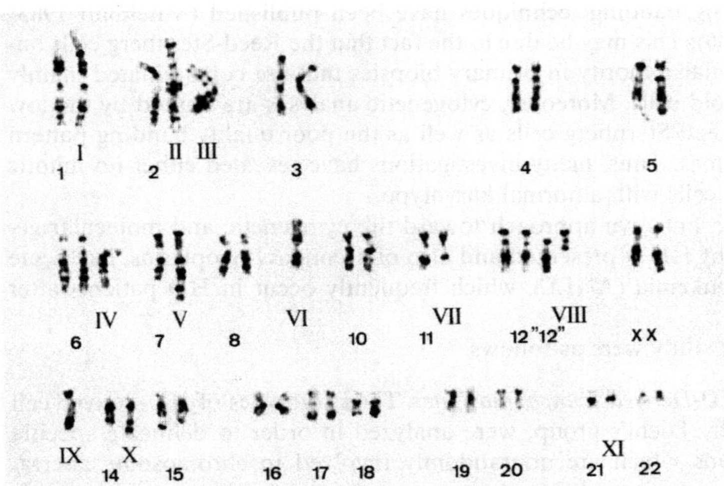

Fig. 1. Representative G-banded karyotype of HD-derived cell line L 428. Marker chromosomes are designated by Roman numerals *I–XI*

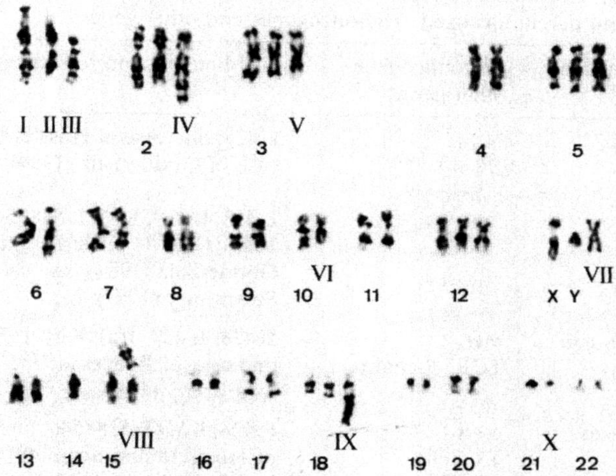

Fig. 2. Representative G-banded karyotype of HD-derived cell line L 439. Marker chromosomes are designated by Roman numerals *I–X*

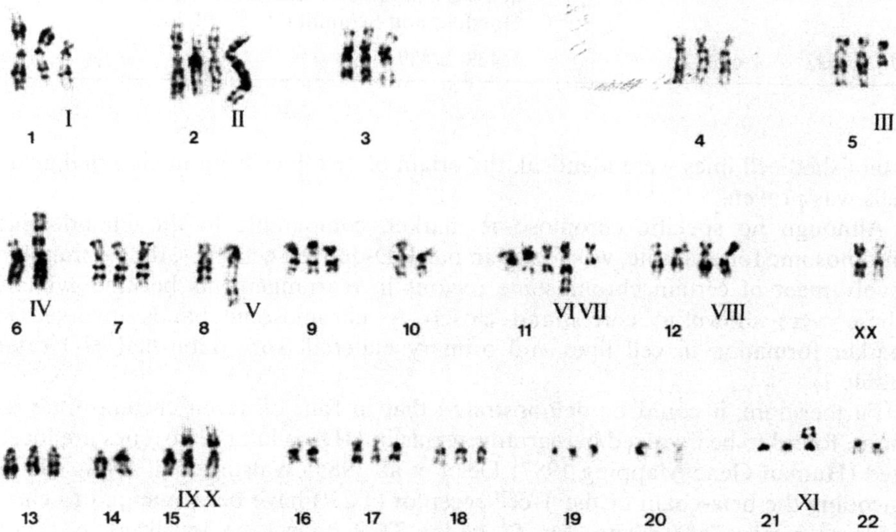

Fig. 3. Representative G-banded karyotype of HD-derived cell line L 540. Marker chromosomes are designated by Roman numerals I–XI

Table 1. Chromosome regions nonrandomly involved in marker formation in HD-derived cells and herein localized protooncogenes and other genes

Chromosome region	Protooncogene/ other genes	Cell line/patient, references
1p21-p22	N-*ras* *Blym* 1 L-*myc*	L 428, one case of Hossfeld and Schmidt (1978), one case of Gödde-Salz (1984)
2q33		L 428, L 439, KM/L 538/540, (DEV, Poppema et al. 1985), (KM-H 2, Kamesaki et al. 1986), one case of Gödde-Salz (1984), one case of Lawler and Swansbury (1987)
7q11.2-q36	*met* TCR, β-chain	L 428, L 439 (fra), KM/L 538/540, L 591, KM-H 2, one case of Reeves (1973), one case of Lawler and Swansbury (1987)
11q21-q23	c-*ets* 1 T3, delta-, epsilon-chains	L 428, KM/L 538/540, one case of Reeves, one case of Hossfeld and Schmidt (1978), one case of Lawler and Swansbury (1987)
14q32	Ig, heavy chain	L 428, L 591, DEV, S 95 (Zech et al. 1976), two cases of Reeves, two cases of Hossfeld and Schmidt (1978), one case of Fukuhara and Rowley (1978), one case of Reeves and Pickup (1980), PI 200, one case of Gödde-Salz (1984)
15p12	rRNA 3	L 439, KM/L 538/540, one case of Reeves, one case of Fleischmann and Krizsa (1977), one case of Hossfeld and Schmidt (1978), PI 200
21q21-q22	c-*ets* 2	L 428, L 439

established cell lines were identical, the origin of cell lines from in vivo malignant cells was proven.

Although no specific chromosome marker, comparable to the Philadelphia chromosome for example, was found in our HD-derived cell lines, the nonrandom involvement of certain chromosome regions in rearrangements became evident. These were shown to correspond closely to chromosome bands involved in marker formation in cell lines and primary material from published HD cases (Table 1).

Furthermore, it could be demonstrated that in four of seven chromosome regions, found to be involved in rearrangements in HD, cellular oncogenes are localized (Human Gene Mapping 1987; Dean et al. 1985; Watson et al. 1986). Genes encoding the beta-chain of the T-cell receptor (TCR) have been mapped to chromosome region *7q35*. Genes for T3 of the TCR have been localized in *11q23*. *14q32* represents the locus of the heavy chain genes of immunoglobulin and genes for the ribosomal RNA, which constitutes the nucleoli, have been found in *15p12* (Human Gene Mapping 1987). Tesch (1986, personal communication), from Diehl's group, demonstrated the expression of genes for the alpha-chain of TCR in L 540. In addition to his results we found an abnormal restriction fragment, different to the germ line configuration, after Southern blot analysis (Fig. 4; Fonatsch

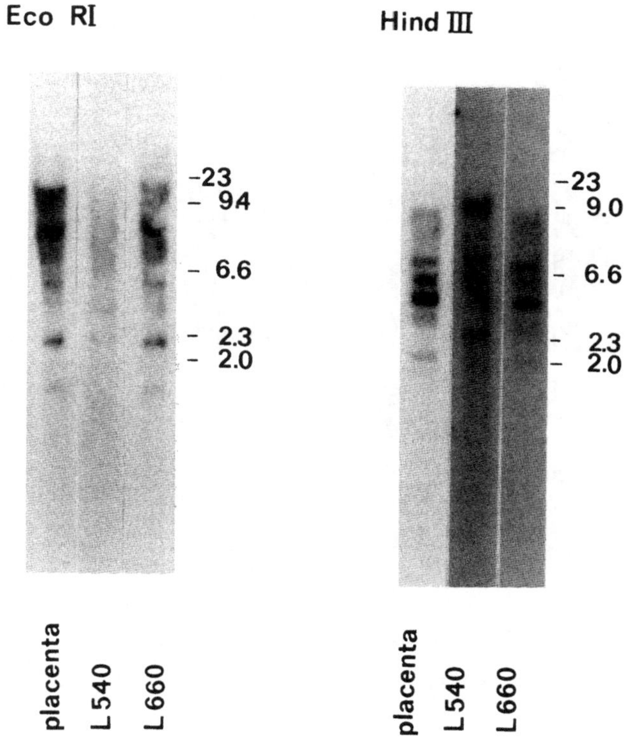

Fig. 4. Southern blot analysis of lymphoma lines L 660 and L 540 DNA and placental DNA using a T-cell receptor alpha, J alpha specific probe *pY 14*

and Gradl 1987). This indicated a rearrangement of alpha-chain genes in L 540. In situ hybridization, by the use of the same DNA probe, revealed a significant number of silver grains in *14q11*, the band to which the TCR alpha-chain genes have been mapped (Fig. 5 a). However, a remarkable labeling was also recognized on marker chromosome IX - a chromosome 15 with a rearrangement in the short arm (Figs. 3, 5 b). Evidently, in L 540 part of chromosome 14, containing genes for TCR alpha, is translocated to the short arm of chromosome 15, where genes for the ribosomal RNA are located (Fig. 6).

A second interesting gene translocation in L 540 has been detected by in situ hybridization using a DNA probe from the *met* oncogene. In addition to the labeling in *7q21-31*, a chromosome region to which *met* oncogene has been mapped (Dean et al. 1985), we found an abundance of silver grains on marker chromosome XI (Fig. 7).

This marker is composed of a chromosome 21 and part of the long arm of chromosome 7 (Fig. 8). As in marker IX with the rearranged TCR alpha-chain genes, again in the marker XI a chromosome region is involved in the translocation, which carries genes for the ribosomal RNA, the so-called nucleolus organizer region (NOR). In both markers, positive silver (Ag. NOR) staining was proof of the activity of these regions in nucleolus organization (Fig. 9). Thus, cell line L 540 is

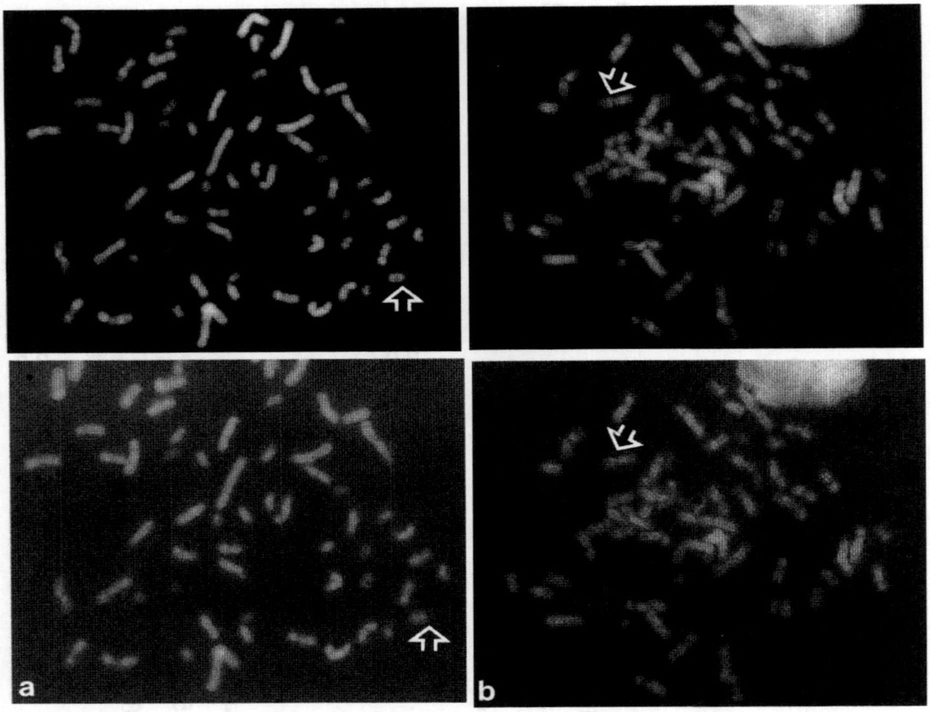

Fig. 5 a, b. In situ hybridization of metaphases from the cell line L 540 using a TCR alpha-specific cDNA probe. *Upper row:* quinacrine mustard-stained chromosomes in incident ultraviolet light; *lower row:* the same metaphases with the grains made visible by simultaneously admitting visible light to the field. **a** *Arrow* indicates chromosome region *14q11* where TCR alpha-chain genes have been mapped. Seventeen out of 294 grains (5.7%) in 80 metaphases were found in this region. **b** *Arrow* indicates the centromere region of marker chromosome IX. Twelve out of 294 grains (4%) in 80 metaphases were found in this region

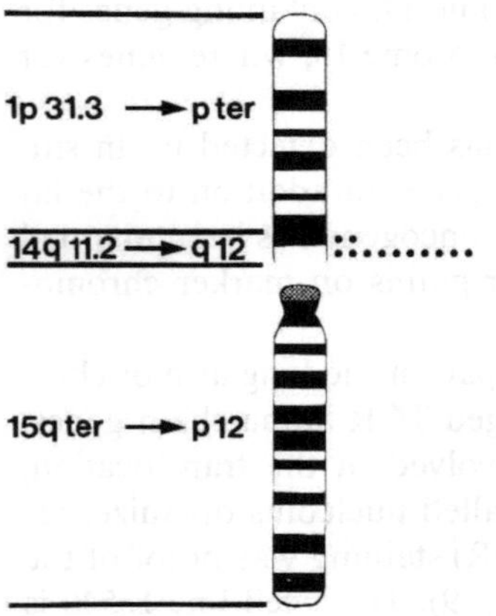

Fig. 6. Idiogram of marker chromosome IX, which can be interpreted as being composed of a chromosome 15, part of chromosome 14 and part of the short arm of chromosome 1. The distribution of silver grains (12/294) in 80 cells is shown

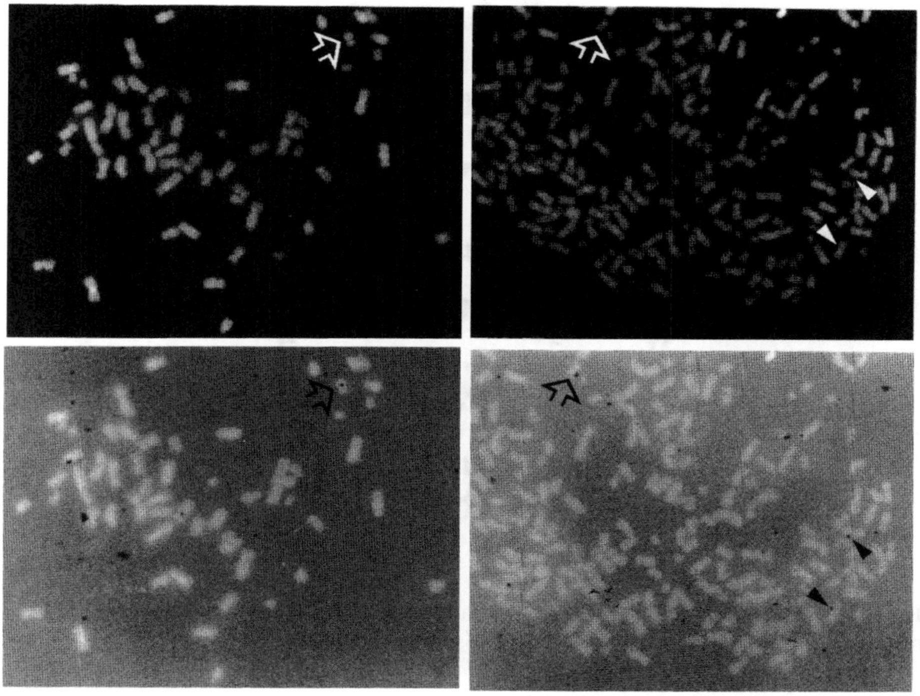

Fig. 7. In situ hybridization of partial metaphases from L 540 using a *met*-oncogene-specific DNA probe. *Upper row:* quinacrine mustard-stained chromosomes in incident ultraviolet light; *lower row:* the same metaphases with the grains made visible by simultaneously admitting visible light to the field. *Arrow* (▲) indicates chromosome region *7q21-31* where the *met* oncogene has been mapped. *Arrow* (⇧) indicates the marker chromosome XI. Twelve out of 67 grains (17.9%) in 17 metaphases were found in *7q21-31* and 9 out of 67 grains (13.4%) were located in the short arm of marker XI

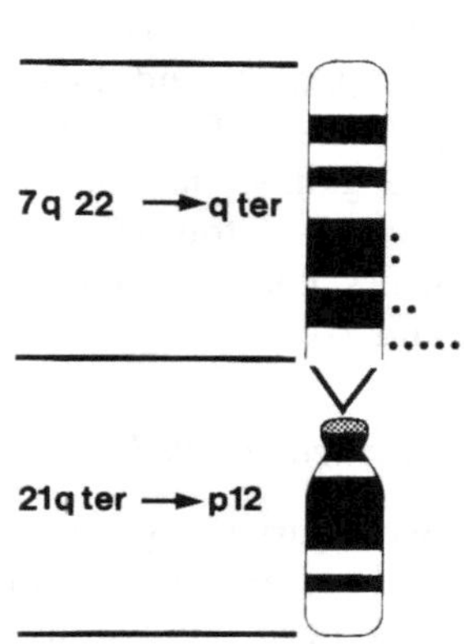

Fig. 8. Idiogram of marker chromosome XI, which can be interpreted as being composed of a chromosome 21 and a segment of a chromosome 7 *(7q22→qter)*. The distribution of silver grains (9/67) in 17 cells is shown

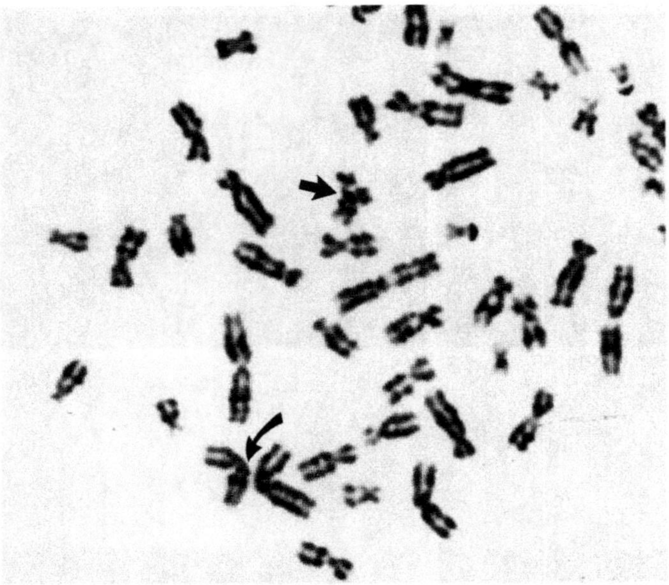

Fig. 9. Partial metaphase of L 540 after G-banding and Ag. NOR staining. Silver-positive regions in marker chromosomes IX (↙) and XI (↑) are indicated by *arrows*

characterized by two chromosomal rearrangements involving regions which contain active ribosomal DNA.

It can be speculated that the spatial connection of genes encoding the alpha-chain of TCR and *met* oncogene to active genes for ribosomal RNA provides the control of expression of TCR alpha and/or *met* oncogene by rDNA-specific promoter or enhancer sequences. Expression of *met* oncogene in L 540 has been demonstrated by Tesch (personal communication).

The significance of rDNA rearrangements in HD-derived cells is furthermore underlined by the detection of aberrations of nucleolus organizer regions bearing chromosomes in HD-derived cell lines L 428 and L 439 (Fonatsch et al. 1986), as well as in Jones' line Co (Jones et al. 1985; Fonatsch unpublished) and in Kamesaki's line KM-H2 (Kamesaki et al. 1986) and in primary material of one of our HD patients and three patients of Hossfeld and Schmidt (1978). If future investigations confirm these preliminary observations, it could be concluded that in Hodgkin's lymphoma genes, active in nucleolus organization, play a role in the control of transcription of normally *silent* genes, as for example of protooncogenes, comparable to the role of immunoglobulin and TCR chain genes in B-cell and T-cell malignancies.

Investigations in Lymphoblastoid Cell Lines (LCLs) from HD Patients

Secondary malignancies in the form of acute nonlymphocytic leukemias (ANLLs) and non-Hodgkin's lymphomas occur frequently – in up to 10% – in patients who have been successfully treated for HD (Anonymous 1985). The secondary leukemias are characterized by specific chromosome abnormalities, mainly concerning chromosomes 5, 7, 17, and 11 (De Braekeleer 1986; Pedersen-Bjergaard and Philip 1987b). The question arose whether a genetically determined chromosome instability, possibly caused by a repair deficiency, is inherent in patients with HD. Elevated rates of spontaneous chromosome aberrations and of sister chromatid exchanges, as have been observed in T- and B-lymphocytes of untreated HD patients, can be taken as an indication for a chromosome instability (Fonatsch et al. 1981 and unpublished observations).

Proceeding from these findings, we tested whether in lymphoblastoid cell lines from patients with HD *specific* chromosome aberrations can be induced in vitro by those cytostatics that are commonly used in HD therapy - and whether in vitro induced chromosome aberrations correspond to those found in secondary leukemias after HD. Furthermore, we tried to clarify whether these aberrations are specific for certain *agents* and whether they occur *only* in HD-derived LCLs or in control LCLs as well.

The LCLs from patients with HD and healthy controls have been established by Kirchner in Hannover and were treated in vitro by several cytostatic drugs in long-term and short-term experiments. The first results of our long-term experiments are shown in Table 2. In L 55, derived from an HD patient, 3 out of 100 untreated metaphases show sporadic anomalies in chromosomes 2, 11, and X. After treatment with activated cyclophosphamide over 3 months, 7 out of 50 metaphases revealed chromosome aberrations, 6 of them showing an identical clonal transloca-

Table 2. Cytogenetics of an LCL from a patient with HD (long-term experiments)

Cell line, in vitro treatment	Number of metaphases analyzed	Number of metaphases with aberrations	Clonal aberrations	Number of metaphases with clonal aberrations
L 55 untreated	100	3	–	–
L 55 CP activated Five courses (10–40 μg/ml/2 h) 25.10. 85 to 27.01. 86	50	7	*t(1;17)(q23;p13)*	6
L 55 BLM Five courses (50–100 μg/ml/2 h) 02.08. 85 until 27.01. 86	100	31	*t(3;5)(q27/29;q31),* *t(7;9)(p15;q34),* *1q+(q42/44)(1)* *t(3;5)(q27/29;q31),* *t(1;11)(q21/23;q23/25)(7)* Additional anomalies to *t(3;5)* and *t(1;11)*: *t(16;?)(q22/24;?)(2)* and *t(16;?)(q22/24;?),* *t(2;10)(q13/14;q11.2),* *dup(7)(q34→q36)(1)*	11

CP, cyclophosphamide; BLM, bleomycin.

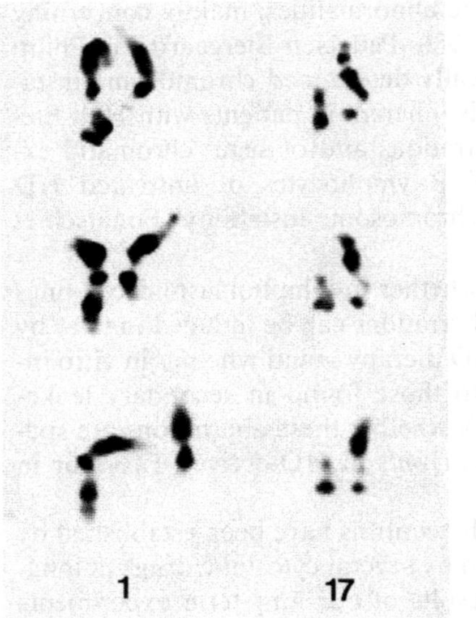

Fig. 10. Clonal chromosome translocation *t(1;17)(q21/23;p13)* in an HD-derived LCL after long-term in vitro treatment with cyclophosphamide. In each pair the normal homologue is *on the left*, the translocated chromosome *on the right*

tion between the long arm of chromosome 1 and the short arm of chromosome 17 (Fig. 10). The chromosome regions *1q21/23* as well as *17p13* are often involved in structural aberrations in ANLL and non-Hodgkin's lymphomas (Mitelman 1985). An identical translocation has been described in ANLL of FAB type M4 and in preleukemia (Mitelman 1985). *17p13* is known to be frequently rearranged in secondary leukemias (Pedersen-Bjergaard and Philip 1987 a).

After bleomycin treatment over 6 months, 31% of the metaphases showed structural aberrations (Table 2): 20 cells had sporadic anomalies, but 11 belonged to a clone characterized by a translocation *t(3;5)(q27/29;q31)* and a duplication in *1q* or a translocation of this duplicated *1q* to *11q23/25* (Fig. 11 a, b).

All the regions whose aberrations led to clonal proliferation in L 55 are well-known and consistent breakpoints in acute nonlymphocytic leukemias, myelodysplastic syndromes (MDS), and non-Hodgkin's (non-HD) lymphomas (Mitelman 1985; Table 3). Moreover, protooncogenes and genes for growth factors and cell differentiation are localized in these regions (Human Gene Mapping 1987; Pedersen-Bjergaard and Philip 1987 a).

Other chromosome segments which were found to be frequently rearranged in long-term as well as in short-term experiments with HD-derived LCLs are the long arm of chromosome 7 and different bands in chromosome 2 and *16q22/24* (Table 3). These bands are also affected in malignancies which can occur after HD treatment.

Data obtained from control LCLs (derived from healthy donors) in short-term experiments suggest that bleomycin may induce aberrations in the long arms of chromosomes 5, 3, and 11 (Table 3). Therefore, these chromosome regions should be defined as agent-specific rather than as disease-specific hot spots. But these preliminary results remain to be confirmed by long-term experiments. Summarizing our data, it becomes evident that the percentage of abnormal metaphases is

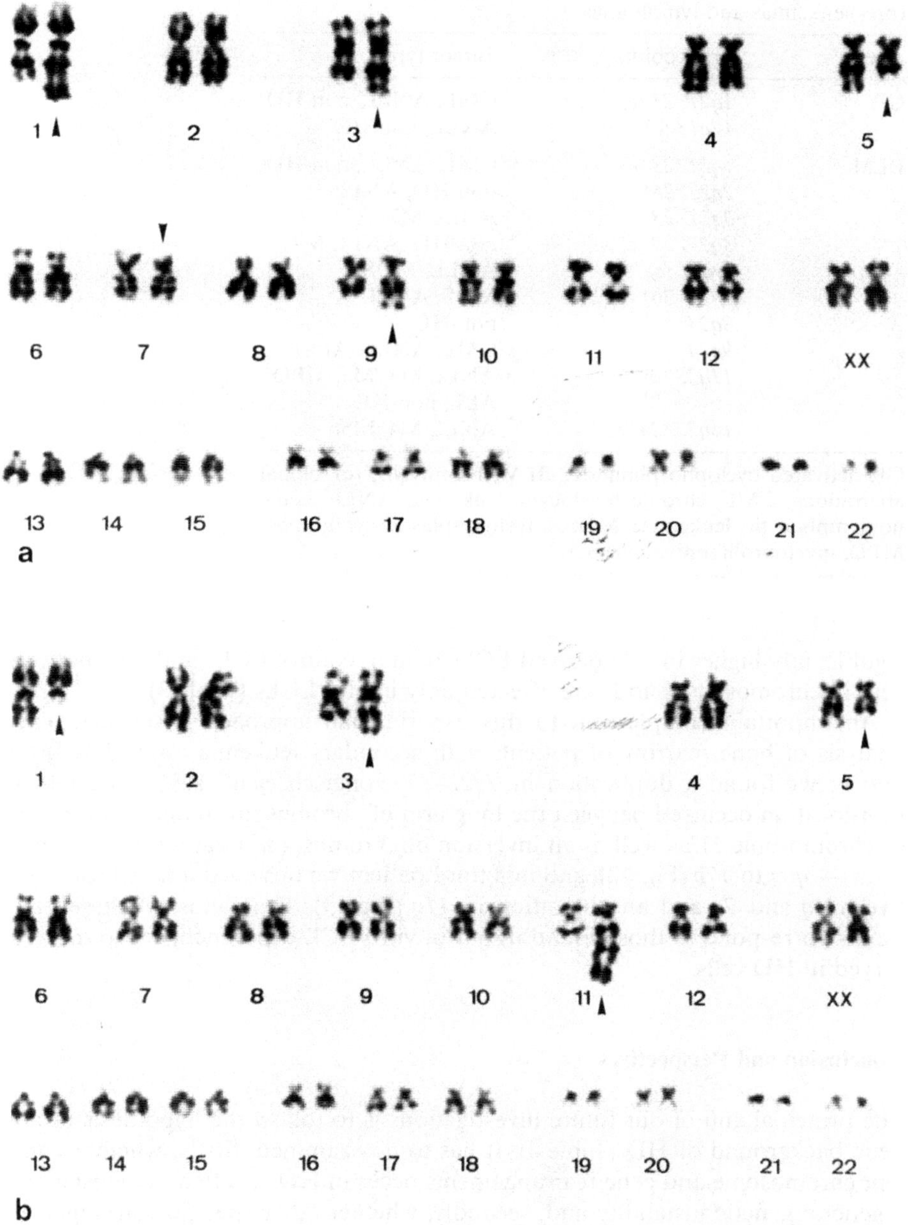

Fig. 11 a, b. Clonal chromosome aberrations in an HD-derived LCL after long-term in vitro treatment with bleomycin. **a** Karyotype, showing translocations *t(3;5)(q27/29;q31)* and *t(7;9)(p15;q34)* as well as a duplication in the long arm of chromosome 1. **b** Karyotype, showing the same translocation *t(3;5)* as in **a** as well as a translocation of the duplicated *1q* to *11q23/25*

Table 3. Breakpoints frequently found in LCLs after in vitro treatment and consistently involved in marker formation in (pre-)leukemias and lymphomas

Agent	Breakpoints	Tumor type
CP	*1q21/23 (c)*[a]	CML, ANLL, non-HD
	17p13 (c)[a]	ANLL, non-HD
BLM	*1q21/23 (c)*	CML, ANLL, non-HD
	2p23/25[a]	non-HD, ANLL
	3q21/23	ANLL M7
	3q27/29 (c)[a]	non-HD, ANLL M7
	5rq31 (c)	ANLL, MDS
	7q22–36[a]	MDS, ANLL
	8q24[a]	non-HD
	9q34[a]	CML, ANLL, ALL
	11q23 (c)	ANLL M4, M5, MPD, ALL, non-HD
	16q22/24 (c)[a]	ANLL M4, M5b

CP, activated cyclophosphamide; BLM, bleomycin; (c), clonal aberrations, CML, chronic myelocytic leukemia; ANLL, acute nonlymphocytic leukemia; MDS, myelodysplastic syndromes; MPD, myeloproliferative diseases.
[a] Chromosome bands only affected in HD-derived LCLs.

significantly higher in HD-derived LCLs than in control LCLs and, secondly, that certain chromosome bands are affected only in HD LCLs (Table 3).

An important complement to this experimental approach is the cytogenetic analysis of bone marrow of patients with secondary leukemia after HD: In one patient we found a duplication in *7q22–32* (Fonatsch et al. 1986), in another a translocation occurred between the long arm of chromosome 6 and the short arm of chromosome 21 as well as an inversion of chromosome 17 and translocation of *8q24 → qter* to *17p* (Fig. 12), and in a third patient we observed a translocation between *2q* and *7q* and an aberration in *11q* (Fig. 13). Thus, most of these breakpoints correspond to those found in our in vitro LCL experiments or to those observed in HD cells.

Conclusion and Perspectives

The principal aim of our future investigations is to follow the hypothesis of a genetic background of HD (Table 4). It has to be examined, firstly, whether consistent chromosome and gene rearrangements occur in HD as a first manifestation of a general genetic instability and, secondly, whether other specific rearrangements are induced by therapy leading to secondary malignancies in different organ systems and representing a second manifestation of this putative genetically determined chromosome vulnerability.

A genetic component in the etiology of HD is discussed in accounts of the familial cases, the increased chromosome instability observed in lymphocytes of

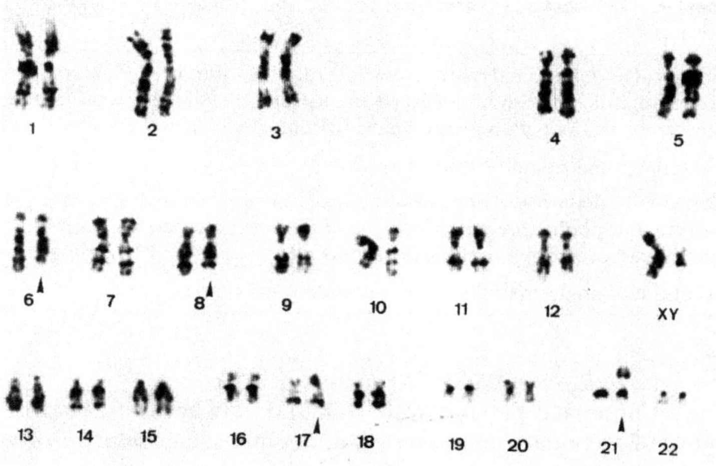

Fig. 12. G-banded karyotype of a bone marrow metaphase from a patient with secondary ANLL (FAB type M4) after HD, showing a translocation *t(6;21)(q23;p12)*, a pericentric inversion of a chromosome 17 – *inv(17)(p13q21.3)* –, as well as a translocation of *8q24→qter* to the short arm of the inverted chromosome 17

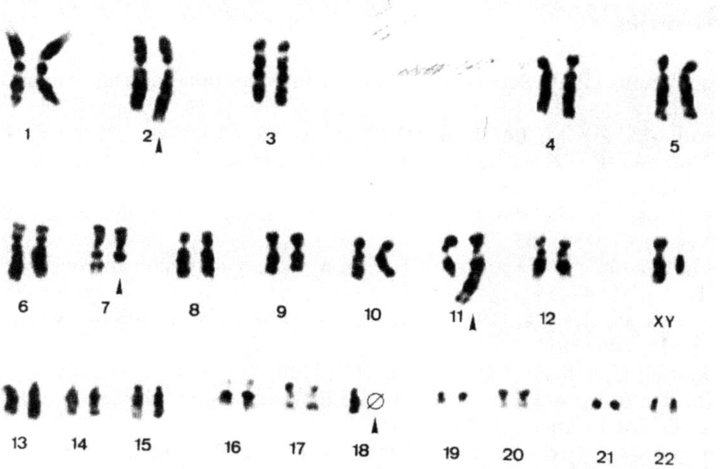

Fig. 13. G-banded karyotype of a bone marrow metaphase from a patient with secondary ANLL (FAB type M4) after HD, showing a translocation *t(2;7)(q33;q22)* and a translocation of a chromosome 18 to *11q23/25*

Table 4. Hypothesis of a common genetic background of Hodgkin's disease and secondary neoplasia

Genetically determined chromosome instability (repair deficiency?) (=putative first hit), predominantly in "Sternberg-Reed precursor cells," leads to an increased susceptibility to exogenous or endogenous mutagenic/carcinogenic agents (=putative second hit):
First malignant manifestation: Hodgkin's disease
Genetically determined chromosome instability (=putative first hit), predominantly in myeloid (or lymphoid) precursor cells, leads to an increased vulnerability by cytostatic agents and/or radiotherapy (=putative second hit):
Second malignant manifestation: secondary neoplasias

several untreated patients with HD, and, last not least, because of the high incidence of secondary neoplasias. But, doubtless, a genetic factor would only be part of the great mosaic picture which makes up the heterogeneous disease entity of Hodgkin's lymphoma.

Acknowledgments. We are grateful to Ms. Ulrike Kolbus, Ms. Ulrike Koop, and Ms. Nicole Erbe for excellent technical assistance and to Ms. Marianne Kock for skilful secretarial help.

References

Anonymous (1985) Second malignancies in lymphoma patients. Lancet II: 1163-1164 (editorial)

Dean M, Park M, Le Beau MM, Robins TS, Diaz MO, Rowley JD, Blair DG, Vande Woude GF (1985) The human *met* oncogene is related to the tyrosine kinase oncogenes. Nature 318: 385-388

De Braekeleer M (1986) Cytogenetic studies in secondary leukemia: statistical analysis. Oncology 43: 358-363

Fleischmann T, Krizsa F (1977) Chromosomes in malignant lymphomas. Haematologia Budap 11: 47-55

Fonatsch C, Gradl G (1987) Chromosomenaberrationen und maligne Lymphome. Onkologie 10: 206-216

Fonatsch C, Schaadt M, Kirchner HH, Diehl V (1980) A possible correlation between the degree of karyotype aberrations and the rate of sister chromatid exchanges in lymphoma lines. Int J Cancer 26: 749-756

Fonatsch C, Kirchner HH, Schaadt M, Günzel M, Diehl V (1981) Sister chromatid exchange in lymphoma lines and lymphoblastoid cell lines before and after heterotransplantation into nude mice. Int J Cancer 28: 441-445

Fonatsch C, Diehl V, Schaadt M, Burrichter H, Kirchner HH (1986) Cytogenetic investigations in Hodgkin's disease: I. Involvement of specific chromosomes in marker formation. Cancer Genet Cytogenet 20: 39-52

Fonatsch C, Gradl G, Ragoussis J, Ziegler A (1987) Assignment of the TCP1 locus to the long arm of human chromosome 6 by in situ hybridization. Cytogenet Cell Genet 45: 109-112

Fukuhara S, Rowley JD (1978) Chromosome 14 translocations in non-Burkitt lymphomas. Int J Cancer 22: 14-21

Gödde-Salz E (1984) Chromosomenaberrationen bei Lymphomen und Lymphom-ähnlichen Erkrankungen. Thieme, Stuttgart

Hossfeld DK, Schmidt CG (1978) Chromosome findings in effusions from patients with Hodgkin's disease. Int J Cancer 21: 147-156

Human Gene Mapping 9 (1987) Ninth International Workshop on Human Gene Mapping. Frézal J, Klinger HP (eds) Cytogenet Cell Genet 46: 1-4

Jones DB, Scott CS, Wright DH, Stein H, Beverley PCL, Payne SV, Crawford DH (1985) Phenotypic analysis of an established cell line derived from a patient with Hodgkin's disease (HD). Hematol Oncol 3: 133-145

Kamesaki H, Fukuhara S, Tatsumi E, Uchino H, Yamabe H, Miwa H, Shirakawa S, Hatanaka M, Honjo T (1986) Cytochemical, immunologic, chromosomal, and molecular genetic analysis of a novel cell line derived from Hodgkin's disease. Blood 68: 285-292

Lawler SD, Swansbury GJ (1987) Cytogenetic studies in Hodgkin's disease. In: Selby P, McElwain TJ (eds) Hodgkin's disease. Blackwell Scientific, Oxford, pp 43-67

Mitelman F (1985) Catalog of chromosome aberrations in cancer, 2nd edn. Alan R Liss, New York

Pedersen-Bjergaard J, Philip P (1987a) Chromosome characteristics of therapy-related acute non-lymphocytic leukemia and preleukemia: possible implications for pathogenesis of the disease. Leuk Res 11: 315-318

Pedersen-Bjergaard J, Philip P (1987b) Cytogenetic characteristics of therapy-related acute nonlymphocytic leukaemia, preleukaemia and acute myeloproliferative syndrome: correlation with clinical data for 61 consecutive cases. Br J Haematol 66: 199-207

Poppema S, de Jong B, Atmosoerodjo J, Idenburg V, Visser L, de Ley L (1985) Morphologic, immunologic, enzyme histochemical and chromosomal analysis of a cell line derived from Hodgkin's disease. Cancer 55: 683-690

Reeves BR (1973) Cytogenetics of malignant lymphomas. Studies utilising a Giemsa-banding technique. Hum Genet 20: 231-250

Reeves BR, Pickup VL (1980) The chromosome changes in non-Burkitt lymphomas. Hum Genet 53: 349-355

Watson DK, McWilliams-Smith MJ, Kozak C, Reeves R, Gearhart J, Nunn MF, Nash W, Fowle III JR, Duesberg P, Papas TS, O'Brien SJ (1986) Conserved chromosomal positions of dual domains of the *ets* protooncogene in cats, mice, and humans. Proc Natl Acad Sci USA 83: 1792-1796

Zech L, Haglund U, Nilsson K, Klein G (1976) Characteristic chromosomal abnormalities in biopsies and lymphoid cell lines from patients with Burkitt and non-Burkitt lymphomas. Int J Cancer 17: 47-56

Hodgkin's Disease Derived Cell Lines

Biology of Hodgkin Cell Lines

M. Schaadt, H. Burrichter, M. Pfreundschuh, E. Schell-Frederick, H. Tesch, C. Fonatsch, H. Stein, and V. Diehl

Medizinische Klinik I, Universität Köln, Joseph-Stelzmann-Straße 9, 5000 Köln, FRG

Introduction

Studies to characterize Hodgkin and Reed-Sternberg cells in histological sections have yielded many conflicting data over the years. Even with progress in immunohistological technology, the histogenetic derivation of these cells remains unclear. Functional and genetic investigations are hampered by the scarcity of tumor cells in Hodgkin's disease biopsy samples and the admixture of a variety of nonmalignant cells.

The establishment of in vitro cell lines may help to overcome these limitations. Tumor cell lines serve as a constantly available source of clonally derived material for the characterization of the tumor cell and open up the possibility of investigating functional properties of these cells.

There are several reports of short- or long-term cultures from Hodgkin's disease tissue, but many could not be maintained in vitro over a longer period, have not been sufficiently characterized, or have turned out not to be Hodgkin derived. Since 1980, our group and other investigators have reported on cell lines that have been claimed to be Hodgkin specific (Table 1). These lines are still in culture and available for further investigation. In this contribution we summarize the investigations performed with four Hodgkin-derived cell lines established in our laboratory.

Table 1. Hodgkin-derived cell lines

Cell line	Reference
KM-H2	12
HDLM-2	6
DEV	19
CO	11
SU/RH-HD1	17
L 538, L 540, L 591	5
L 428	21

Recent Results in Cancer Research. Vol. 117

Table 2. Hodgkin-derived cell lines

	L 428[a]	L 439	L 538/L 540	L 591
Histology	NS	NS	NS	NS
Stage	IVB	IVB	IVB	IVB
Material	Pleural effusion	Ascites	Peripheral blood/ bone marrow	Pleural effusion

NS, Nodular sclerosing subtype.
[a] Sublines: L 428 KS, L 428 KSA.

Results and Discussion

Cell Lines

Originally, we established five cell lines (Table 2). L 439 died after 7 months, while the other four are still in culture. L 538 and L 540 are identical; they are derived from the same patient, one from bone marrow and the other from the peripheral blood. All five lines originate from patients with histologically proven Hodgkin's disease (HD), all of whom showed the nodular sclerosing subtype. In addition, we have two sublines of L 428. L 428 KS is a calf serum-adapted variant, and L 428 KSA has been induced by tetradecanylphorbolacetate (TPA) treatment. In contrast to all other lines which grow as suspension cultures, L 428 KSA is plastic adherent and grows as a monolayer.

Cytogenetics

The malignant origin of the cell lines can be shown by the demonstration of chromosomal aberrations. Each line exhibits a monoclonal pattern of chromosomal abnormalities, but the aberrations differ between the lines. The chromosome region found to be most frequently involved was the long arm of chromosome 7 from *q22* to *q36* (L 428, L 439, L 591). Other regions involved in the formation of marker chromosomes were *1p22, 2q33, 11q21/23, 14q32, 15p12,* and *21q21*. However, there was no Hodgkin-specific marker chromosome [9].

Heterotransplantation

After transplantation into nude mice, L 538/540 was the strongest line to induce tumors (Table 3). L 428 was also tumorigenic, but only after intracranial inoculation or after being embedded in a fibrin clot. This rather weak tumorigenicity has also been observed with other tumor cell lines from Hodgkin's disease (H. Kamesaki, personal communication), as well as from other hematological malignancies [14]. In contrast to the other lines, L 591 did not grow in mice. This is a really surprising finding, since in our hands even Epstein-Barr virus (EBV)-transformed lymphoblastoid cell lines are capable of inducing tumors after intracranial inoculation [20].

Table 3. Hodgkin-derived cell lines: heterotransplantation into nude mice

Inoculation	L 428	L 538/L 540	L 591
Intracranial	5/6[a]	24/25	0/8
Subcutaneous (suspension)	0/13[b]	23/25	0/7

[a] Positive attempts.
[b] 1/9 positive in fibrin clot.

Marker profile

The four cell lines differ in their expression of most antigens and coincide only in the expression of a few markers usually found on activated cells. They all express HLA-DR, the interleukin 2 receptor, and react with Hodgkin-associated antibodies (CD30) produced against L 428. The following reactions were seen:

- L 428 shows only a weak reaction with B 4 (CD 19) in some cells and is negative for other markers of the B-cell, T-cell, and macrophage lineage listed in Table 4.

Table 4. Hodgkin-derived cell lines: antibody staining

	L 428	L 538/L 540	L 591
Anti-B-cell			
Ig	–	–	+(IgA_λ)
To 15/CD 22	–	–	+
B 1/CD 20	–	–	+
B 4/CD 19	(+)	–	+
EBNA	–	–	+
Anti-T-cell			
T 1/CD 5	–	–	–
T 2/CD 7	–	–	–
T 3/CD 3	–	–	–
T 11/CD 2	–	+	+
T 4/CD 4	–	+	–
T 8/CD 8	–	–	–
Antimacrophage			
Antilysozyme	–	–	–
MO 1/CD 11 b	–	+	(+)
MO 2/CD 14	–	+	(+)
Ki-M 1	–	(+)	–
Various			
Ki-1, HS-1, 2/CD30	+	+	+
Anti TAC/CD 25	(+)	+	+
HLA-DR	+	+	+
Anti-TdT	–	–	–
CALLA/CD 10	–	–	–

- L 538/540 exhibits some T-cell antigens (T 4/CD 4 and T 11/CD 2), as well as some macrophage-associated antigens. However, antilysozyme staining was negative.
- L 591 is positive for B-cell markers, in particular for immunoglobulin production and expression of the EBV-associated nuclear antigen (EBNA). In addition, the line expresses T 11/CD 2 and shows weak reactions with two macrophage-related antibodies.

With the exception of L 591, which expresses major markers of the B-cell line, the assignment of the other lines to a defined cell lineage remains open after phenotypical analysis. However, markers of activated cells are expressed in common. Since the reactivity of most monoclonal antibodies is defined with resting cells, the characterization of activated cells by phenotypical analysis remains crucial.

The other Hodgkin cell lines so far reported also express one or more activation antigens and, in addition, T- or B-cell-related antigens. Only KM-H2 [12] is devoid of most other markers comparable to the features of L 428. In contrast to the other HD lines, SU/RH-HD 1 [17] has been reported to exhibit major macrophage markers (nonspecific esterase, phagocytosis).

Molecular Biology

Since immunophenotyping did not clarify the cellular origin of the in vitro lines, we expected more evidence from the hybridization experiments. In fact, all four lines show rearranged genes coding either for immunoglobulins or for the T-cell receptor [7] (Table 5):

- *L 428* cells have one Ig heavy-chain allele rearranged to C_γ and transcribed into RNA, while the second is deleted. In addition, a κ-light-chain rearrangement has been shown. These data support derivation from the B-cell lineage.

Table 5. Hodgkin cell lines: compilation of DNA and RNA hybridization results

	J_H	C_μ	C_γ	J_κ	C_κ	C_λ	TcR_α	TcR_β	TcR_γ
L 428									
DNA	r/−	−/−	r/−	gl/gl/r[a]	gl/gl/−[a]	gl/gl	gl	gl	gl/gl
RNA	n.d.	−	+	n.d.	−	−	+[b]	−	−
L 540									
DNA	gl/gl	gl/gl	gl/gl	gl/gl	gl/gl	gl/gl	r	r	r/r
RNA	n.d.	−	−	n.d.	−	−	+	−	−
L 591									
DNA	r/−	−/−	−/?	gl/gl	gl/gl	gl/r	n.d.	n.d.	n.d.
RNA	n.d.	−	−	n.d.	+	+	n.d.	n.d.	n.d.

r, allele rearranged; gl, allele germline; −, not detectable, negative; +, positive; n.d., not done.

[a] L 428 cells have three chromosomes #2.

[b] Short transcript.

- *L 540* exhibits rearrangements of genes coding for the α-, β-, and γ-chain of the T-cell receptor and, in addition, α-chain transcripts. These are strong arguments for assigning this cell line to the T-cell lineage.
- In *L 591*, genes for heavy chains and light chains are rearranged with transcription into RNA. In addition, we demonstrated the gene product, IgA λ. There is no doubt about the B-cell nature of this line.

The incomplete expression of the B-cell receptor genes in L 428, or the T-cell receptor genes in L 538/540, may point to an early differentiation stage of these cells. However, this is in contrast to the finding of immunological markers of activated cells. Falk et al. [7] have proposed a virus infection of immature cells as an explanation for these contradictory results. In the cell lines in question (L 428, L 538/540), however, virus-specific proteins have not been detected. Thus, the state of differentiation of the cell lines remains unclear.

In the HD cell lines reported from other laboratories, rearrangements of genes coding for immunoglobulin heavy chains (DEV, KM-H2) or for the T-cell receptor (CO, HDLM-2) have also been demonstrated. We have no hybridization results for the line SU/RH-Hd 1.

Genetic expression of the B- and T-cell receptor has been obtained not only from clearly defined lymphocytes but also from myeloid-leukemia cells in up to 10% of cases [4]. But, since all cell lines which are accepted as Hodgkin derived show gene rearrangements typical of B or T cells, we have good evidence for the lymphocytic derivation of these in vitro cells.

Anti-HD Monoclonal Antibodies

A number of monoclonal antibodies have been developed against the HD cell lines. Some of them are well characterized. The antibody Ki-1 was produced by Schwab et al. [23] in Kiel and has been characterized by Stein et al. [24]. HRS-1 and HRS-2 antibodies were developed in our laboratory [18]. All three antibodies are very similar in their reactivity. They stain H and SR cells in HD biopsy samples and neoplastic cells in some non-Hodgkin's lymphomas. The antigen (CD30) has been characterized as a 120-kD protein which is expressed on activated cells such as activated lymphocytes or macrophages. The three antibodies seem not to detect the same epitope, as shown by blocking tests; therefore, a combined application of all three monoclonal antibodies may be of diagnostic and therapeutic value.

Functional Studies

Immunohistological techniques allow the demonstration of antigens in histological sections from biopsy material. For functional tests, however, single cell suspensions or cell lines are needed in most cases. L 428 could be shown to act as an antigen-presenting cell and is a potent stimulator of the mixed lymphocyte reaction [8].

We and other groups have found that our HD lines produce a variety of biological mediators. The production of colony-stimulating factors (GM-CSF) was demonstrated. In the murine assay, 80%–90% of the colonies were positive for peroxidase staining, indicating predominantly the release of g-CSF [3]. The production of Interleukin 1 by the HD cell lines was detected in an human IL-1 assay as well as in the standard mouse thymocyte assay [13].

It is well known that T-lymphocyte-function is defective in HD patients. Bieber reported an E-rosette inhibiting lipoprotein isolated from HD-spleens [1]. Our observations show that the cell lines release a factor suppressing the rosette-forming capacity of about 50% of T cells without blocking the reaction with the OKT 11 antibody. After passage over a Biogel A column, we found peaks of activity in the range of 12.5, 25, 50, and 100 kD. One might speculate that the rosette-inhibiting factor (RIF) interferes with the alternate pathway of T-cell activation via the CD2-receptor. This will be investigated after further purification of the factor.

Since all Hodgkin cell lines from our laboratory originated from patients with the nodular sclerosing subtype of HD, the question arises whether a fibroblast-activating factor is released by the tumor cells. We demonstrated an activity in the supernatant of the cell lines which stimulates thymidine uptake of 3-T-3 fibroblasts in a dose-dependent fashion. This is in line with the findings of Newcom et al. [16], who described a β-transforming growth factor released by L 428.

Another cytokine, Hodgkin-derived leukocyte factor (HDLF), partially purified and characterized by our group from 428 KSA, inhibits random and directed migration of normal human neutrophils. Compared controls, preincubation of neutrophils with supernatant of L 428, 428 KSA, and L 540 reduces spontaneous migration as well as directed migration to different chemoattractants. This factor appears to be different from previously described neutrophil migration inhibitory factors [22]. Additional mediators have been identified in the supernatant of other HD lines.

These biological factors are not useful for clarifying the histogenetic derivation of Hodgkin and Reed-Sternberg cells, since a variety of cell types have been shown to release one or more biological mediators. But the presence of active factors in the supernatant of HD cell lines supports the hypothesis that Hodgkin tumor cells can interfere with the immunological and hematopoetic regulation network. This may possibly account for the particular histological picture of Hodgkin's lymphomas surrounding, with a majority of nonmalignant cells a minority of tumor cells.

Conclusions and Prospects

Based on molecular biological studies, the HD cell lines established by our group, as well as those from other investigators, can be assigned to the B-lymphocytic (L 591, L 428, DEV, KM-H 2) or T-lymphocytic lineage (L 538, L 540, CO, HDLM-2). Immunostaining results were not unequivocal but did not argue in general against a lymphocytic derivation of these cells. Only one cell line (Su/RH-HD 1) exhibits major phenotypical and functional properties of the macrophage lineage. However, molecular genetic data from this particular cell line are not available. To prove the nonlymphocytic origin of these cells, the lack of rearrangements of genes coding for B- or T-cell receptors should at least be demonstrated.

Since all but one of the in vitro lines have been established from the nodular sclerosing subtype, conclusions can be drawn only for that histological entity. Moreover, one has to bear in mind that the artifical in vitro system might have growth conditions particularly advantageous for lymphocytes. Thus, we conclude from the analysis of cell lines that malignant transformation of lymphatic cells can give rise to nodular sclerosing Hodgkin's disease. The state of differentiation of in vitro Hodgkin cells is unclear, since the incomplete expression of B- and T-cell receptors contrasts with the demonstration of activation antigens on the membrane. Whether Hodgkin tumor cells may originate from other cell types such as macrophages or reticulum cells remains open.

Only a few tumor cell lines from Hodgkin's disease, predominantly of the nodular sclerosing type, have been established in several hundred attempts. Further development of tissue cultures and the use of new generations of immunodeficient animals for xenotransplantation might open up new prospects for the experimental propagation of Hodgkin and Reed-Sternberg cells. In addition, it seems reasonable to go back to single cell preparations from primary tumor material. New separation techniques such as counterflow centrifugation in combination with different steps of density sedimentation should allow the enrichment of Hodgkin and Reed-Sternberg cells to such an extent that reproducible cytogenetic and molecular genetic analysis would become possible.

The particular histological picture of Hodgkin's disease might be due to the interaction of tumor cells with reactive cells of the immune system via biological mediators. The release of CSF, IL-1, Hodgkin-derived growth factor, Hodgkin-derived leukocyte factor, and possibly other hitherto unknown mediators could account for the cytological composition and the architecture of Hodgkin's lymphomas and, moreover, for some typical clinical features. IL-1 is known to be one of the most powerful pyrogens. RIF is capable of interacting with CD 2, which has recently been shown to play a role in an alternative pathway of T-cell activation.

Release of biological mediators is not specific for Hodgkin tumor cells. In addition, each lymphokine does not act in isolation, but is part of the framework of regulation of cell proliferation and function. Thus, specificity does not imply the production of a particular mediator, but the release of a specific combination of biological factors. Differences in composition may be responsible for different histological and clinical features.

Since production of proteins is highly dependent on sophisticated growth conditions, artificial systems such as in vitro tissue cultures or xenotransplantation in mice are inadequate tools to investigate details of this system. Interest must go back to primary tumor tissue. In the last decade, we have obtained a lot of information from tissue culture data concerning the origin and function of Hodgin tumor cells. Now, the hypotheses based on these investigations have to be verified and specified in primary tissue.

Acknowledgements. This work was supported by grants from the Deutsche Forschungsgemeinschaft and Stiftung Deutsche Krebshilfe.

References

1. Bieber MM, King DP, Strober S, Kaplan HS (1979) Characterization of an E-rosette inhibitor in the serum of patients with Hodgkin's disease as a glycolipid. Clin Res 27: 81 A
2. Breitmeyer JB (1987) Lymphocyte activation; how T cells communicate. Nature 329: 760-761
3. Burrichter H, Heit W, Schaadt M, Kirchner HH, Diehl V (1983) Production of colony stimulating factors by Hodgkin cell lines. Int J Cancer 31: 269-274
4. Cheng G, Minden M, Toyonaga B, Mak T, McCullogh A (1986) T-cell receptor and immunoglobulin gene rearrangements in acute myelobastic leukemia. J Exp Med 163: 414-424
5. Diehl V, Kirchner HH, Burrichter H, Stein H, Fonatsch C, Gerdes J, Schaadt M, Heit W, Uchanska-Ziegler B, Ziegler A, Heintz H, Sueno K (1982) Characteristics of Hodgkin's disease-derived cell lines. Cancer Treat Rep 66: 615-632
6. Drexler HG, Gaedicke G, Lok MS, Diehl V, Minowada J (1986) Hodgkin's disease derived cell lines HDLM-2 and L 428. Comparison of morphology, immunological and isoenzyme profiles. Leuk Res 10: 487-500
7. Falk M, Tesch H, Stein H, Diehl V, Jones D, Fonatsch C, Bornkamm G (1987) Phenotype versus immunoglobulin and T-cell receptor genotype of Hodgkin derived cell lines: activation of immature lymphoid cells in Hodgkin's disease. Int J Cancer 40: 262-269
8. Fischer RI, Bostick-Bruton F, Sander DN, Scala G, Diehl V (1983) Neoplastic cells obtained from Hodgkin's disease are potent stimulators of human primary mixed lymphocyte cultures. J Immunol 130: 2666-2670
9. Fonatsch C, Diehl V, Schaadt M, Burrichter H, Kirchner HH (1986) Cytogenetic investigations in Hodgkin's disease: I. Involvement of specific chromosomes in marker formation. Cancer Genet Cytogenet 20: 39-52
10. Giovanella BC, Nilsson K, Zeich L, Yim O, Klein G, Stehlin JS (1979) Growth of diploid, Epstein-Barr-virus-carrying human lymphoblastoid cell lines heterotransplanted into nude mice under immunologically privileged conditions. Int J Cancer 24: 103-113
11. Jones DB, Scott CS, Wright DH, Stein H, Beverly PCL, Payne SV, Crawford DH (1985) Phenotype analysis of an established cell line derived from a patient with Hodgkin's disease (HD). Hematol Oncol 3: 133-145
12. Kamesaki H, Fukuhara S, Tatsumi E, Uchino H, Yamabe H, Miwa H, Shiragawa S, Hatanaka M, Honjo T (1986) Cytochemical, immunologic, chromosomal and molecular genetic analysis of a novel cell line derived from Hodgkin's disease. Blood 68: 285-292
13. Kortmann C, Burrichter H, Jahn G, Diehl V, Peter HH (1984) Interleukin-1 activity constitutively generated by Hodgkin derived cell lines. I. Measurement in a human lymphocyte costimulator assay. Immunobiology 166: 318-333
14. Lozzio BB, Lozzio CB, Machado E (1976) Human myelogenous (Ph^1+) leukemia cell-line; transplantation into athymic mice. J N C I 56: 627-628
15. Meuer S, Hussey R, Fabbi M, Fox D, Acuto O, Fitzgerald K, Hodgdon J, Protentis J, Schlossman S, Reinherz E (1984) An alternative pathway of T-cell activation: a functional role for the 50 kd T 11 sheep erythrocyte receptor protein. Cell 36: 897-906
16. Newcom S, Kadin ME, Ansari AA (1988) Production of transforming growth factor beta activity by Ki-1 positive lymphoma cells and analysis of its role in the regulation of Ki-1 positive lymphoma growth. Am J Pathol 131 (3): 569-577
17. Olsson L, Behnke O, Pleibel N, D'Amore F, Werdelin O, Fry KE, Kaplan HS (1984) Establishment and characterization of a clonal gaint cell line from a patient with Hodgkin's disease. J N C I 73: 809-830
18. Pfreundschuh M, Mommerz E, Meissner M, Feller A, Hassa R, Krueger HR, Diehl V (1988) Hodgkin and Reed-Sternberg cells associated monoclonal antibodies HRS-1 and HRS-2 react with activated cells of lymphoid and monocytoid origin. Anticancer Res 8: 217-224

19. Poppema S, De Jong B, Atmosoerodjo J, Idenburg V, Visser L, De Ley L (1985) Morphologic, immunologic, enzyme histochemical and chromosomal analysis of a cell line derived from Hodgkin's disease. Evidence for a B-cell origin of Sternberg-Reed cells. Cancer 55: 683-690
20. Schaadt M, Kirchner H, Fonatsch C, Diehl V (1979) Intracranial heterotransplantation of human hematopoetic cells in "nude"-mice. Int J Cancer 23: 751-561
21. Schaadt M, Diehl V, Stein H, Fonatsch C, Kirchner H (1980) Two neoplastic cell lines with unique features derived from Hodgkin's disease. Int J Cancer 26: 723-731
22. Schell-Frederick E, Radtke H, Sommer H, Helbing I, Burrichter H, Schaadt M, Diehl V (1988) Inhibition of human neutrophil migration by supernatants from Hodgkin's disease derived cell lines. Eur Clin Invest 18: 290-296
23. Schwab U, Stein H, Gerdes J, Lemke H, Kirchner HH, Schaadt M, Diehl V (1982) Production of a monoclonal antibody specific for Hodgkin- and Sternberg-Reed-cells of Hodgkin's lymphoma and a subset of normal lymphoid cells. Nature 299: 65-67
24. Stein H, Gerdes J, Schwab U, Lemke H, Mason DY, Ziegler A, Schienle W, Diehl V (1982) Identification of Hodgkin and Sternberg-Reed cells as a unique cell type derived from a newly detected small-cell population. Int J Cancer 30: 445-459

Phenotypic and Genotypic Analysis of Two Cell Lines Derived from Hodgkin's Disease Tissue Biopsies

D. B. Jones, A. J. W. Furley, J. Gerdes, M. F. Greaves, H. Stein, and D. H. Wright

Department of Patholgogy, Southampton General Hospital, University of Southamptom, Tremona Road, Southamptom SO6 4XY, Great Britain

Introduction

Although the demonstration of Reed-Sternberg cells (RS) and their mononuclear counterparts (MC) is essential for a histological diagnosis of Hodgkin's disease (HD), the origin of these cell types remains obscure (Jones 1986).

The development of techniques for cell phenotyping in both frozen and paraffin sections (Jones 1986; Norton and Isaacson 1985) has succeeded only in further demonstrating the complexity of the phenotypic pattern exhibited by RS and MC. Further, CD30 antibodies (Stein et al. 1982), initially thought to identify a unique lineage for both RS and MC in HD, clearly identify activated lymphocytes. Cell lines derived from biopsy tissue from patients with HD might enable the cell of origin of the RS and MC populations to be determined and we briefly report phenotypic and genotypic observations obtained from two cell lines derived in our laboratory.

Materials and Methods

Conditions of Culture. The conditions of culture employed for the generation of the two cell lines are described elsewhere (Jones et al. 1985). Briefly, unfractionated, dispersed HD lymph node biopsies were maintained in RPMI 1640 with 10% fetal calf serum added. No other growth supplements were employed.

Antibodies Employed for Phenotypic Analysis. The antibodies employed for the characterisation of the two cell lines are identified in Tables 1–4. For details of these reagents readers are referred to Jones et al. (1985), Stein and Gerdes (1986) and McMichael (1987). Antibody binding was demonstrated either by flourescence-activated cell sorting (FACS) or by immunohistochemical staining of acetone-fixed, cytocentrifuge preparations (Gerdes et al. 1986).

Genotypic Analysis. Immunoglobulin super family gene rearrangements and gene expression were demonstrated by Southern or Northern blot analysis of DNA or RNA extracted from viable cells.

Recent Results in Cancer Research. Vol. 117

Origin of Biopsies

Fresh biopsy material from both patients (Co and Ho) was submitted to the Pathology Department at Southampton General Hospital for phenotypic analysis. Full histological examination was made of formalin-fixed biopsy material obtained from the same lymph node and established a diagnosis of Hodgkin's disease, nodular sclerosis type, for both cases. Karyotypic analysis undertaken at the Wessex Regional Cytogenetics Laboratory has demonstrated that both cell lines are aneuploid with no consistent shared chromosomal abnormality.

Results

Phenotypic Analysis. Tables 1 and 2 present the results of antibody characterisation studies of both Ho and Co. Both cell lines lack phenotypic markers characteristic of the B-cell lineage and immunoglobulin light and heavy chains. Ho shows surface reactivity with CD3 reagents, together with the presence of surface CD4 and CD7. Co also exhibits surface CD7; CD3 cannot be demonstrated on the cell surface but is present in the cytoplasm. Both cell lines are positive when stained with CD30 antibodies. Interestingly, the B-cell blast markers, BB1 and BB2, are present on Co cells, whilst BB2 is also present on the surface of Ho.

Genotypic Analysis. Full and detailed results of the genotypic analysis of both Co and Ho have been partially published (Falk et al. 1987) and will be the subject of a subsequent publication. Briefly (Tables 3, 4), both Co and Ho exhibit a germline pattern for immunoglobulin heavy chain genes. The T-cell receptor (TCR) gamma

Table 1. Phenotypic analysis of two HD cell lines

Specificity	Co	Ho
CD3	c	+
WT31	−	+
TR	+	+
CD1	−	−
CD2	−	−
CD4	−	+
CD8	−	−
CD7	+	+
CD11c (p150:95)	−	−
my9	−	−
my7	−	−
CD15	+	−

c, cytoplasmic; +, positive; −, negative. Compilation of FACS data and cytoprep staining.

Table 2. Phenotypic analysis of two HD cells lines

Specificity	Co	Ho
Ki-1	+	+
Ki-24	–	+
Ki-27	–	–
BERH2	+	+
CD25	–	–
OKT10	–	(+)
MHC class II	–	+
Tu35	–	+
BB1	+	–
BB2	+	+

Both lines negative for:
CD19, 22, 23, k, λ, GMAD, EBNA.

Table 3. Gene rearrangement in two HD cell lines

Gene		Co	Ho
IgH		G	G
TCR gamma		R	R
TCR	C beta 1	D/D	R/R
	C beta 2	R/R	G/G

R, rearranged; G, germline; D, deleted.

Table 4. Gene expression in two HD cell lines

Gene		Co	Ho
IgH		–	–
CD3	Delta	+	nd
	Epsilon	+	nd
TCR	1.3K	+	+
	1.0K	–	–

+, mRNA present; –, mRNA absent; nd, not done.

chain is rearranged in both cases and rearrangements of TCR beta are also present. These studies are consistent with the phenotypic data presented above.

TCR message is present in both cell lines in its 1.3-kd mature form but the encoded CD3 molecule is not inserted into the membrane in Co.

Discussion

From the phenotypic and genotypic data presented it is clear that both Co and Ho exhibit at least some of the characteristics of T-cells. However, Co, in particular, shows an abnormal T-cell phenotype.

Phenotypic investigations of RS and MC show heterogeneity in staining pattern (Doreen et al. 1984; Abdul Azziz et al. 1983; Stein et al. 1982). Recent biopsy studies (Griesser et al. 1987; O'Connor 1986) also show that Hodgkin's tissue biopsies demonstrate gene rearrangements consistent either with the T- or B-cell lineages. In many biopsies it has not been possible to demonstrate clonal rearrangements of either immunoglobulin or TCR genes. Genotypic investigations of HD cell lines show similar heterogeneity (Falk et al. 1987; Drexler et al. 1987; Poppema et al. 1985; Stein and Gerdes 1986). The data presented in this communication add to our observations concerning the lineage heterogeneity of cell lines derived from HD biopsies.

It is difficult to interpret the lineage data available for HD-derived cell lines, including the histiocytic line of Olsson et al. (1984), in terms of a common origin for the neoplastic population in HD. Immunopathological investigation strongly suggests that HD represents a heterogeneous entity (Abdul Azziz et al. 1983; Stein and Gerdes 1986; Jones 1986) and the genotypic data available (O'Connor 1986, Griesser et al. 1987) are not inconsistent with HD representing an initial polyclonal event with the subsequent appearance of a clonal population. In the evaluation of genotypic investigations, however, it is always possible that a low tumour cell number, often a feature of HD, may result in the failure to detect a monoclonal population on Southern blot hybridisation. In conclusion, the data presented on the two cell lines, Co and Ho, when considered in parallel with the information existing on cell lines from other laboratories, are insufficient to attribute a particular lineage to RS and MC, but suggest an origin within the lymphocyte lineage.

Acknowledgements. We acknowledge the Cancer Research Campaign for funding and Miss Julie T. Foster for secretarial assistance.

References

Abdul-Azziz Z, Mason DY, Stein H, Gatter KC, Nash JRG (1983) An immunohistochemical study of the cellular constituents of Hodgkin's disease using a monoclonal antibody panel. Histopathology 8: 1-25

Doreen HS, Habeshaw JA, Stansfeld AG, Wrigley PMF, Lister TA (1984) Characteristics of Sternberg-Reed and related cells in Hodgkin's disease. Br J Cancer 49: 465-476

Drexler HG, Norton J, Leber BF, Yaxley J, Tatsumi E, Hoffbrand AV, Minowada J (1987) Leukaemia (in press)

Falk MH, Stein H, Diehl V, Jones DB, Bornkamm GW (1987) Immunoglobulin gene rearrangements in two of four cell lines derived from patients with Hodgkin's disease. Int J Cancer 40: 262-269

Gerdes J, Schwarting R, Stein H (1986) High proliferative activity of Reed-Sternberg associated Ki-1 positive cells in normal lymphoid tissue. J Clin Pathol: 993-997

Griesser H, Feller AC, Mak TW, Lennert K (1987) Clonal rearrangements of T-cell receptor and immunoglobulin genes and immunophenotypic antigen expression in different subclasses of Hodgkin's disease. Int J Cancer 40: 157-160

Jones DB (1986) The histogenesis of the Reed-Sternberg cells and its mononuclear counterparts. J Pathol 151: 191-195

Jones DB, Scott CS, Wright DH, Stein H, Beverley PCL, Payne SV, Crawford DH (1985) Phenotypic analysis of an established cell line derived from a patient with Hodgkin's disease. Hematol Oncol 3: 133-145

McMichael AJ et al. (eds) (1987) Leucocyte typing III. Oxford University Press, Oxford.
Norton AJ, Isaacson PG (1985) Granulocyte and HLA-D region specific monoclonal antibodies in the diagnosis of Hodgkin's disease. J Clin Pathol 38: 1241-1246
O'Connor N (1986) Genotypic analysis of lymph node biopsies. J Pathol 151: 185-190
Olsson L, Bernke O, Pleibel N, D'Amore F, Werdelin O, Fry KE, Kaplan HS (1984) Establishment and characterisation of a cloned giant cell line from a patient with Hodgkin's disease. JNCI 73: 809-830
Poppema S, de Jong B, Atmosoerodjo J, Idenberg V, Visser L, de Ley L (1985) Morphologic, immunologic, enzymohistochemical and chromosomal analysis of a cell line derived from Hodgkin's disease. Cancer 55: 683-690
Stein H, Gerdes J (1986) Phenotypical and genotypical marker in malignant lymphomas: cellular origin of Hodgkin and Sternberg-Reed cells and implications on the classification of T-cell and B-cell lymphomas. Verh Dtsch Ges Pathol 70: 127-151
Stein H, Gerdes J, Schwab U, Lemke H, Mason DY, Ziegler A, Schjenle W, Diehl V (1982) Identification of Hodgkin and Reed-Sternberg cells as a unique cell type derived from a newly-detected small cell population. Int J Cancer 30: 445-459

The Typical Reed-Sternberg Phenotype and I g Gene Rearrangement of Hodgkin's Disease Derived Cell Line ZO Indicating a B-Cell Origin

S. Poppema, L. Visser, B. de Jong, M. Brinker, J. Atmosoerodjo, and W. Timens

Department of Laboratory Medicine, Cross Cancer Institute, Edmonton, Canada and the Departments of Pathology and Human Genetics, University of Groningen, Groningen, The Netherlands

Introduction

Hodgkin's disease differs from non-Hodgkin's lymphomas by the presence of reactive lymphocytes, histiocytes, plasma cells, fibroblasts, and eosinophils in addition to the abnormal, so-called Reed-Sternberg cells and their variants. Usually, Reed-Sternberg cells constitute only a minor population, whereas there is a majority of small reactive lymphocytes. Non-Hodgkin's lymphomas have been demonstrated by immunological or gene analysis approaches to be monoclonal populations of B- or T-lymphocyte-derived cells. In Hodgkin's disease neither the cell of origin nor the monoclonal origin of Reed-Sternberg cells has been established. One approach to analyze the origin and nature of Reed-Sternberg cells is the establishment of cell lines derived from tissues or fluids involved in Hodgkin's disease. Ideally, one should be able to demonstrate identical membrane and cytoplasmic markers, chromosomal abnormalities, and gene rearrangements in Reed-Sternberg cells in tissue sections and in the in vitro counterparts. Here we will describe the establishment and characterization of cell line ZO, derived from a pericardial effusion in a patient with the nodular sclerosis type of Hodgkin's disease. In addition we will describe the preparation of three new antibodies against this cell line and report their staining patterns and those of other anti-Reed-Sternberg cell reagents on Hodgkin's and non-Hodgkin's cell lines and Hodgkin's and non-Hodgkin's lymphomas.

Characterization of the Patient

A 26-year-old female was admitted to hospital because of dyspnea and a large supraclavicular lymph node. On further examination she was found to have a large mediastinal mass. A lymph node biopsy was performed and diagnosed as Hodgkin's disease, nodular sclerosis type with many lacunar Reed-Sternberg cells and areas of necrosis. The patient was treated with six courses of MOPP (mustine, oncovin, procarbazine, and prednisone), but complete remission was not achieved and 7 months later she was readmitted to the hospital because of pleural and peri-

Recent Results in Cancer Research. Vol. 117

cardial fluid. Cytological examination showed groups of mesothelial cells, several macrophages, few lymphocytes, many eosinophils, and a large number of cells with the cytological features of Reed-Sternberg cells and mononuclear variants. A volume of 250 ml pericardial fluid was used to perform immunological, enzyme histochemical, cytogenetic, and molecular biological analysis and to establish the cell line. The patient was treated with chemotherapy but no remission could be achieved and she died 3 months later.

Establishment of the Cell Line

The pericardial fluid cells were cultured at a density of 10^6 cells/ml in RPMI1640 with 20% fetal calf serum (FCS, Flow Laboratories, UK) and growth of sheets of cells in suspension could be observed after 2 weeks. At passage 10 after approximately 1 month of culturing the cells stopped growing. Subsequently we added interleukin 2, containing supernatant (Lymphocult-T-LF, Biotest Diagnostics, Frankfurt, FRG) at 15 units/ml tissue culture medium to part of the culture flasks,

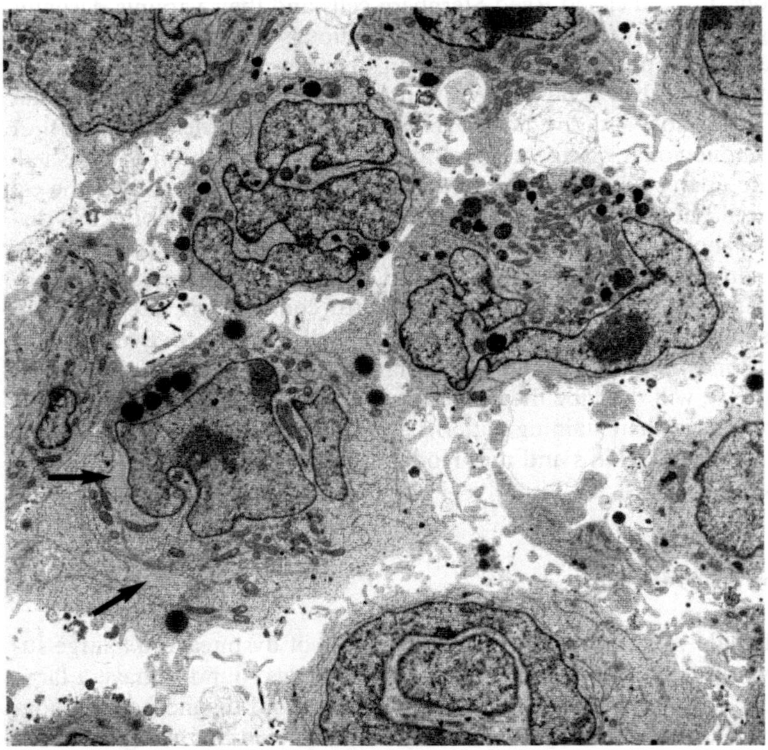

Fig. 1. Transmission electron microscopy of the ZO cell line, demonstrating lobated nuclei with huge nucleoli and a cytoplasm containing polyribosomes, some strands of RER, fat globules, dense bodies, and microfilaments *(arrows)*

resulting in a continuation of growth, whereas the cells without this addition died. However, continuous addition of Lymphocult also appeared to be disadvantageous and presently we culture this cell line (ZO) under careful observation in medium with 20% FCS and add Lymphocult when the cell growth slows down at approximately monthly intervals.

Morphological Studies

The cell line grows in large sheets in suspension, and the cells have small villous projections as can be well seen with scanning electron microscopy. With transmission electron microscopy the cultured cells have lobulated nuclei with huge nucleoli and a cytoplasm with several strands of endoplasmic reticulum, ribosomes, many microfilaments, several fat droplets, and some dense vesicles (Fig. 1).

Immunological Analysis

The results of the immunophenotypic analysis of the cell line and the Reed-Sternberg cells in the pericardial fluid and in frozen tissue sections are summarized in Table 1. It can be seen that ZO cells have the typical Reed-Sternberg cells immunophenotype without reactivity with anti-common leukocyte antigens (200 k, 190 k) and B-cell and T-cell markers and with reactivity for markers like Ki-1, Leu M1, HLA class II, and interleukin 2 receptors. In addition, the ZO cells ex-

Table 1. Immunophenotype Reed-Sternberg cells in lymph node, pericardial fluid, and ZO cell line

Reagent	Lymph node	Pericardial fluid	ZO cell line
Immunoglobulin	–	–	–
Ki-1	+	+	+
LeuM1	+	+	+
SR7	+	+	+
HLA class I	+	+	+
HLA class II	+	+	+
Anti IL2 receptor	+	+	+
CD45 (CLA)	–	–	–
MB1	–	–	–
MT1	–	–	–
CD19	–	–	–
CD20	–	–	–
CD21	–	–	–
CD2	–	–	–
CD3	–	–	–
CD7	–	–	–
CD4	?	?	cytoplasmic +
CD8	–	–	–
CD11c (p150, 95)	–	–	–

press strong receptors for T-lymphocytes as demonstrated by lymphocyte-rosetting techniques.

Cytogenetic Analysis

Cytogenetic analysis was performed on the pericardial fluid cells and on the established cell line. In the pericardial fluid cells a few cells had karyotype 47, XX, +g, and in the established cell line all cells had karyotype 53, XX, iso 1q, t(1;13), iso 2p, −3, 4q−, 6q−, t(7;17), −15, 16p+, 16q−, +17, +20, +5 markers.

Molecular Biological Analysis

Ig gene and T-cell receptor gene analysis were performed on the pericardial fluid cells and on the established cell line. The results are shown in Fig. 2, and indicate that the pericardial fluid contained no detectable clonal population, whereas the cell line, despite the absence of immunoglobulin production and B-cell membrane markers, has clonal rearrangements of the immunoglobulin heavy chain gene, as shown by hybridization with the JH probe and of the kappa light chain gene, as shown with a C kappa probe. Hybridization with the C lambda probe shows an identical germ line pattern in both the pericardial fluid and the cell line (Fig. 2). No rearrangements were found with the probes for the T-cell-receptor beta- and gamma-chain genes.

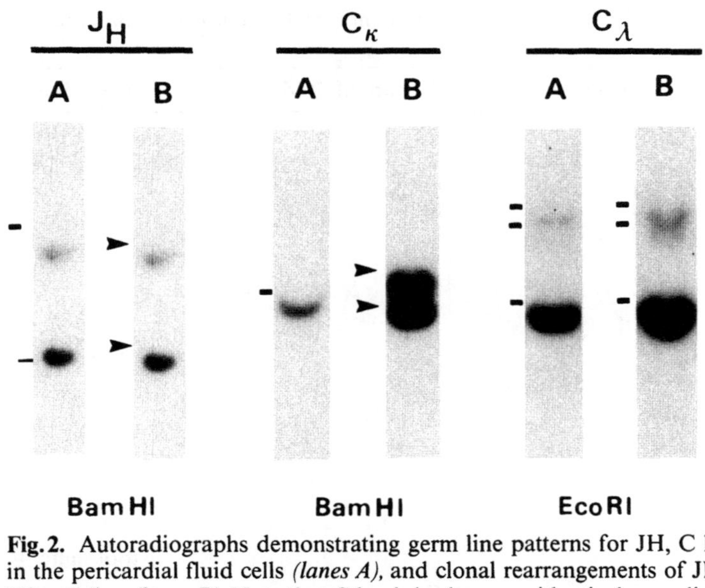

Fig. 2. Autoradiographs demonstrating germ line patterns for JH, C kappa, and C lambda in the pericardial fluid cells *(lanes A)*, and clonal rearrangements of JH and C kappa in the ZO cell line *(lanes B)*. Note that C lambda shows an identical germ line pattern in both materials

Growth in Nude Mice

At passage 16 the cell line was injected subcutaneously into nude mice. After 1 month the mice showed signs of disease although no local subcutaneous disease was evident. At autopsy the mice had large hepatic tumors. Morphological, ultrastructural, and immunohistological analyses of the tumor gave similar results to in the original cell line.

Production of Monoclonal Antibodies

Balb/c mice were immunized intraperitoneally with cells of the ZO cell line. Four days after intravenous boosting the spleen was removed and the spleen cells were hybridized with SP20 myeloma cells according to standard procedures. Screening of supernatants of the resulting clones was performed on frozen tissue sections of a lymph node involved in the nodular sclerosis type of Hodgkin's disease of another patient with Hodgkin's disease. Clones with Reed-Sternberg cell reactivity were selected and recloned. Here we describe the staining results of three of these clones on normal lymphoid tissue, Hodgkin's cell lines, non-Hodgkin's cell lines, on Hodgkin's and non-Hodgkin lymphomas, including cases of so called Ki-1-positive lymphoma. The staining results are summarized in Table 2 and illustrated in Fig. 3.

Comparison of ZO with Other Hodgkin's Cell Lines

We compared the immunocytological features and the gene analysis results of ZO with those of some other Hodgkin's cell lines as given by the authors or performed in our laboratory. These results are summarized in Table 3. The findings indicate that cell lines ZO, L428 (Schaadt et al. 1980), and KM-H2 (Kamesaki et al. 1986) have typical Reed-Sternberg cell phenotypes and immunoglobulin gene rearrangements indicating a possible B-cell origin. Our cell line DEV (Poppema et al. 1985; Timens et al. 1985) is also reactive with some pan-B-cell reagents and expresses alpha-2 immunoglobulin heavy chains, resulting in a B-cell phenotype that is less frequently encountered in the nodular sclerosis and mixed-type subtypes of Hodgkin's disease, but is a typical feature in the lymphocyte predominance type of Hodgkin's disease. Finally, cell lines L540 (Diehl et al. 1981) and CO (Jones et al. 1985) lack the typical immunophenotype of Reed-Sternberg cells, express several T-cell markers, and do have T-cell receptor gene rearrangements, indicating a T-cell origin of these cell lines.

Comparison of ZO with Reed-Sternberg Cells

The immunological marker analysis of the ZO cell line gives results that are identical to the markers of Reed-Sternberg cells and mononuclear variants. An interesting finding is the cytoplasmic reactivity with CD4 antibodies; a feature sometimes

Table 2. Staining results of monoclonal antibodies produced against ZO cells

Clone	Zo	R-S cells	L&H cells	Ki-1	B-IB	T-IB	React. LN
200-1E11	+	+	+	–	+	–	–
200-1G11	+	+	–	–	–	–	–
200-4C2	+	+	+	–	+	–	–

R-S, Reed-Sternberg; L&H, L&H-type Reed-Sternberg; Ki-1, non-Hodgkin's lymphomas with Ki-1 positivity (five cases); B-IB, B-immunoblastic non-Hodgkin's lymphomas (five cases); T-IB, T-immunoblastic non-Hodgkin's lymphomas (five cases); react. LN, reactive lymph node cells, including myeloid cells.

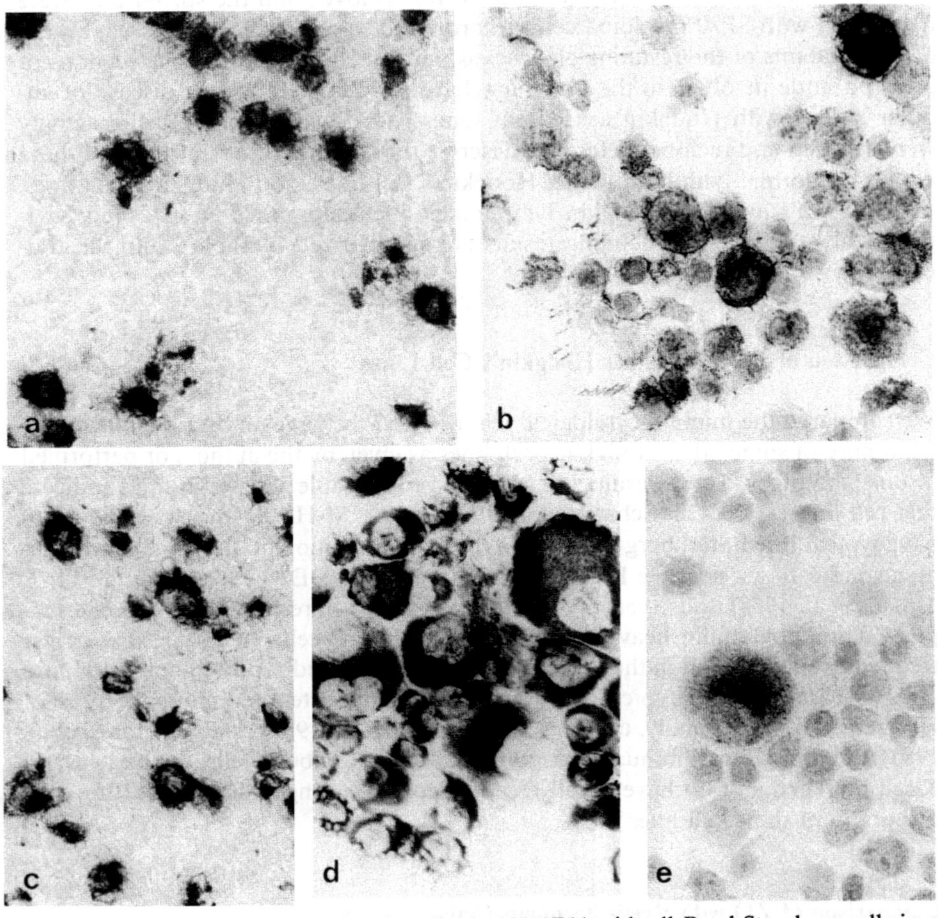

Fig. 3 a–e. Reactivity of monoclonal antibody 200-1E11 with all Reed-Sternberg cells in a frozen tissue section of the nodular sclerosis subtype of Hodgkin's disease (**n**), and on a cytospin of Hodgkin's cell line L428 (**n**). Reactivity of monoclonal antibody 200-1G11 with all Reed-Sternberg cells on tissue section (**n**), and on cytospins of ZO cell line (**n**), and L428 cell line (**n**). Note that only a minority of cells of L428 are stained by this reagent

Table 3. Immunophenotypic and genotypic analysis of Hodgkin's and other cell lines

Reagent	Hodgkin's						EBV$^+$		Non-Hodgkin's	
	ZO	L428	KM-H2	L540	CO	DEV	DUS	CAT	SCHI	VER
Immunoglobuline	−	−	−	−	−	+	+	+	+	+
Ki-1	+	+	+	+	−	+	+	+	+	+
Leu M1	+	+	+	−	−	+	+	+	−	−
HLA class II	+	+	+	+	−	−	+	+	+	+
anti IL2r	+	+	+	+	+	+	+	+	+	+
CLA (200K)	−	+		+	+	+	+	+	+	+
MB1 (200K)	−	+		+	+	+	+	+	+	+
MT1 (190K)	−	−		+	+	−	−	−	−	−
CD20 (B1)	−	−	−	−	−	+	+	+	+	+
CD22 (leu14)	−	−		−	−	+	+	+	−	+
CD21 (B2)	−	−	−	−	−	−	+	+	−	−
CD10 (CALLA)	−	−	−	−	−	−	−	−	−	−
CD11c (p150, 95)	−	−	−	−	−	−	−	−	−	−
CD2 (leu5)	−	−	−	+	−	−	−	−	−	−
CD7 (WT1)	−	−	−	−	−	−	−	−	−	−
CD1 (OKT6)	−	−	−	−	−	−	−	−	−	−
CD5 (leu1)	−	−	−	+	+	−	−	−	−	−
CD3 (OKT3)	−	−	−	+	+	−	−	−	−	−
CD4 (leu3)	(+)	−	−	−	−	−	−	−	−	−
CD8 (OKT8)	−	−	−	−	−	−	−	−	−	−
p19 (HTLV1)	−	−	−	−	−	−	−	−	−	−
EBNA	−	−	−	−	−	−	+	+	−	−
Receptor for T cells	+	+	+	+	−	−	−	−	−	−
200-1E11	+	+		+	−	+	+	+	+	+
200-1G11	+	+/−		−	−	−	−	−	−	−
200-4C2	+	+		+	−	+	+	+	+	+
Gene analysis										
Ig heavy chain	+	+	+	−	−	+	+	+	+	+
Ig light chain	+	+		−	−	+	+	+	+	+
TCR beta chain	−	−		+	+	−	−	−	−	−
Conclusion	B	B	B	T	T	B	B	B	B	B

observed in Reed-Sternberg cells in tissue sections and thought to be the result of endocytosis of shedded T-lymphocyte-derived CD4 antigen. However, the presence of CD4 antigen in these cultured cells indicates endogenous CD4 antigen production. CD4 positivity has been demonstrated in some EB virus and HTLV1 virus-infected lymphoblastoid B-cell lines. The monoclonal antibodies produced after immunization with ZO cells are remarkably specific for Reed-Sternberg cells in tissues involved in Hodgkin's disease and show no reactivity at all in normal lymphoid tissues and in peripheral blood specimens. Antibodies 200-1E11 and 200- 4C2 react with B-immunoblastic lymphoma cells and also with other cell lines of B-cell origin. No reactivity was found with T-immunoblastic lymphomas and with so-called Ki-1-positive lymphomas. Antibody 200-1G11 appears to be only reactive with cell lines ZO and L428 and with typical and lacunar-type Reed-Sternberg cells, but not with so-called L&H type Reed-Sternberg cells. The results indicate that cell line ZO and also L428 are cell lines with a typical immunophenotype of Reed-Sternberg cells.

Conclusion

Based on the results of these studies we conclude that ZO cells are of mature B-lymphocyte origin since the cell line has clonal heavy chain as well as kappa light chain gene rearrangements. In addition, this cell line has a typical Reed-Sternberg cell immunophenotype. Therefore, we conclude that Reed-Sternberg cells may be of B-cell origin. The results with the ZO cell line are also supported by our findings with the DEV cell line and to a certain extent the results published on the L428 and KM-H2 cell lines. Further arguments for a possible B-cell origin of Reed-Sternberg cells are the presence of clonal Ig gene rearrangements in Hodgkin's tissues with the predominance of Reed-Sternberg cells (Brinker et al. 1987), as well as a membranous staining pattern with pan-B-cell antibodies of CD20 found on Reed-Sternberg cells in all cases of nodular lymphocyte predominant type and some cases of nodular sclerosis and mixed-type Hodgkin's disease.

References

Brinker MGL, Poppema S, Buys CHCM, Timens W, Osinga J, Visser L (1987) Clonal immunoglobulin gene rearrangements in tissues involved by Hodgkin's disease. Blood 70: 186-191

Diehl V, Kirchner HH, Schaadt M, Fonatsch C, Stein H, Gerdes J, Boie C (1981) Hodgkin's disease: establishment and characterization of four in vitro cell lines. J Cancer Res Clin Oncol 101: 111-124

Jones DB, Scott CS, Wright DH, Stein H, Beverly PCL, Payne SV, Crawfords DH (1985) Phenotypic analysis of an established cell line derived from a patient with Hodgkin's disease. Hematol Oncol 3: 133-145

Kamesaki H, Fukuhara S, Tatsumi E, Uchino H, Yamabe H, Miwa H, Shikakawa S, Hatanaka M, Honju T (1986) Cytochemical, immunologic, chromosomal and molecular genetic analysis of a novel cell line derived from Hodgkin's disease. Blood 68: 285-292

Poppema S, de Jong B, Atmosoerodjo J, Idenburg V, Visser L, de Leij L (1985) Morphological, immunological, enzymehistochemical and chromosomal analysis of a cell line derived from Hodgkin's disease. Cancer 55: 683-690

Schaadt M, Diehl V, Stein H, Fonatsch C, Kirchner HH (1980) Two neoplastic cell lines with unique features derived from Hodgkin's disease. Int J Cancer 26: 723-731

Timens W, Brinker M, Osinga J, Buys CHCM, Visser L, Halie MR, Poppema S (1985) Ig gene rearrangement in Hodgkin's disease cell line. Exp Hematol 13: 421

Characterization of Hodgkin's Disease Derived Cell Line HDLM-2

H. G. Drexler, S. M. Gignac, A. V. Hoffbrand, B. F. Leber, J. Norton, M. S. Lok, and J. Minowada

Department of Haematology, Royal Free Hospital, School of Medicine, London NW3 2QG, Great Britain

Introduction

The origin of the cells which are pathognomonic for Hodgkin's disease (HD), the Hodgkin (H) and Reed-Sternberg (RS) cells, remains controversial (Drexler et al. 1987; Jones 1987). The analysis of H-RS cells is hampered by the scarcity of the neoplastic cells and contamination with bystander cells. The event of improved tissue culture methodology for the establishment of immortalized cell lines has greatly enhanced the possibilities of studying neoplastic hematopoietic cells. The advantages are the monoclonality of the cell population and the unlimited access to cells which retain their phenotypic characteristics in long-term culture.

Recently, a number of cell lines have been established from tissues or pleural effusions from patients with HD (Jones et al. 1985; Poppema et al. 1985; Diehl et al. 1982; Kamesaki et al. 1986; Olsson et al. 1984). The in vitro cultured cells presumably represent the in vivo H-RS cells having identical or very similar characteristic features.

Since 1982 we have cultured a cell line established from the pleural effusion of a patient with HD, termed HDLM-2 (Drexler et al. 1986). In this report we summarize the characteristics of this cell line and attempt to compare it with other known HD-derived cell lines.

Materials and Methods

Culture Conditions

Mycoplasma-free cultures were grown in RPMI 1640 medium supplemented with 10% heat-inactivated fetal calf serum plus penicillin/streptomycin and L-glutamine at 37 °C in a 5% CO_2 atmosphere without the addition of any growth factors.

Recent Results in Cancer Research. Vol. 117

Morphological and Cytochemical Analysis

The morphological and cytochemical features of the cells were examined on cytospin slide preparations stained with May-Grünwald-Giemsa and the respective standard cytochemical stains.

Functional Studies

Phagocytosis was tested using latex beads. Antigen presentation, nitroblue tetrazolium (NBT) reduction, NK cell activity, and colony-stimulating factor activity were examined with standard assays.

Immunophenotyping

Expression of surface and intracytoplasmic markers was examined by indirect immunofluorescence in suspension and on fixed cytospin slide preparations, respectively. Cells were analyzed by immunofluorescence microscopy and by flow cytometry (EPICS V).

Gene Rearrangement Studies

DNA extraction, Southern blot analysis, and hybridization with J_H, TCR, β-chain, and γ-chain probes were carried out according to established procedures.

Results

Origin

Three cell lines (HDLM-1, -2, and -3) were established from the pleural effusion of a 74-year-old male with nodular-sclerosing HD of stage IV. These cell lines show very similar or identical features. The results below refer mainly to the analysis of HDLM-2.

Establishment, Maintenance, and Cell Kinetics

In 1982, the cell lines were initiated in RPMI 1640 medium at Roswell Park Memorial Institute, Buffalo, New York. The cells can be cultured in RPMI 1640 medium supplemented with 5% or 10% fetal calf serum. The cultures grow as free-floating, single cells in suspension. The cell doubling time ranges from 3 to 5 days at a cell density of $0.1-0.2 \times 10^6$ cells/ml.

HDLM cell lines are negative for the Epstein-Barr virus nuclear antigen.

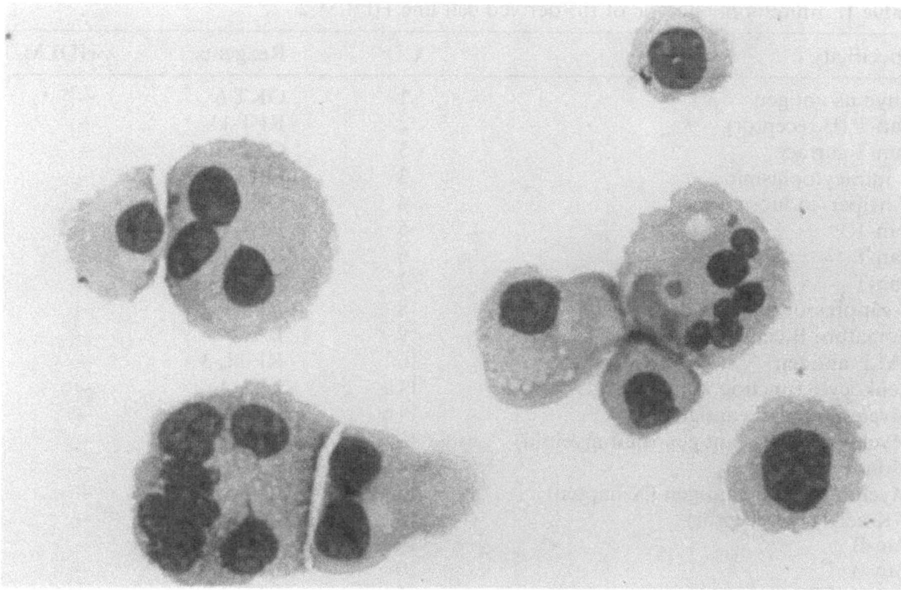

Fig. 1. Cytospin slide preparation of HDLM-2 cells stained with May-Grünwald-Giemsa, ×500

Morphology

A marked heterogeneity in cell size, number of nuclei, and structure of cytoplasm can be seen in HDLM-2: mono-, bi-, and multinucleated cells are found with cell sizes ranging from about 10 to 100 μm or more. The predominant cell type is a relatively small (10-20 μm), round, mononucleated cell with smooth surface, one to three prominent nucleoli, and a moderate amount of basophilic cytoplasm with occasional fine vacuoles. A medium-sized cell (20-50 μm) is mono- or binucleated, round or polygonal with lighter, foamy-appearing cytoplasm. Another variant is a very large to giant cell (accounting for about 5% of the cell population) of 50-100 μm diameter with multiple nuclei and an abundance of cytoplasm. Eighty percent of the cells are mononucleated; 5%-10% of the cells are giant, but mononucleated cells; 20% of the cells are multinucleated (two to ten nuclei) (Fig. 1).

Cytochemistry and Enzyme Marker Analysis

HDLM-2 is negative for peroxidase, alkaline phosphatase, and naphthol-AS-D-chloroacetate esterase staining, but does show positivity for acid phosphatase and nonspecific esterase (α-naphthylacetate esterase) activities.

Using isoenzyme marker analysis on horizontal polyacrylamide isoelectric focusing gels, HDLM-2 cells display the tartrate-resistant acid phosphatase isoenzyme, but are negative for an esterase isoenzyme which is specific for monocytes (Drexler et al. 1986; Scott et al. 1988).

Table 1. Immunophenotype of HD-derived cell line HDLM-2

Specificity	CD[a]	Reagents	HDLM-2
Thymus antigen	1	OKT-6	–[b]
Pan-T (E-receptor)	2	RFT-11	+
Pan-T surface	3	OKT-3	–
intracytoplasmic	3	OKT-3	–
T-helper/inducer	4	OKT-4	–
Pan-T	5	RFT-1	–
Pan-T	6	RFT-12	–
Pan-T	7	RFT-2	–
T-suppressor/cytotoxic	8	RFT-8	–
Immature B cells	9	BA-2	–
cALL antigen	10	RFAL-3	–
Leukocyte function antigen	11 a	LFA-1	–
Myelomonocytic antigen	11 b	Mo1	–
Myelomonocytic antigen (pan-myeloid)	13	MCS-2	–
Monocytic antigen	14	Leu-M3	–
Myelomonocytic antigen (X-hapten)	15	Leu-M1	+
NK cells (Fc-receptor)	16	Leu-11 b	–
Pan-B	19	B4	–
Pan-B	20	B1	–
Pan-B (C3d-receptor)	21	RFB-6	–
Pan-B	22	RFB-4	–
B cells	24	BA-1	–
Activated cells (IL-2 receptor)	25	Tac	+
Activated cells	30	Ki-1	+
Myelomonocytic antigen	33	MY9	–
Stem cell antigen	34	MY10	–
Immunoglobulin surface		IgM	–
intracytoplasmic		IgM	–
Terminal deoxynucleotidyl transferase		TdT	–
HLA-A,B,C		W6/32-1	+
Monomorphic HLA-DP		B7/21	+
Monomorphic HLA-DQ		Tü22	+
Monomorphic HLA-DR		RFDR-1	+
Common HLA-DR/DQ/DP		RFDR-2	+
Glycophorin A		LON R10	–
RS-cell marker		HeFi-1	+
NK cells		Leu-7	–
NK cells		Leu-19	–
Nature B-cell marker		FMC-7	–
Transferrin-receptor (proliferation)		B3/25	+
Nuclear proliferation antigen		Ki-67	+

[a] As proposed by the Workshops on Human Leukocyte Differentiation Antigens (Paris, France, 1982; Boston, MA, USA, 1984; Oxford, UK, 1986).

[b] +, 80% or more cells were positive; –, negative.

Immunophenotype

HDLM-2 cells are positive for markers of activation (IL-2 receptor and HLA class I and II antigens) and of proliferation (transferrin-receptor and Ki-67 nuclear antigen) which are generally present on H-RS cells (Jones 1987) (Table 1). HDLM-2 cells express the "H-RS cells associated" Ki-1 and HeFi-1 antigens which presumably are markers of activation (Hecht et al. 1985; Schwab

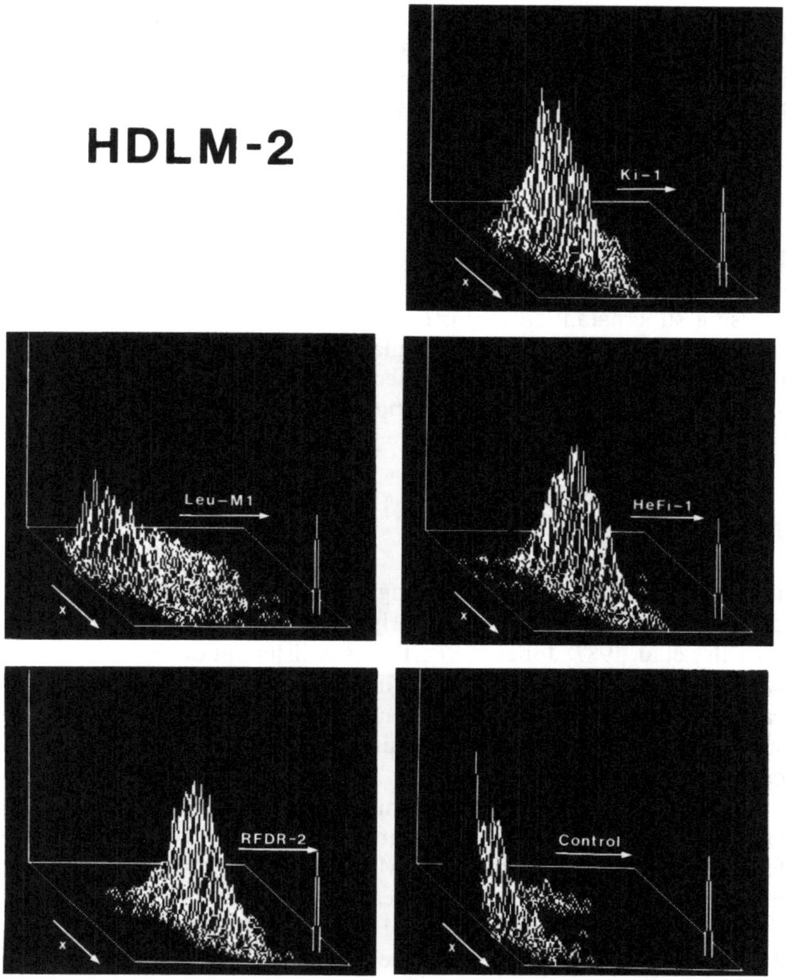

Fig. 2. Three-dimensional flow cytometry histogram of HDLM-2 cells stained with the McAbs Ki-1 (CD30), Leu-M1 (CD15), HeFi-1 ("H-RS cell-associated" marker produced against L-428), and RFDR-2 (HLA-DR class II antigen); an irrelevant McAb of IgG class was used for background staining. *x-axis*, cell size on linear scale; *y-axis*, positivity for fluorescence staining on logarithmic scale; *z-axis*, cell number on linear scale. Note that small, medium, and giant cells are equally positive for the respective markers

et al. 1982) (Fig. 2). The CD15 McAbs (Leu-M1, MCS-1, and others) also label HDLM-2. Leu-M1 has been described as "a marker" for H-RS cells; however, it has been found on cells from other lineages as well (Drexler et al. 1987). Except for CD2 (sheep erythrocyte receptor), all markers of the T-, B-, myeloid, monocytic, or NK cell lineages were negative on HDLM-2 (cell surface and intracytoplasmic stainings).

Gene Status

Whereas the immunoglobulin heavy chain genes show germline configuration, T-cell receptor β- and γ-chain genes both show biallelic rearrangement (Drexler et al. 1988).

Functional Studies

HDLM-2 cells are negative in phagocytosis, antigen presentation, NK activity, colony-stimulation factor activity, and NBT reduction assays.

Using supernatant from unstimulated HDLM-2 cultures, differentiation could be induced in a number of leukemia cell lines (Drexler et al. 1986). Whereas the phorbol ester TPA was not very effective in inducing differentiation (Drexler et al. 1986), extracellular matrix was reported to promote differentiation of HDLM-1 (Hsu et al. 1987).

Discussion

Despite extensive efforts employing a multitude of techniques from a variety of disciplines, a consensus on the origin of H-RS cells has not yet been reached (Drexler et al. 1987; Jones 1987). The establishment of cell lines which have identical or very similar phenotypic profiles as in vivo H-RS cells (and which therefore can be regarded operationally as in vitro representatives of H-RS cells) enabled studies to be performed which require large numbers of cells and monoclonal, uncontaminated cell populations.

In this study we report on an immortalized cell line established from the pleural effusion of a patient with HD. Using a multiparameter analysis approach we attempted to characterize the HDLM-2 cell line and to detect any distinctive features indicative of a known cell lineage.

On morphological examination, HDLM-2 cultures contain a majority of undistinguished mono- or binucleated cells and conspicuous giant cells with multiple nuclei. Cytochemical, enzymological, and functional data speak against a monocyte/macrophage origin of HDLM-2 cells. Except for CD2 no cell lineage-specific or -associated immunological markers were detected on HDLM-2 cells. Genotypic analysis, however, demonstrated that the cells have rearranged T-cell β- and γ-receptor genes with germline immunoglobulin heavy chain genes, which is a pattern typical of monoclonal T-cell proliferation (Sangster et al. 1986).

A comparison of HDLM-2 cells with in vivo H-RS cells is virtually impossible as nearly every cell of the hematopoietic system has been proposed as the normal counterpart to H-RS cells (Drexler et al. 1987; Jones 1987). Nevertheless, HDLM-2 and H-RS cells share the following characteristic features: a heterogeneous morphological picture with giant, multinucleated cells besides mono- and binucleated variants; positivity for immunological activation markers (IL-2 receptor and HLA class II antigens) and for markers associated with, but not specific for, H-RS cells such as Ki-1 (produced against L-428 and the so far best marker of H-RS cells), HeFi-1 (equally produced against L-428), and CD15 McAbs (Leu-M1); and negativity for a number of cell lineage-specific markers.

A comparison of the data on six HD-derived cell lines (CO, DEV, HDLM-2, KM-H2, L-428, SU-RH/HD-1) indicates that they display similarities in morphology and cytochemistry, but each cell line differs in at least one respect from all the rest. A common, albeit negative, denominator is the fact that the cells cannot easily be assigned to a given cell lineage. In other words, it appears that each cell line is unique within the group of HD-derived cell lines and with regard to other hematopoietic cell lines.

The above data seem to indicate two major conclusions: (1) HDLM-2 is a unique cell line when compared with a panel of more than 100 leukemia-lymphoma cell lines of acknowledged cell lineage origin; but HDLM-2 cells share positive and negative markers with in vivo H-RS cells and can therefore be regarded as in vitro representatives of H-RS cells and (2) extrapolation of the information gained on HDLM-2 cells to the in vivo situation would suggest a T-cell origin of H-RS cells. The HD-derived cell line CO is also of T-cell lineage whereas a B-cell origin has been reported for the cell lines L-428, KM-H2, and DEV (Jones 1987; Diehl et al. 1982; Kamesaki et al. 1986; Poppema et al. 1985; Drexler et al. 1988). Taken together, these studies on cell lines strongly support the hypothesis of a lymphoid origin of H-RS cells. It should be noted, however, that another HD-derived cell line, SU-RH/HD-1, showed characteristics of monocytic/macrophage-like cells (Olsson et al. 1984).

Summary

The cell line HDLM-2 was established from the pleural effusion of a patient with Hodgkin's disease. Here, we describe the morphological, cytochemical, enzymological, immunological, molecular biological, and functional characteristics of the cell line. The results of this multiparameter profile show that HDLM-2 is different from other well-studied leukemia-lymphoma cell lines including other Hodgkin's disease derived cell lines. HDLM-2 cultures contain mainly mono- or binucleated cells, but also prominent giant cells with two to ten nuclei. HDLM-2 cells do not express an immunophenotype characteristic of a given cell lineage. However, the cells are positive for Ki-1, HeFi-1, Leu-M1, Tac, and HLA class II markers. Cytochemical, enzymological, and functional data are equally inconclusive, but are definitely not compatible with a monocyte/macrophage profile. Analysis of the gene status documents that T-cell receptor β- and γ-chain genes are rearranged while immunoglobulin heavy chain genes are in germline configuration. The combined results indicate a T-cell origin of HDLM-2 cells. The evidence available from this and other established Hodgkin's disease derived cell lines suggests a lymphoid origin of Hodgkin and Reed-Sternberg cells.

References

Diehl V, Kirchner HH, Burrichter H, Stein H, Fonatsch C, Gerdes J, Schaadt M, Heit W, Uchanska-Ziegler B, Ziegler A, Heintz H, Sueno K (1982) Characteristics of Hodgkin's disease-derived cell lines. Cancer Treat Rep 66: 615-632

Drexler HG, Gaedicke G, Lok MS, Diehl V, Minowada J (1986) Hodgkin's disease derived cell lines HDLM-2 and L-428: comparison of morphology, immunological and isoenzyme profiles. Leuk Res 10: 487-500

Drexler HG, Amlot PL, Minowada J (1987) Hodgkin's disease derived cell lines - conflicting clues for the origin of Hodgkin's disease? Leukemia 1: 629-637

Drexler HG, Leber BF, Norton J, Yaxley J, Tatsumi E, Hoffbrand AV, Minowada J (1988) Genotypes and immunophenotypes of Hodgkin's disease derived cell lines. Leukemia 2: 371-376

Hecht TT, Longo DL, Cossman J, Bolen JB, Hsu SM, Israel M, Fisher RI (1985) Production and characterization of a monoclonal antibody that binds Reed-Sternberg cells. J Immunol 134: 4231-4236

Hsu SM, Zhao X, Hsu PL, Lok MS (1987) Extracellular matrix does not induce the proliferation, but promotes the differentiation of Hodgkin's cell line HDLM-1. Am J Pathol 127: 9-14

Jones DB (1987) The histogenesis of the Reed-Sternberg cell and its mononuclear counterparts. J Pathol 151: 191-196

Jones DB, Scott CS, Wright DH, Stein H, Beverley PCL, Payne SV, Crawford DH (1985) Phenotype analysis of an established cell line derived from a patient with Hodgkin's disease (HD). Hematol Oncol 3: 133-145

Kamesaki H, Fukuhara S, Tatsumi E, Uchino H, Yamabe H, Miwa H, Shiragawa S, Hatanaka M, Honjo T (1986) Cytochemical, immunologic, chromosomal, and molecular genetic analysis of a novel cell line derived from Hodgkin's disease. Blood 68: 285-292

Olsson L, Behnke O, Pleibel N, D'Amore F, Werdelin O, Fry KE, Kaplan HS (1984) Establishment and characterization of a cloned giant cell line from a patient with Hodgkin's disease. J N C I 73: 809-830

Poppema S, De Jong B, Atmosoerodjo J, Idenburg V, Visser L, De Ley L (1985) Morphologic, immunologic, enzymehistochemical and chromosomal analysis of a cell line derived from Hodgkin's disease. Evidence for a B-cell origin of Sternberg-Reed cells. Cancer 55: 683-690

Sangster RN, Minowada J, Suciu-Foca N, Minden M, Mak TW (1986) Rearrangement and expression of the α, β and γ chain T cell receptor genes in human thymic leukemia cells and functional T cells. J Exp Med 163: 1491-1508

Schwab U, Stein H, Gerdes J, Lemke H, Kirchner HH, Schaadt M, Diehl V (1982) Production of a monoclonal antibody specific for Hodgkin- and Sternberg-Reed cells of Hodgkin's lymphoma and a subset of normal lymphoid cells. Nature 299: 65-67

Scott CS, Stark AN, Jones DB, Minowada J, Roberts BE, Drexler HG (1988) Quantitative and qualitative enzyme studies of Hodgkin's disease derived cell lines. Leukemia 2: 447-452

A New Hypothesis on the Cellular Origin of Reed-Sternberg and Hodgkin Cells Based on the Immunological and Molecular Genetic Analysis of the KM-H2 Line

H. Kamesaki, S. Fukuhara, H. Uchino, and T. Nosaka

First Division of Internal Medicine, Faculty of Medicine, Kyoto University, Kyoto, Japan

Introduction

Previously we reported a novel cell line, KM-H2, established from the pleural effusion of a patient who was initially diagnosed as having Hodgkin's of mixed cellular type (Kamesaki et al. 1986). We describe here the results of further investigation of this line (ultrastructural, immunological, and molecular genetic studies) and discuss its cellular origin. Based on these analyses of the KM-H2 line, as well as those of other cell lines derived from Reed-Sternberg and Hodgkin cells (e.g., L428, L540), we propose a new hypothesis for the cellular origin of Hodgkin's disease.

Materials and Methods

As most of the methods used in the analysis of the KM-H2 line were described in our previous report (Kamesaki et al. 1986), we describe only the new methods used in our further studies.

Electron Microscopy

Cells were fixed in 2.5% glutaraldehyde in 0.1 *M* cacodylate buffer, pH 7.4 at 4 °C for 2 h. After washing, the cells were postfixed in 1% Os_3O_4 in the same buffer at 4 °C for 2 h, dehydrated in graded ethanols, and embedded in an epoxy resin. Ultrathin sections were examined under an electron microscope. Counterstaining with uranyl acetate and lead citrate was used with these sections. Detection of myeloperoxidase by electron microscope was performed by the modified method of Graham and Karnovsky (1966). Detection of platelet peroxidase was done by the modified method of Breton-Gorius and Guichard (1972).

Recent Results in Cancer Research. Vol. 117

Immunological Markers

In addition to the antibodies described in our previous report, B2 (CD21), B4 (CD19), and L26 (Ishii et al. 1984) were employed as anti-B-cell monoclonal antibodies. Antigens defined by these antibodies were detected by indirect immunofluorescence.

Molecular Genetic Analysis

Southern Blot Analysis. In addition to the probes used in our previous report, the *Hin*dIII-*Eco*RI fragment of the joining region of the T-cell receptor gamma chain gene and the *Xba*I-*Bgl*II fragment of the variable region of the immunoglobulin heavy chain gene were used as Jr_I and V_{HI} probes, respectively.

Northern Blot Analysis. Total cellular RNA was extracted by the guanidinium isothiocyanate technique, and the poly(A)-containing fraction was purified by oligodeoxythymidylic acid cellulose chromatography, then 2 μg (poly(A)+ RNAs was denatured in 50% formamide at 60 °C, and electrophoresed through 1% agarose with 6% formaldehyde and transferred to Gene Screen (New England Nuclear). Blots were hybridized at 42 °C in 50% formamide, 3 × Denhardt's solution, 5 × SSC (SSC = 0.15 *M* NaCl, 0.015 *M* tri-sodium citrate), 1% SDS, 200 μg/ml denatured salmon sperm DNA, and radiolabeled human DNA probe. Hybridized blots were washed finally in 0.2 × SSC, 0.5% SDS at room temperature and autoradiographed.

The mixture of VH_I (V_{266BL}, *XBa*I-*Bgl*II), VH_{II} ($V_{CE\text{-}1}$, *Hha*I-*Hha*I), VH_{III} ($V_{HB\text{-}26}$, *Sau*3AI-*Hha*I), and VH_{IV} ($V_{71\text{-}2}$, *Rsa*I-*Hha*I) was used as probes.

Studies by Short-Term Culture

The KM-H2 cells were cultured at the concentration of 5×10^5/ml in RPMI 1640 for 24 h with or without TPA (1 ng/ml). After centrifugation at 400 g for 15 min, the supernatant fluid of the culture medium was tested for IL1-alpha and IL1-beta. They were also cultured at the concentration of 5×10^5/ml in Eagle's minium essential medium supplemented with 10% fetal calf serum. After 24 h culture with or without lipopolysaccharide (*Escherichia coli* 026:B6, Sigma) the culture medium was centrifuged at 400 g for 15 min and the supernatant fluid was tested for tumor necrosis factor. The assay of IL1-alpha, IL1-beta, and tumor necrosis factor was performed by ELISA (Hayashi et al. 1985).

Results

As some of the morphological, cytochemical, immunological, and molecular genomic findings of the KM-H2 line are described in our previous report (Kamesaki et al. 1986), we describe mainly the new findings obtained by further investigation.

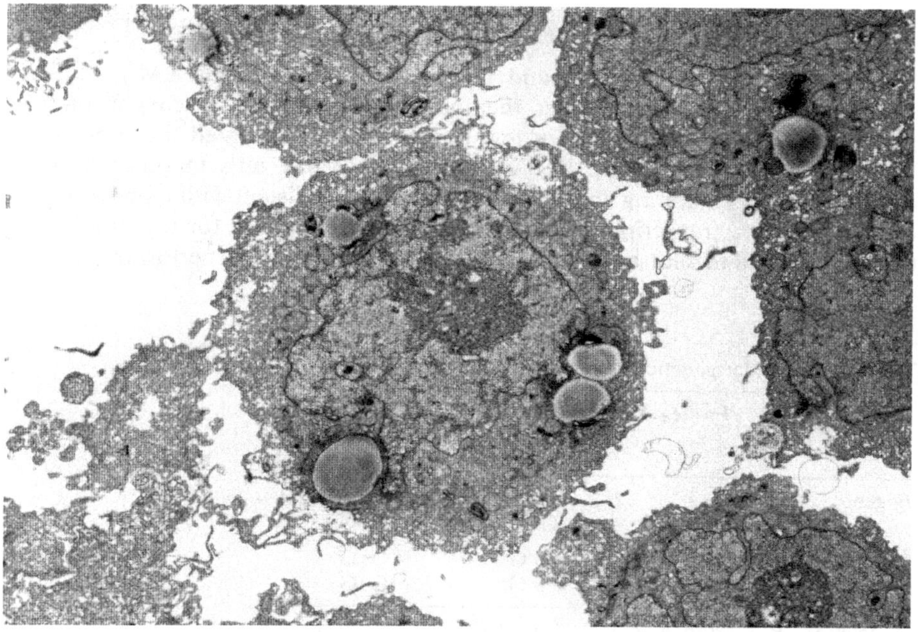

Fig. 1. Ultrastructure of KM-H2 cells, × 5500

Morphology and Cytochemistry

The ultrastructure of the KM-H2 cells is shown in Fig. 1. Their surface was irregular with several projections. The nuclear outline was frequently irregular. The nuclei frequently contained a highly developed nucleolus of regular outline and usually displayed little condensation of heterochromatin. The cytoplasm was notable for its lack of any structures resembling monocyte granules, but frequently contained one or more lipid droplets. Birbeck granules were not observed. Both myeloperoxidase and platelet peroxidase were negative.

Immunological Marker Analysis

As our previous investigation disclosed the immunoglobulin (*Ig*) heavy chain gene rearrangements, we carried out a further study using anti-B-cell monoclonal antibodies. The KM-H2 cells expressed B2 (CD21) and L26 antigens, which are usually expressed on sIg(+) mature B cells, but lacked B1 (CD20) and B4 (CD19) antigens. They were negative for J5 (CD10) and TdT, which are usually expressed on immature B cells.

Molecular Genomic Analysis

Southern blot analyses of the *Ig* and T-cell receptor genes in the KM-H2 line are summarized in Table 1. Moreover, the analysis using VH_I as a heavy-chain gene probe disclosed the deletion of a few fragments and suggested the presence of VDJ recombinations (Fig. 2). Because the KM-H2 line fails to produce heavy chain in spite of exhibiting an *Ig* gene organization consistent with a pre-B cell, we asked whether a transcriptional defect could be responsible for this finding. To this end, Northern blot analysis using $VH_{I\text{-}IV}$ as probes was performed, and no

Table 1. Molecular genetic findings

	Probes	Restriction enzyme	
		*Bam*HI	*Eco*RI
Ig gene	J_H	D/R	
	J_K	G/G	G/G
TcR gene	$C\beta_1$	G/G	G/G
	$J\gamma_1$	G/G	G/G

D, deletion; R, rearrangement; G, germ line; TcR, T-cell receptor.

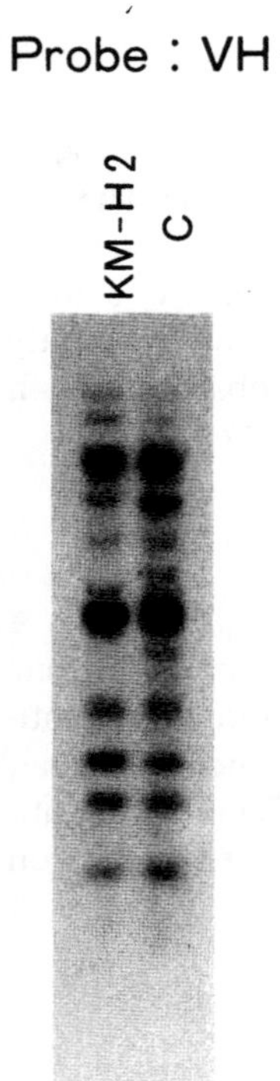

Fig. 2. Southern blot analysis of the KM-H2 line using *VHI* as a probe. A few fragments are deleted in comparison with the control (*C*, AML cells)

transcripts of the heavy-chain gene were detectable (data not shown). These results suggest that a lack of *Ig* gene transcription accounts for the absence of *Ig* heavy-chain production by this line.

Studies by Short-Term Culture

Production of Biologically Active Substances. The supernatants of culture medium of the KM-H2 cells contained 0.24±0.01 U/ml tumor necrosis factor (TNF) after 24 h culture. Moreover, its production increased to 0.44±0.03 U/ml when these cells were stimulated with LPS. We were unable to detect IL1-alpha or IL1-beta in the supernatant fluid after 24 h culture. However, it contained 66±30 pg/ml IL1-alpha when the KM-H2 cells were stimulated with TPA.

Phenotypic Changes Induced by Interferon Gamma. Interferon gamma is known to be an important regulator of the monocyte/macrophage system. Therefore, we investigated the response of the KM-H2 line to it, using the expression of Fcr receptors, HLA-DR antigens, and Tac (CD25) antigens as markers. While the U937 line (a cell line derived from the monocyte/macrophage system) increased the expression of these markers in response to interferon gamma, it showed no noticeable changes in their expression.

Discussion

From our previous report, it is clear that the KM-H2 line is derived from the Reed-Sternberg (RS) and Hodgkin (H) cells. Therefore, it may provide us with important insights into the nature of the RS and H cells to determine its cellular origin.

Our observations on the cellular origin of the KM-H2 line can be summarized as follows: (1) The KM-H2 cells expressed neither OKT6 (CD1A) antigens nor S100 protein. They had no Birbeck granules. Thus, there was no evidence for their derivation from interdigitating reticulum cells or Langerhans cells. (2) The KM-H2 cells were negative for specific monocyte/macrophage markers such as lysozyme, *My4* (CD14), or *My7* (CD13). Moreover, they showed neither phagocytosis nor response to interferon gamma. Therefore, it seems unlikely that they are derived from the monocyte/macrophage system although a small amount of tumor necrosis factor was secreted from them. (3) The KM-H2 line did not express T-cell antigens such as Leu1 (CD5), OKT3 (CD3), WT1 (CD7), and OKT11 (CD2) antigens. Southern blot analysis showed that its T-cell receptor beta- and gamma-chain genes were in germ line configuration. Thus, it seems reasonable to exclude its origin from T-cell lineage. (4) The J_H region of the immunoglobulin (*Ig*) gene in the KM-H2 line was rearranged, while the J_k region was in germ line configuration. The transcripts of the heavy-chain gene or cytoplasmic u chains were not detected. The KM-H2 cells were also positive for markers specifically expressed on sIg(+) mature B cells (L26, CD21), although they were negative for markers expressed on immature B cells (CD10, TdT). These findings suggested their relation to B-cell lineage, though their differentiation stage could not be determined because they were genotypically undifferentiated but phenotypically differentiated.

These findings, described in Ishii et al. (1984), are not peculiar to the KM-H2 line. The L428 line also has similar features (Stein and Gerdes 1986), although its *Ig* genes were at a more differentiated stage than those of the KM-H2 line (a functional heavy-chain gene without the formation of a functional light-chain gene). Moreover, a recent report[1] by Weiss et al. (1986) suggests that at least some of the RS and H cells may have similar genotypic and phenotypic characteristics (three of eight cases showed only the heavy-chain rearrangements and one case showed the heavy-chain and kappa-chain gene rearrangements without the production of monoclonal Ig).

It seems clear that these abortive *Ig* gene rearrangements in these cell lines are formed through the normal *Ig* gene recombination process rather than through the neoplastic process, if we judge from their *Ig* gene recombination pattern. Therefore, if we abandon the current concept that B-cell differentiation is determined by the maturation stage of Ig synthesis system, these observations suggest that at least some of the RS and H cells are derived from an aberrant (or rare, but still normal) B-cell population with the abortive or aberrant *Ig* gene rearrangements which somehow survived and differentiated in the other aspects of cellular functions.

The nature of the HDLM 2 cell line (a RS and H cell line with the common phenotype that has only T-cell receptor gene rearrangements) may also indicate the presence of the RS and H cells derived from an aberrant T-cell population with the abortive or aberrant T-cell receptor rearrangements which somehow survived and differentiated in the other aspects of cellular functions.

Recently, Stein et al. (1985) have proposed the hypothesis that the RS and H cells originate from the acitvated B or T cells. Although this hypothesis can explain several features of the SR and H cells (expression of Ki-1 and Tac antigens or production of biologically active products etc.), it cannot explain why most of the SR and H cells lack sIg, cIg, or CD3 (T-cell receptor-associated antigen) because these proteins are usually not lost through activation of B- or T-lymphocytes by viruses or lectins [e.g., Epstein-Barr (EB) viruses, HTLV I viruses, or phytohemagglutinin (PHA)]. Moreover, it cannot explain the molecular genomic findings of the KM-H2 or L428 line as well as those findings reported by Weiss et al. (1986).

Therefore, we propose a new hypothesis that the majority of the SR and H cells are derived from aberrant (or rare but still normal) lymphocytes with defective expression of antigen receptors (Ig or T-cell receptors) through an intrinsic mechanism (e.g., abortive or aberrant rearrangements of antigen receptor genes) which somehow differentiated in the other aspects of cellular functions (Fig. 3). Although this cell population may appear heterogeneous as it includes the cells with various maturation stages of the *Ig* or T-cell receptor genes, we should note the following points: (1) However heterogeneous the maturation stages of the antigen receptor genes may be, there is no difference among these cells in the point that they lack

---

[1] As they did not perform the control experiments using DNA extracted from nonlymphatic cells, the "lambda chain gene rearrangements" in their report might be due to DNA polymorphism.

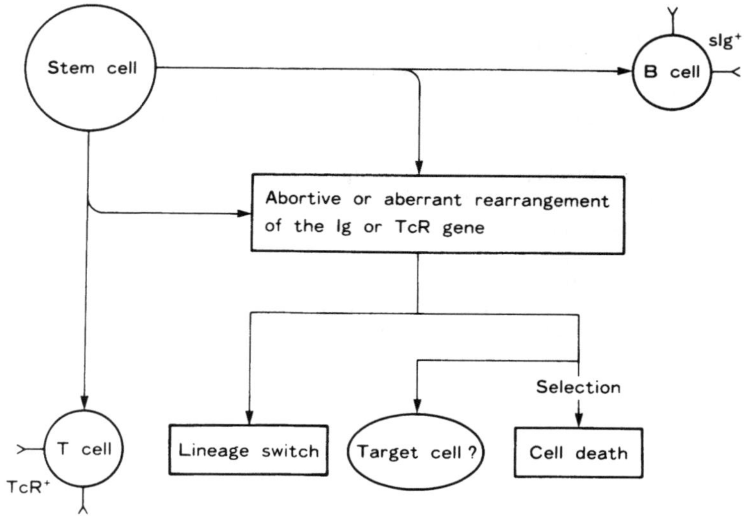

Fig. 3. Although the mechanisms of negative selection of non-functional B- or T-lymphocytes are not well understood, the analysis of KM-H2 and other cell lines suggests that the majority of Reed-Sternberg and Hodgkin cells might be derived from "abortive" lymphocytes which somehow survived and differentiated in cell functions other than the antigen recognition pathway

the functional antigen receptors. (2) It is now clear that B and T cells have many features in common. Both use identical proliferative systems (e.g., IL2 - IL2 receptor system, IL4 - IL4 receptor system) and secrete the same biologically active substances (IL1, etc.). Therefore, these aberrant cells may have similar phenotypic and functional properties in spite of their apparent heterogeneity.

Acknowledgment. We thank Drs. H. Miwa, Y. Arita, Y. Ohno, H. Yamabe, Y. Mizutani, and M. Hatanaka for their kind advice and cooperation. We are grateful to Dr. T. Honjo for his generous gift of probes. We also thank S. Yoshida for his excellent technical assistance.

References

Breton-Gorius J, Guichard J (1972) Ultrastructural localization of peroxidase activity in human platelets and megakaryocytes. Am J Pathol 66: 277-294

Graham RC, Karnovsky MJ (1966) The early stages of absorption of injected horseradish peroxidase in the proximal tubules of mouse kidney; ultrastructural cytochemistry by a new technique. J Histochem Cytochem 14: 291-302

Hayashi H, Kiyota T, Sakamoto H, Seto M (1985) An enzyme-linked immunosorbent assay for recombinant human tumor necrosis factor using monoclonal antibody. In: Ishigami J (ed) Recent advances in chemotherapy. University of Tokyo Press, Tokyo, pp 820-821

Ishii Y, Takami T, Yuasa H, Takei T, Kikuchi K (1984) Two distinct antigen systems in human B lymphocytes: identification of cell surface and intracellular antigens using monoclonal antibodies. Clin Exp Immunol 58: 183-192

Kamesaki H, Fukuhara S, Tatsumi E, Uchino H, Yamabe H, Miwa H, Shirakawa S, Hatanaka M, Honjo T (1986) Cytochemical, immunologic, chromosomal, and molecular genetic analysis of a novel cell line derived from Hodgkin's disease. Blood 68: 285-292

Stein H, Gerdes J (1986) Phenotypical and genotypical marker in malignant lymphomas: cellular origin of Hodgkin and Sternberg-Reed cells and implications on the classification of T-cell and B-cell lymphomas. Verh Dtsch Ges Pathol 70: 127-151

Stein H, Mason DY, Gerdes J, O'Connor N, Wainscoat J, Pellesen G, Gatter K, Falini B, Delsol G, Lemke H, Schwarting R, Lennert K (1985) The expression of the Hodgkin's disease associated antigen Ki-1 in reactive and neoplastic lymphoid tissue: evidence that Reed-Sternberg cells and histiocytic malignancies are derived from activated lymphoid cells. Blood 66: 848-858

Weiss LM, Strickler JG, Hu E, Warnke RA, Sklar J (1986) Immunoglobulin gene rearrangements in Hodgkin's disease. Hum Pathol 17: 1009-1014

*A Marker and Putative Pathoantigen of Hodgkin's Cells**

E. Paietta, R. J. Stockert, A. Morell, J. Racevskis, and P. H. Wiernik

Department of Oncology, Montefiore Medical Center and Departments of Medicine and Biochemistry, and Liver Research Center, Albert Einstein Cancer Center, Bronx, New York, NY, 10467, USA

Introduction

Hodgkin's disease has become a curable disease thanks to the sensitivitiy of its abnormal cells to available treatment strategies. However, the origin of these malignant cells, the typical giant, multinucleated Reed-Sternberg cells and their mononuclear variants, is still a subject of debate as is the early natural history of the disease (Bonadonna and Santoro 1985; Kaplan 1980). Furthermore, the success of modern therapy tends to blur the influence of classic prognostic factors such as histological type and immunodeficiency in Hodgkin's disease (Bonadonna and Santoro 1985). Hence, fundamental questions such as the cause-effect relationship of defective immune responsiveness in the pathogenesis of Hodgkin's disease remain unresolved. The early proposition that Hodgkin's disease might represent an autoimmune process, involving an interaction between the neoplastic cells and normal lymphocytes (Kaplan 1980; Zwitter 1984), was based upon the unique manifestation of the disease that it is initially confined to lymphoid tissue, and the observation that the few abnormal cells found in involved organs are surrounded by normal lymphoid cells. These lymphocytes, predominantly T4-helper cells (Archibald 1973), have the antigenic profile of activated cells ($T10^+$, $HLA\text{-}DR^+$, interleukin-2 $receptor^+$) (Pizzolo et al. 1984; Poppema et al. 1982), and, presumably as a result of this activation (Landolfi and Cook 1986), express hyposialylated surface glycans (Aisenberg and Wilkes 1982; Dorreen et al. 1984). A preexisting lymphocyte activation in vivo may, at least partially, explain the impaired lymphocyte reactivity in Hodgkin's disease as measured by reduced mitogen-induced and mixed lymphocyte culture responses of T-lymphocytes and abnormal immunoglobulin synthesis of B-lymphocytes after mitogenic stimulation (Romagnani et al. 1985). The postulate that lymphocyte activation may result from a reaction against unknown antigens on the abnormal Hodgkin's cells, possibly associated with the major histocompatibility (MHC) class II antigens (Romagnani et al. 1986), was

* Supported in part by grant P30CA13330 awarded by the NCI, DHHS, by a grant from the Chemotherapy Foundation, Inc., by a Henry and Lillian Stratton Foundation grant, and by the NIH grants DK-32972-04 and DK-34668.

supported by the finding of all 3 MHC class II molecules (HLA-DR, SB, DS) (Fisher et al. 1985) on cells cultured from a patient with advanced Hodgkin's disease and maintained in vitro, line L428 (Diehl et al. 1982). As additional features typical for cell types capable of expressing antigen to T-lymphocytes or acting as accessory cells for their activation (including dendritic cells and macrophages (Unanue and Allen 1987), L428 cells are able to present soluble antigen to T cells (Fisher et al. 1985), to serve as potent stimulators of the primary allogeneic mixed lymphocyte reaction (Fisher et al. 1983), to function as accessory cells for mitogen-induced human T-cell proliferative responses (Fisher et al. 1984), and to secrete interleukin-1 (Kortmann et al. 1984). However, this lymphokine was found not to be involved in the accessory cell function of L428 cells and neither was the HLA-DR molecule (Fisher et al. 1984), suggesting an as yet unknown specific molecule on the surface of Hodgkin's cells with which lymphocytes may be reacting.

Using the above Hodgkin's (HD) cell line and its variants (Diehl et al. 1982), we have recently obtained evidence for the existence and identity of such a hypothetical recognition site and for a true receptor-ligand interaction between Hodgkin's cells and lymphocytes (Paietta 1986a, b, 1987). We have found this "lymphocyte-receptor" to be a galactose-specific lectin, termed the HD lectin, which immunologically and functionally resembles the galactophilic asialoglycoprotein receptor of hepatocytes, the hepatic-binding protein (HBP) (Stockert 1983). The detection of the HD-lectin in HD-involved lymph nodes, spleens, and bone marrows suggests its potential physiological significance in the pathogenesis of Hodgkin's disease.

Results and Discussion

Functional Similarities Between the HD Lectin and HBP

The Hodgkin's cell lines (HD cells) designated L428, its variants (L428KS and KSA), and L540 have been provided to us by Dr. V. Diehl, University of Cologne, FRG. The variant L428KS has spontaneously arisen upon adaptation of the parental line to calf serum. L428KSA, the only line growing as adherent monolayer, was established by treating L428 cells with 12-tetradecanoylphorbol-13-acetate (10^{-8}–10^{-6} *M*) for 3 weeks. These lines vary in the percentage of cells positive for certain antigens (Diehl et al. 1982; Paietta and Wiernik 1985; Paietta et al. 1986a, b), and show differences in their highly abnormal karyotypes (Fonatsch et al. 1986, our own unpublished data). By indirect immunofluorescence, these cells lack a marker profile for a specific hemopoietic cell lineage including dendritic cells and monocytes (Diehl et al. 1982; Paietta and Wiernik 1985). Notably, the reactivity of these cells with myeloid-specific antibodies is limited to those recognizing the X-hapten (e.g., Leu-M1, VIM-D5), thus arguing against a granulocytic origin of these cells (Paietta and Wiernik 1985; Paietta et al. 1986a).

Recognition of the X-hapten, the 3-fucosyl-*N*-acetyllactosamine carbohydrate structure, on myeloid leukemic cells has been reported to be enhanced upon desialylation, and on lymphoblasts to occur de novo following neuraminidase treatment (Paietta et al. 1986a; Stockinger et al. 1984; Tabilio et al. 1984; Tetteroo et al. 1984). We found that treating HD cells with neuraminidase decreased the reactivity of anti-X-hapten antibodies by up to 50% (Paietta et al. 1986a).

The other characteristic feature of HD cells is the expression of high levels of ectosialyltransferase activity measured as the incorporation of radioactive sialic acid into endogenous, neuraminidase-desialylated substrate(s) (Paietta et al. 1986a). The level of ectosialyltransferase activity in HD cells correlated positively with the neuraminidase sensitivity of VIM-D5 antigenicity.

The neuraminidase-induced loss of VIM-D5 recognizition could be modulated by means known to alter the physiological requirements for sialyltransferase activity: (a) chelation of Ca^{2+} and high concentrations of exogenous galactose inhibited transferase activity and prevented the loss of VIM-D5 reactivity and (b) resialylation of the desialylated HD-cell surface with cytidine 5-monophosphate sialic acid restored preneuraminidase VIM-D5 antigenicity. Trypsinization of the neuraminidase-treated cells before incubation with the radioactive substrate inhibited incorporation of [^{14}C]sialic acid by 84%. These data support the surface localization of the enzyme in HD cells.

Thus, a galactophilic lectin is proposed to exist on the surface of HD cells that exhibits sialyltransferase activity. There is another well-characterized lectin which exhibits similar binding characteristics as proposed for the HD lectin, the asialoglycoprotein receptor or hepatic-binding protein of hepatocytes (HBP): HBP recognizes the oligosaccharide moiety of glycoproteins only when the terminal sialic acid residue has been removed (Stockert 1983); and the lectin activity of HBP is reversibly lost following neuraminidase treatment due to the binding of its own exposed galactosyl residues (Hudgin et al. 1974; Paulson et al. 1977; Stockert et al. 1977). This autoinhibition is mechanistically equivalent to that proposed for the loss of VIM-D5 antigenicity on HD cells. However, sialyltransferase activity as a possible physiological function of HBP has never been demonstrated, at least in the purified preparations tested (Hudgin et al. 1974).

Antigenic Similarities Between the HD Lectin and HBP

The putative galactophilic binding protein on the surface of HD cells appeared to be immunologically related to HBP in that polyclonal antiserum against rat HBP demonstrated cross-reactivity with the HD lectin (Paietta et al. 1986a, b, 1987). Presence of anti-HBP antiserum inhibited the neuraminidase-induced loss of X-hapten recognition on the surface of HD cells. Anti-HBP antiserum absorbed onto and eluted from the surface of HD cells was able to inhibit the lectin activity of HBP as defined by the binding of asialoorosomucoid (ASOR) to hepatocytes. The efficacy to inhibit ligand binding to HBP as exerted by antibody absorbed by HD cells was equivalent to a 1:200 dilution of unabsorbed anti-HBP antiserum. The absorption of anti-HBP-reactive material was specific to HD cells and their surface lectin, since non-HD cells lacking such lectin, e.g., the leukemic cell line HL-60, did not absorb HBP-inhibitory material from anti-HBP antiserum.

In immunoblot analysis of HD-cell proteins, polyclonal and monoclonal anti-HBP antibodies recognizing either the carbohydrate-binding domain or the polypeptide cytoplasmic portion of rat or human HBP all recognized the HD-lectin as a single band of approximately 55 Kd. This band was HD prominent both in vitro and in vivo (Fig. 1). While reactive or non-HD-involved tissues (e.g., thymus, lymph nodes) were essentially negative for the 55-Kd protein as recognized by an-

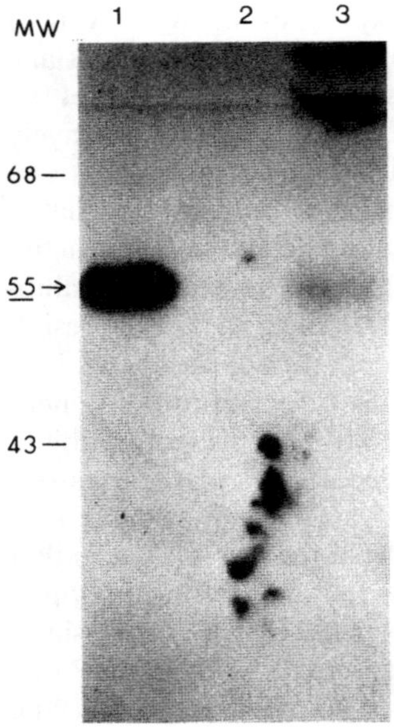

Fig. 1. Immunoblotting of proteins from L428KSA cells (*lane 1*) and from a lymph node involved with nodular sclerosing Hodgkin's disease (*lane 3*). *Lane 2* shows the lack of the 55-kd protein recognition in tissue from a reactive lymph node. *Numbers* indicate mobility of molecular weight markers (*MW*). The *arrow* points at the 55-kd HD protein

ti-HBP antiserum in immunoblotting or enzyme-linked immunosorbent assay, the protein was demonstrable in HD-involved lymph nodes, spleen, and bone marrow as well as in tissues heavily involved by leukemic monocytes and in a case of true histiocytic lymphoma. A protein of identical molecular mass was detected by two monoclonal antibodies recognizing the X-hapten (VIM-D5 and 1G10) both on the cell surface and, to an even larger extent, in lysates of HD cells. This strongly suggested that the surface HD-lectin as recognized by anti-HBP antiserum and the anchorage protein for the X-hapten in HD cells appeared to be identical (Paietta et al. 1986).

Expression of Sialyltransferase Activity by the 55-kd HD Protein

When HD-cell lysates were affinity chromatographed by incubation with ASOR-linked agarose beads, sialyltransferase activity was lost and the 55-kd HD protein could be recovered from the affinity matrix through elution with EDTA, which also recovered sialyltransferase activity. Demonstration of sialyltransferase activity in the EDTA eluate was dependent upon its reconstitution into a cellular milieu (Paietta et al. 1986b, 1987).

Regulation of the Expression of the HD Protein and of HD Cell Surface Sialyltransferase Activity by Culture Conditions Known To Regulate Expression of Human HBP

Expression of HBP in the human hepatoma cell line HepG2 is decreased by 60%–80% when the cells are grown in dialyzed fetal bovine serum (Collins et al.

1987). When L428KSA cells were grown in dialyzed fetal bovine serum, a progressive decrease in enzyme activity was seen, to 51% of control activity in complete serum at day 12 and 24% at day 18 of culture. In parallel, expression of the HD protein as recognized in immunoblotting by anti-HBP antiserum was gradually decreased (Paietta et al. 1987).

Thus, in addition to unique functional and immunological similarities between the hepatic and the HD-cell galactose lectin, a common regulatory mechanism for the expression of these two proteins can be postulated.

Lymphocyte-Agglutinating Ability of HD Cells

The HD-cell lectin, as HBP, the first lectin of mammalian origin described (Stockert et al. 1974), has the ability to agglutinate erythrocytes of A or B type, expressing *N*-acetyl-α-D-galactosamine and α-D-galactose as antigenic structures, but not erythrocytes of 0-type (Paietta et al. 1986a). Furthermore, as purified HBP (Novogrodsky and Ashwell 1977), we found HD cells to agglutinate lymphocytes in an HD-lectin-dependent fashion (Paietta et al. 1986b). Peripheral blood lymphocytes were isolated from patients who had received interleukin-2 for 5 days. Immunological phenotyping of the mononuclear cell population revealed less than 2% B-lymphocytes, greater than 80% T-lymphocytes (in the majority $T4^+$ cells), and approximately 15% monocytes. On average, 15% of the cells expressed the Tac antigen (3%-5% normal range) and 70% carried the T10 antigen, reflecting their activation due to the administration of interleukin-2 in vivo. Agglutination of these lymphocytes by addition of peanut agglutinin (50 µg/ml) suggested an asialo state of their surface membrane (Novogrodsky et al. 1975). Hence, this cell population showed the surface antigen pattern characteristically described for lymphocytes which closely interact with Reed-Sternberg cells in vivo and in vitro (Dorreen et al. 1984; Poppema et al. 1982). This macrophage-depleted (by adherence to plastic) lymphocyte population was mixed with HD cells at a ratio of approximately 20:1 and allowed to settle for 24 h at 4 °C, before rosette formation was evaluated. At that time, 72% of HD cells had formed rosettes with three to ten or more lymphocytes attached to the cell surface (Fig. 2). Addition of 200 m*M* galactose after rosette formation was completed substantially reversed the effect (greater than 70% of the rosettes were disrupted). When rosette formation was attempted in the presence of anti-HBP antiserum, only 29% of the HD cells formed rosettes. Thus, sugar specifically recognized by the HD protein as well as antiserum against the HD protein counteracted rosette formation. Further support for an involvement of the HD protein in the lymphocyte-HD cell interaction came from the use of HD cells which had been grown in dialyzed serum and showed a substantial decrease in the expression of the HD protein, as evaluated by immunoblotting, as well as in ectosialyltransferase activity (20% of control activity). Twenty-nine percent of these cells formed rosettes with lymphocytes. A cell line not expressing the HD protein, such as the promyelocytic leukemia cell line HL-60, did not react with these lymphocytes. Agglutination by HD cells resulted in the resialylation of lymphocyte membrane asialoglycans, further implying a dual function of the HD protein as galactose-specific lectin and ectosialyltransferase (Paietta et al. 1987). Purified

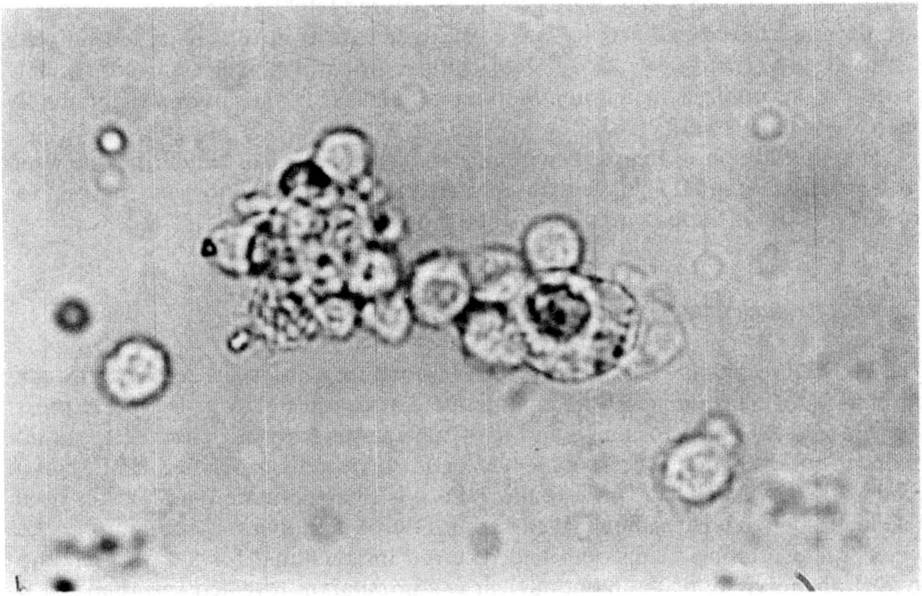

Fig. 2. HD cell-lymphocyte rosette formation. Two large HD cells are shown with agglutinating lymphocytes

HBP has been shown not only to agglutinate desialylated lymphocytes but also to cause their blastogenic transformation (Novogrodsky and Ashwell 1977) and to induce these cells to mediate mitogen-induced cellular cytotoxicity (Vierling et al. 1978). We have recent evidence that agglutination via the HD lectin, analogous to HBP, is followed by activation of agglutinated lymphocytes (Paietta et al. 1987).

Summary

A galactose-specific lectin, recently described by our laboratory, is immunologically demonstrable on the surface of neoplastic cells derived from patients with Hodgkin's disease. This Hodgkin's lectin is shown to be functionally and antigenically related to the galactose-*N*-acetylgalactosamine-specific lectin of the hepatocyte (HBP). Poly- and monoclonal antibodies against either the cytoplasmic tail or the cell-surface binding site of HBP recognize the Hodgkin's lectin as a 55 Kd protein. Expression of the 55 Kd antigen appears to be restricted to Hodgkin's disease involved tissues and cells of the monocyte/macrophage lineage. The putative identification of the Hodgkin's lectin as an ectosialyltransferase unique to Hodgkin's cells is supported by inhibition of enzymatic activity by anti-HBP antibodies. Cultured Hodgkin's cells, in analogy to purified HBP, agglutinate T-lymphocytes mediated by the Hodgkin's lectin. This cell-to-cell interaction results in the incorporation of sialic acid into lymphocyte surface asialoglycans as well as in the stimulation of lymphocyte proliferation. The function of the Hodgkin's lectin as lymphocyte agglutinant in vitro suggests its role as an immunomodulator contributing to the immunodeficiencies associated with Hodgkin's disease.

Acknowledgments: We are very grateful to Dr. V. Diehl and his collegues for providing us with the Hodgkin's cell lines. The skillful technical help of Ms. D. Thompson and M. McManus is greatly appreciated.

References

Aisenberg AC, Wilkes BM (1982) Lymph node T cells in Hodgkin's disease: analysis of suspensions with monoclonal antibody and rosetting technique. Blood 59: 522-527

Archibald RB (1973) Qualitative ultrastructural analysis of in vivo lymphocyte-Reed-Sternberg cell interactions in Hodgkin's disease. N C I Monogr 36: 239-245

Bonadonna G, Santoro A (1985) Clinical evolution and treatment of Hodgkin's disease. In: Wiernik PH, Canellos GP, Kyle RA, Schiffer CA (eds) Neoplastic diseases of the blood. Churchill Livingstone, New York, pp 789-826

Collins JC, Paietta E, Green R, Morell AG, Stockert RJ (1988) Biotin-dependent expression of the asialoglycoprotein receptor in $HepG_2$. J Biol Chem 263: 11280-11283

Diehl V, Kirchner HH, Burrichter H, Stein H, Fonatsch C, Gerdes J, Schaadt M, Heit W, Uchanska-Ziegler B, Ziegler A, Heintz F, Sueno K (1982) Characteristics of Hodgkin's disease-derived cell lines. Cancer Treat Rep 66: 615-632

Dorreen MS, Habeshaw JA, Stansfeld AG, Wrigley PFM, Lister TA (1984) Characteristics of Sternberg-Reed, and related cells in Hodgkin's disease. An immunohistological study. Br J Cancer 49: 465-476

Fisher RI, Bostick-Bruton F, Sauder DN, Scala G, Diehl V (1983) Neoplastic cells obtained from Hodgkin's disease are potent stimulators of human primary mixed lymphocyte cultures. J Immunol 130: 2666-2670

Fisher RI, Bates SE, Bostick-Bruton F, Tuteja N, Diehl V (1984) Neoplastic cells obtained from Hodgkin's disease function as accessory cells for mitogen-induced human T cell proliferative responses. J Immunol 132: 2672-2677

Fisher RI, Cossman J, Diehl V, Volkmann DJ (1985) Antigen presentation by Hodgkin's disease cells. J Immunol 135: 3568-3571

Fonatsch C, Diehl V, Schaadt M, Burrichter H, Kirchner HH (1986) Cytogenetic investigations in Hodgkin's disease: I. Involvement of specific chromosomes in marker formation. Cancer Genet Cytogenet 20: 39-52

Hudgin RL, Ashwell G (1974) Studies on the role of glycosyltransferases in the hepatic binding for asialoglycoproteins. J Biol Chem 249: 7369-7372

Hudgin RL, Pricer WE Jr, Ashwell G, Stockert RJ, Morell AG (1974) The isolation and properties of a rabbit liver binding protein specific for asialoglycoproteins. J Biol Chem 249: 5536-5543

Kaplan HS (1980) Hodgkin's disease: unfolding concepts concerning its nature, management and prognosis. Cancer 45: 2439-2474

Kortmann C, Burrichter H, Monner D, Jahn G, Diehl V, Peter HH (1984) Interleukin-1-like activity constitutively generated by Hodgkin derived cell lines. I. Measurement in a human lymphocyte co-stimulator assay. Immunobiology 166: 318-333

Landolfi NF, Cook RG (1986) Activated T-lymphocytes express class 1 molecules which are hyposialylated compared to other lymphocyte populations. Mol Immunol 23: 297-309

Novogrodsky A, Ashwell G (1977) Lymphocyte mitogenesis induced by a mammalian liver protein that specifically binds desialylated glycoproteins. Proc Natl Acad Sci USA 74: 676-678

Novogrodsky A, Lotan R, Ravid A, Sharon N (1975) Peanut agglutinin, a new mitogen that binds to galactosyl sites exposed after neuraminidase treatment. J Immunol 115: 1243-1248

Paietta E, Wiernik PH (1985) Hodgkin's cell lines: immunologically incompatible with any defined hemopoietic cell type. Clin Res 33: 384

Paietta E, Stockert RJ, Morell AG, Diehl V, Wiernik PH (1986a) Lectin activity as a marker for Hodgkin disease cells. Proc Natl Acad Sci USA 83: 3451-3455

Paietta E, Stockert RJ, Morell AG, Diehl V, Wiernik PH (1986b) Unique antigen of cultured Hodgkin's cells. A putative sialyltransferase. J Clin Invest 78: 349-354

Paietta E, Hubbard AL, Wiernik PH, Diehl V, Stockert RJ (1987) Hodgkin's cell lectin: an ectosialyltransferase and lymphocyte agglutinant related to the hepatic asialoglycoprotein receptor. Cancer Res 47: 2461-2467

Paulson JC, Hill RL, Tanabe T, Ashwell G (1977) Reactivation of asialo-rabbit liver binding protein by resialylation with β-D-galactosidase α 2→6 sialyltransferase. J Biol Chem 252: 8624-8628

Pizzolo G, Chilosi M, Semenzato G, Caligaris-Cappio F, Fiore-Donati L, Perona G, Janossy G (1984) Immunohistological analysis of Tac antigen expression in tissues involved by Hodgkin's disease. Br J Cancer 50: 415-417

Poppema S, Bhan AK, Reinherz EL, Posner MR, Schlossman SF (1982) In situ immunologic characterization of cellular constituents in lymph nodes and spleen involved by Hodgkin's disease. Blood 59: 226-232

Romagnani S, Rossi Ferrini PL, Ricci M (1985) The immune derangement in Hodgkin's disease. Semin Hematol 22: 41-55

Romagnani S, Maggi E, Parronchi P, Macchia D, Del Prete GF, Rossi Ferrini PL, Ricci M, Moretta L (1986) Clonal analysis of T lymphocytes in spleen from patients with Hodgkin's disease. Frequent occurrence of unusual T4-positive cells which co-express cytolytic activity and production of interleukin-2. Int J Cancer 37: 343-349

Stockert RJ (1983) Asialoglycoproteins: hepatic clearance and degradation of serum proteins. In: Cuatrecasas P, Roth TF (eds) Receptor-mediated endocytosis, chap 7. Chapman and Hall, London, pp 169-185 (Receptors and recognition, series B, vol 15)

Stockert RJ, Morell AG, Scheinberg IH (1974) Mammalian hepatic lectin. Science 186: 365-366

Stockert RJ, Morell AG, Scheinberg H (1977) Hepatic binding protein: the protective role of its sialic acid residues. Science 197: 667-668

Stockinger H, Majdic O, Liszka K, Aberer W, Bettelheim P, Lutz D, Knapp W (1984) Desialylation exposes myeloid antigens on acute lymphoblastic leukemia cells. J N C I 73: 7-11

Tabilio A, Del Canizo MC, Henri A, Guichard J, Mannoni P, Civin CI, Testa U, Rochant H, Vainchenker W, Breton-Gorius J (1984) Expression of SSEA-1 antigen (3-fucosyl-*N*-acetyllactosamine) on normal and leukaemic human haemopoietic cells: modulation by neuraminidase treatment. Br J Haematol 58: 697-710

Tetteroo PAT, Mulder A, Lansdorp PM, Zola H, Baker DA, Visser FJ, von dem Borne AEGKr (1984) Myeloid-associated antigen 3-α-fucosyl-*N*-acetyllactosamine (FAL): location on various granulocyte membrane glycoproteins and masking upon monocyte differentiation. Eur J Immunol 14: 1089-1095

Unanue ER, Allen PM (1987) The basis for the immunoregulatory role of macrophages and other accessory cells. Science 236: 551-557

Vierling JM, Steer CJ, Hickman JW, James SP, Jones EA (1978) Cell-mediated cytotoxicity of desialylated human lymphocytes induced by a mitogenic mammalian liver protein. Gastroenterology 75: 456-461

Zwitter M (1984) Hodgkin's disease: five steps from autoimmunity to cancer. Med Hypotheses 15: 399-413

Development in Diagnostic Procedures

Radiolabeled Monoclonal Antibodies Against Reed-Sternberg Cells for in Vivo Imaging of Hodgkin's Disease by Immunoscintigraphy

P. Carde, M. Pfreundschuh, L. da Costa, L. Manil, J. D. Lumbroso, B. Caillou, C. Parmentier, J. C. Saccavini, and V. Diehl

Institut Gustave-Roussy, 39 Rue Camille Desmoulins, 94805 Villejuif Cedex, France

Introduction

Although it is characterized by its propensity for dissemination, Hodgkin's disease (HD) can be cured by locoregional irradiation (Carde et al. 1988; Kaplan 1980; Tubiana et al. 1985). Since intensive chemotherapy and/or extended irradiation carry the risk of considerable acute and long-term toxicity (Longo et al. 1982), limiting treatment as much as possible is a valid goal. However, such a therapeutic strategy must be based on accurate staging procedures that reflect the actual extent of disease. Such staging procedures have long been relying on exploratory laparotomy and splenectomy (Castellino et al. 1984; Kadin et al. 1971; Tubiana et al. 1985). Although criticized (Bergsagel et al. 1982; Gomez et al. 1983; Lacher 1983) because of its morbidity, the sensitivity and specificity of surgical staging has not been reached by computerized tomography of the abdomen (Castellino et al. 1984) or high-dose gallium scanning (Blackwell et al. 1986) with lymphangiography. Even sophisticated combinations of prognostic factors (Tubiana et al. 1985) identified by multivariate analyses of large series are unable to predict an infradiaphragmatic involvement in 10%-30% of cases, depending on the combination of factors that is used.

Other important diagnostic issues are the recognition of lung, bone, pleural and chest wall involvement; after treatment, the interpretation of residual masses, present in up to 40% of stage IIIB-IV patients in the European Organization for Research and Treatment of Cancer (EORTC) series, is a major problem, one-fifth of these patients having equivocal visceral disease.

The development of monoclonal antibodies (Mab) against Hodgkin's derived cell lines (Pfreundschuh et al. 1988) made it possible to investigate the value of immunoscintigraphy of Hodgkin's disease. We now report the results of a pilot study that has been conducted in six patients with active Hodgkin's disease using the whole HRS-1 Mab. In five cases, the Mab labeled with ^{131}I was detected with linear scintigraphy. In one case, ^{123}I labeling allowed a three-dimensional reconstruction of radiolabeled Mab distribution through the utilization of tomoscintigraphy (Mach et al. 1981).

Recent Results in Cancer Research. Vol. 117

Patients, Materials, and Methods

Production, and Radiolabeling of the HRS-1 Mab

The Hodgkin's derived cell line 428 (Diehl et al. 1982, 1985) was used to immunize Balb/c mice to obtain immune spleen cells, which were fused with mouse myeloma cells NS-1. One hybridoma, obtained by such a fusion, produced a Mab which shows a restricted reactivity with Hodgkin and Reed-Sternberg cells. The immunoglobulin isotype of HRS-1 as determined by ELISA of HRS-1 is IgG2a. The antigen recognized by HRS-1 is a glycoprotein with a molecular weight of 120 kd (gp 120). The HRS-1 Mab reacts with all the Hodgkin and Reed-Sternberg cell in lymph nodes and spleens from the nodular sclerosis (NS), mixed cellularity (MC), and lymphocyte depletion (LD) specimens and with some of the lymphocyte predominance (LP) subtype (Pfreundschuh et al. 1988). Rare large cells in the perifollicular region of normal lymph nodes and tonsils are also recognized by the Mab. Besides Hodgkin's lymphoma, large-cell anaplastic non-Hodgkin's lymphomas (NHLs) also demonstrate a strong staining in the overwhelming majority of the neoplastic cells. In contrast, cells are only rarely stained in other subtypes of NHLs. Other malignancies (hematopoietic system or solid tumors) are negative. Normal peripheral mononuclear cells become positive after exposure to phytohemagglutinin (PHA) or Epstein-Barr virus (EBV) and after a mixed lymphocyte culture, suggesting that the HRS-1-associated gp 120 is an activation antigen. Blocking experiments suggest that other Mabs such as HRS-2 (Pfreundschuh et al. 1988) and Ki-1 (Stein et al. 1985) recognize different epitopes of the same antigen.

Purification of the HRS-1 Mab. The mouse ascites was purified by protein A Sepharose 4B affinity chromatography (Pharmacia Fine Chemicals, Uppsala, Sweden). The purity was demonstrated by two-dimensional gel electrophoresis (Manil et al. 1986), to be greater than 98%. To remove a possible protein A contaminant for clinical use, the purification was completed by either ion exchange chromatography in DEAE matrix or fast flow column. Protein purity was finally assessed by high-performance liquid chromatography.

Immunoreactivity was tested by radioimmunoassay, using the L428 cell lines as positive control and RAJI and MOLT-4 lines as negative controls.

Labeling of HRS-1. Mab labeling was performed with ^{131}I in five cases and ^{123}I in one case (Oris, Gif-sur-Yvette) by the iodogen method, which provides an 80% labeling efficiency. After removal of free iodine, the Mab was sterilized by filtration through 0.22 μm Millex HA filters (Millipore, Mosheim, France) and tested for sterility and for apyrogenicity (Limulus test) before injection. Patients received 0.5 mg [^{131}I]HRS-1 Mab, labeled with a mean of 70 MBq (1.9 mCi) ^{131}I. One patient received 2 mg HRS-1 Mab labeled with 251 MBq (6.8 mCi) ^{123}I. In addition, a Mab of the same class as HRS-1 (IgG2a), AFO1 (Bellet et al. 1984; Manil et al. 1986), which is directed against alpha-fetoprotein (AFP), was labeled by ^{123}I and used as control in patients injected with [^{131}I]HRS-1 Mab. Two patients received 0.5 mg anti-AFP labeled with 251 and 259 MBq (6.8 and 7 mCi) ^{123}I, respectively.

Table 1. Immunoscintigraphy with iodine-labeled HRS-1: patients' clinical data

Case no.	Age (years)	Sex	Histology	Stage	HRS-1 dose (MG)	Label	Dose (MCi)
1	18	F	NS	IIA	0.5	^{131}I	1.8
2	57	M	MC	IIIA	0.5	^{131}I	1.9
3	19	M	NS	IIA	0.5	^{131}I	1.3
4	59	M	MC	IIA	0.5	^{131}I	1.8
5	13	F	NS	IVA	0.2	^{131}I	1.0
6	36	M	MC	IVB	2.0	^{123}I	6.8

NS, nodular sclerosis; MC, mixed cellularity.

Patients

Six patients with biopsy-proven Hodgkin's disease and no previous allergy were entered into this study after they gave informed consent. The main characteristics of the patients are listed in Table 1. All patients were given 30 daily drops of Lugol's iodide solution for 5 days prior to and after Mab injection. Mab was infused i.v. in 250 ml normal saline over 60 min.

Immunoscintigraphy

Planar scanning was carried out from day 1 up to day 6 after injection in patients who received [^{131}I]HRS-1 Mab. Acquisitions were performed on a CGR gamma camera. Regions of interest (RI) were selected after clinical, radiological, isotopic, or nuclear magnetic resonance (NMR) imaging. Visual and quantitative analysis were performed for immunoscan interpretation.

Tomoscintigraphy was performed with single-photon emission tomography (SPECT). Tomographic reconstruction from collected data was performed on an Infogam (Sopha Medical) nuclear medicine computer with original algorithms without using subtraction techniques (Berche et al. 1978).

A specificity index (Pressman et al. 1967) was estimated in one patient who underwent biopsy of a lymph node and the surrounding soft tissue after he was injected with both the [^{131}I]HRS-1 Mab and the control [^{123}I]anti-AFP Mab. The radioactivities of nodal and soft tissue were both counted with a gamma counter at two windows of energy for ^{131}I (peak 364 keV$\pm$20%) and for ^{123}I (peak 159 keV$\pm$20%).

Pathological control of the HD material with the HRS-1 Mab by immunoperoxidase was performed whenever infixed biopsy material was available.

Results

The clinical data and a summary of immunoscintigraphy (IS) results are given in Tables 1 and 2. All patients tolerated the application of the radiolabeled HRS-1 without any side effects. The results are graded as negative ($-$) or doubtful ($\pm$)

Table 2. Immunoscintigraphy (IS) with iodine-labeled monoclonal antibody HRS-1: results in six patients

Case no.	Known sites of involvement before IS	Radioactive uptake in IS
1	Ln cervical right	Ln cervical right
	Ln mediastinal	Ln mediastinal
2	Ln cervical left	Ln cervical left
	Spleen[a]	Spleen[b]
3	Ln cervical left	Ln cervical left
	Ln axillary right	Ln axillary right
	Ln mediastinal	Ln mediastinal
4	Ln cervical right	Ln cervical right
	Ln cervical left	Ln cervical left
5	Ln cervical right	None
	Ln cervical left	None
	Ln mediastinal	None
6	Bone (anterosuperior iliac crest, L_3 vertebra, right superior humerus)	Bone (anterosuperior iliac crest, L_3 vertebra, right superior humerus)

Ln, lymph nodes.
[a] Negative in sonography and CT scan of abdomen.
[b] Involvement histologically verified.

when radioactive uptake in tumoral areas was absent or identical to surrounding tissues, respectively. Positive tumoral uptake was judged as weak (+) or strong (++).

Of the five patients who received [^{131}I]HRS-1, results were positive in four, and doubtful in the last patient. In the sixth patient, who received [^{123}I]HRS-1, there was also a positive imaging of involved areas.

Imaging of Peripheral Lymph Nodes. Four of five patients with peripheral nodal involvement (cases 1, 2, 3, and 4) had + or ++ scans in these areas.

Imaging of Mediastinal Tumors. There were three patients with mediastinal involvement. The immunoscintigraphic imaging was graded as + in a patient (case 1) with involvement of the upper mediastinum, ++ in a patient (case 3) who received [^{131}I]HRS-1 and [^{123}I] anti-AFP control, and ± in a patient (case 5) with small mediastinal involvement.

Demonstration of Splenic Involvement. In case 2, no clinical or radiological evidence for spleen involvement with Hodgkin's lymphoma could be seen prior to IS. However, IS was performed because a splenic puncture which was done a few weeks prior to the examination demonstrated involvement with Hodgkin's lymphoma. The spleen was not enlarged. IS demonstrated hot spots both in the spleen and in the liver. The latter, however, was not biopsied.

Detection of Bone Involvement. Single-photon emission tomography was performed in case 6, who received [^{123}I]HRS-1. This patient presented with relapsed bony Hodgkin's lymphoma. IS demonstrated three hot spots: one of + + grade in the anterosuperior iliac crest, one of + + grade in the L_3 vertebra, and one of + + grade in the right superior humerus. These hot spots were also visible on $^{99}Tc^{m}$ pyrophosphate planar scintigraphy and by NMR examination. In addition, the left anterosuperior lesion was histologically verified.

Specificity Index. In case 4, who was injected with both the [^{131}I]HRS-1 Mab and the [^{123}I]anti-AFP Mab, a supraclavicular lymph node was biopsied at day 4 and demonstrated a specificity index of 1.3. This same node demonstrated an immunoperoxidase HRS-1 staining on cryostat section of Hodgkin and Reed-Sternberg cells, which were infrequent in these tissue sections.

Discussion

Place of IS in the Staging of Lymphomas Compared with Radiographic and Ultrasound Techniques

Evaluation of Residual Masses. Improvement of the prognosis of lymphoma is due in part to progress in techniques used to detect sites of involvement in Hodgkin's lymphoma. Whereas X-ray films and ultrasound techniques cannot distinguish between residual inactive masses and active disease, IS with specific Mabs might be able to do so, as specific Mabs should not bind to residual masses free of Hodgkin and Reed-Sternberg cells.

Mediastinal/Hilar Nodes. The prognostic significance of mediastinal/hilar nodes has been much debated (Cosset et al. 1984). CT scan undoubtedly improves the design of radiotherapy portals by showing nodes as abnormal when over 15 mm in diameter, particularly if they are located in the retrocardiac area. However, pericardial, chest wall, and pleural involvements are not adequately assessed. Previously unsuspected nodes were detected by IS in the upper mediastinum (case 1) and were confirmed by CT scan; in case 3, mediastinal nodes could be visualized by double labeling, more easily on oblique images; conversely, the contrast was insufficient in case 5, where slightly enlarged retrocardiac nodes were superimposed with the heart shadow.

Spleen. The most important task of any noninvasive diagnostic procedure in Hodgkin's lymphoma, however, is the diagnosis of infradiaphragmatic involvement, which is present in about one-third of Hodgkin's lymphoma and one-half of large-cell NHL cases. Splenic involvement in the absence of splenomegaly is a major prognostic factor (Rosenberg and Kaplan 1985; Tubiana et al. 1985) and is used as a guide in selecting the therapeutic strategy to employ (Bergsagel et al. 1982; Gallez-Marechal et al. 1984). Moreover, a normal-sized spleen may be the only site of infradiaphragmatic involvement in 5%-10% of cases, while one-third of enlarged spleens have no evidence of Hodgkin's lymphoma on histologic examination (Kadin et al. 1971). Ultrasound and CT scan have been shown to be of no value for the detection of splenic involvement in Hodgkin's lymphoma (Castellino et al. 1984) due to their poor sensitivity (33%) and a specificity of 76%, resulting in an accuracy of 58%. In the small series of our pilot study, an increased uptake in the spleen was observed in the immunoscintigrams of one patient; there, splenic involvement could be demonstrated histologically, even though clinical examination and CT had been negative.

Retroperitoneal Lymph Nodes. Retroperitoneal nodes are well detected by lymphography, which demonstrates size and structure alterations presumably due to Hodgkin's lymphoma. Its overall accuracy is better than that of CT scan (Castellino et al. 1984) in very experienced teams. The theoretical advantage of CT scan in showing involved celiac, retrocrural, splenic, renal, and mesenteric nodes concerns mainly NHL. It suffers in any case from its inability to detect architectural modifications in unenlarged nodes and from its poor accuracy for mesenteric nodes. The present series of patients did not provide the opportunity to detect retroperitoneal nodes with IS, as none of the patients in the pilot study had enlarged retroperitoneal nodes (as demonstrated by other techniques). However, the demonstration that nodal uptake was detected in cervical and mediastinal areas suggests that it will be feasible in the retroperitoneum as well.

Liver Involvement. The problems of detecting liver involvement are similar for classical CT and ultrasound and IS. They stem from the fact that hepatic involvement in Hodgkin's lymphoma is characterized by small macroscopic areas of periportal involvement or only microscopic involvement. It is also less frequent (5%) than splenic involvement and almost always associated with the latter. Indeed, IS was positive in the liver in one of the six patients (case 2), where it was also positive in the spleen. The other patients had no increased splenic or hepatic uptake. Patient 2 had histologically proven splenic involvement and thus was at very high risk for liver involvement, although it was not demonstrated by classical CT scan or ultrasound. However, as no biopsies of the liver could be obtained, we do not know whether the uptake of radiolabeled HRS-1 in the liver was unspecific or represented true involvement of Hodgkin's lymphoma.

Place of IS Compared with Classical Scintigraphy

Perivascular and peribronchial infiltrations by Hodgkin's lymphoma may be hardly distinguishable from pneumonia or atelectasis-related abnormalities. Gallium scanning, as a part of lymphoma workup (Anderson et al. 1983), detects many inflammatory processes which may possibly be associated with lymphoma (Castellino et al. 1984; Ford et al. 1987). In addition, [^{67}Ga]citrate scanning concentrates in the colon and some normal organs. Although ^{67}Ga detects Hodgkin's lymphoma more easily than NHL, it fails to recognize not only lesions of less than 2 cm but also some bulky (necrotic) masses. Furthermore, ^{67}Ga scanning is not specific for a given tumor type (Hibi et al. 1987; Tsan 1985). It may, however, be helpful in differentiating residual fibrosis after radiotherapy from residual active or recurrent Hodgkin's lymphoma. Finally, its poor accuracy below the diaphragm (Castellino et al. 1984; Kaplan 1980), where splenic and mainly liver detection is very poor, explains the gloomy prospects for ^{67}Ga scanning development in lymphoma staging.

This is also the case for thallium scintigraphy, which has recently been reported to be valuable (Linde and Basso 1987; Winzelberg et al. 1986), in particular in lymphocyte predominance Hodgkin's lymphoma. However, positive imaging with thallium in a patient with lymphocyte-predominant Hodgkin's lymphoma, which is a rare hypercellular histologic subtype, cannot be extrapolated to other subtypes (Linde and Basso 1987).

Positive fluorine-18 ([^{18}F]-2-fluoro-2 deoxy-D-glucose) imaging has been reported in four of five patients with NHL, whereas only two ^{67}Ga scans were positive. However, fluorodeoxyglucose has no specificity for tumoral tissues (Paul 1987) since it only reflects increased glucose utilization.

Diagnosis of bone involvement is based generally on ^{99}Tc^{m}-labeled pyrophosphate bone scanning. However, ^{99}Tc^{m} scanning is even less specific than conventional X-rays (Skarin 1985), or CT scan, and increased radionuclide uptake is observed in a variety of malignant and benign diseases. Bony hot spots detected by ^{99}Tc^{m} scanning have to be confirmed by conventional radiography, which may fail to do so because it is less sensitive.

Furthermore, an X-ray appearance of osteoplastic or sclerotic lesions may reflect either healed lesions or active disease, a fact which represents a diagnostic dilemma that is not infrequent in Hodgkin's lymphoma. In this regard, the IS results observed in case 6 are particularly informative. The three areas of HRS-1 Mab uptake corresponded with painful zones and areas of increased ^{99}Tc^{m} pyrophosphate uptake although conventional X-ray films were normal. Furthermore, the abnormal anterior iliac crest was confirmed by biopsy as involved with Hodgkin's lymphoma. Moreover, IS clearly demonstrated a paraspinal mass at the L_3 level, related to a periosteal reaction which frequently occurs in Hodgkin's lymphoma. This paraspinal mass was also demonstrated on the CT scan. However, CT scan imaging was unable to differentiate the vertebral lesions at the dorsal level, where they were no longer active after irradiation, from the active lesions at the L_3 level, whereas IS indicated only active vertebral disease. As the interval between the end of vertebral irradiation and IS was short (2 weeks), this observation may indicate that IS is a very early indicator of tumor response. Indeed, the speed of tumor response in Hodgkin's lymphoma (Carde et al. 1983; Somers and Henry-Amar 1987) and in non-Hodgkin's lymphoma (Armitage et al. 1982) is probably of primary prognostic importance. Early indicators for tumor response may help to select in due time those patients who require more intensive or alternative types of treatment.

Place of IS in Lymphoma Imaging

Attempts to detect NHL with IS have been addressed mainly to cutaneous T-cell lymphoma using T101 Mab cells. A first series of 11 patients (Carrasquillo et al. 1986) with ^{111}In-labeled T101 demonstrated an increased Mab concentration in infiltrated erythroderma and nodes, and nonspecific uptake mainly in liver but also in marrow and spleen. Similar results have been reported elsewhere (Dillman et al.

1987). Modulation of the T65 antigen with internalization from the cell surface was interpreted as specific for lymphoma and as a possible explanation for increased and prolonged tumor uptake of the labeled Mab when compared with that observed in solid tumor IS imaging. Other favorable characteristics reported for IS were "good incorporation" of ^{111}In into the Mab; its preserved immunoreactivity after labeling; and the stability of the complex, with minimal ^{111}In translocation from the Mab to transferrin.

In a subsequent study (Carrasquillo et al. 1987) of four patients, [^{131}I]T101 Mab was compared with [^{111}In]T101 Mab. The comparison was to the disadvantage of [^{131}I]Mab, which showed suboptimal uptake in nodes and none in the involved skin. It was suggested that iodinated Mab in vitro and in vivo behave differently, with rapid in vivo deiodination occurring mainly in the liver and spleen. Release of ^{131}I (or ^{125}I) was thought to occur not from loose binding, but from cell release after T65 antigen-Mab internalization. Our own observations with the [^{131}I]HRS-1 Mab indicated a better nodal labeling than with [^{131}I]T101, suggesting a better localization and/or a lesser ^{131}I release of the HRS-1 Mab. In any case, labeling of HRS-1 Mab with ^{111}In should be tested for Hodgkin's lymphoma imaging other than in spleen and liver.

IS imaging has been reported for Hodgkin's disease using a polyclonal [^{131}I] antiferritin IgG (Order et al. 1980). Ferritin is a tumor-associated protein present in a variety of solid tumors. More recently, the same team correlated SPECT using [^{131}I]antiferritin and gallium imaging of lung lesions in Hodgkin's lymphoma (Lenhard et al. 1985). These attempts at Hodgkin's lymphoma imaging with IS were part of a therapeutic trial involving repeated administration of high doses of [^{131}I]antiferritin antibody. In the patients with Hodgkin's lymphoma, the absence of production of autoantibodies was characteristic. Such antibodies were responsible for decreased tumor targeting and antibody half-lives in patients with other tumor types and necessitated "recycling" with antibodies from different animal species in repeated treatments (Klein et al. 1986).

Finally, the disadvantages of each particular method discussed above illustrate the advantages of our approach with the HRS-1 Mab: it is directed with high affinity to an antigen closely associated with the neoplastic cells in Hodgkin's lymphoma and in a subset of large-cell anaplastic NHL. Its monoclonality provides reproducible characteristics of purity and of biological behavior, with the iodogen radioiodination method best preserving the antibody activity (Ghose 1987; Saccavini et al. 1985). Choice of labeling method and of isotopes allowed for good planar imaging with [^{131}I]Mab and SPECT with [^{123}I]Mab. Double labeling with a control Mab suggested specific in vivo imaging and specific (although suboptimal) uptake of HRS-1 in a biopsied node; moreover, a variety of disease localizations in nodes, mediastinum, spleen, liver, and bone were obtained.

Improvement of the imaging techniques should come from the utilization of $F(ab')_2$ fragments, use of higher Mab quantities, and/or use of a cocktail of Mabs directed against different epitopes of the same antigen (Pfreundschuh et al. 1988). Selection of different isotopes (Anderson and Strand 1987; Endo et al. 1987; Saccavini et al. 1985; Vaughan et al. 1987) adapted to specific localizations of Hodgkin's lymphoma, such as ^{111}In for nonhepatosplenic lesions, should be investigated.

Undoubtedly, IS with HRS-1 Mab brings a new functional pattern to the imaging of lymphoma. It will be worth testing IS for posttreatment evaluation of treatment complications (postradiation pleural effusion or pneumonitis, lung opportunistic infections such as pneumocystis carinii) or for the evaluation of residual masses (abdomen, retroperitoneum, mediastinum) with equivocal disease activity.

Provided that the preliminary results reported here can be confirmed in a larger series of patients, the use of HRS-1 and related Mabs may be extended to other diagnostic procedures, such as gadolinium attachment for nuclear magnetic resonance imaging (NMRI) or immunolymphoscintigraphy, as reported in two mycosis fungoides patients injected with [^{111}In]T101 Mab (Keenan et al. 1987). Development of in vitro purging of bone marrow or radioimmunotherapy might be possible if the in vivo imaging techniques with radiolabeled HRS-1 demonstrate sensitive and specific uptake of the antibody by Hodgkin's lymphoma.

Summary

The Hodgkin Reed-Sternberg (HRS-1) monoclonal antibody (Mab) was raised against the L 428 Hodgkin's disease (HD) cell line. The HRS-1 Mab was labeled with radioactive iodine and injected into six patients with Hodgkin's disease of varied histological subtypes for immunoscintigraphic imaging. In five patients, the HRS-1 Mab was labeled with ^{131}I; a control antialpha-fetoprotein (AFP) Mab was injected simultaneously in two of these five cases. Four of five patients had a positive scan (nodal, splenic and hepatic involvements), the results in the fifth patient being equivocal. In the sixth patient, the HRS-1 Mab was labeled with ^{123}I in order to utilize tomoscintigraphy instead of linear scintigraphy. Although the immunoscintigraphy (IS) was performed secondary to effective chemotherapy, images of bony disease were demonstrated. These preliminary results demonstrate that IS with iodine-labeled HRS-1 Mab is feasible and informative in Hodgkin's lymphoma. The real clinical value and the specificity of IS deserves confirmation in a larger series of patients. Several techniques such as the use of Fab or $F(ab')_2$ fragments should further improve the results.

References

Anderson KC, Leonard RCE, Canellos GP (1983) High-dose gallium imaging in malignant lymphoma. Am J Med 75: 327-334

Anderson WT, Strand M (1987) Radiolabeled antibody: iodine versus radiometal chelates. NCI Monogr 3: 149-151

Armitage JO, Dick FR, Corder MP et al. (1982) Predicting therapeutic outcome in patients with diffuse histiocytic lymphoma treated with cyclophosphamide, adriamycin, vincristine and prednisone. Cancer 50: 1695-1702

Bellet DH, Wands JR, Isselbacher KJ, Bohuon C (1984) Serum-fetoprotein levels in human disease: perspective from a highly specific monoclonal radioimmunoassay. Proc Natl Acad Sci USA 81: 3869-3873

Berche C, Aubert B, Bethencourt A, Di Paola R (1978) Three-dimensional reconstruction based on sequential information from a scintillation camera. In: Brill AB, Price RR (eds)

Information processing in medical imaging. Proceedings of the Vth International Conference, Nashville, Vanderbilt University, BCTIC, pp 214-251
Bergsagel DE, Alison RE, Bean HA et al. (1982) Results of treating Hodgkin's disease without a policy of laparotomy staging. Cancer Treat Rep 66: 717-731
Blackwell EA, Joshua DE, McLaughin AF et al. (1986) Early supradiaphragmatic Hodgkin's disease. High dose gallium scanning obviates the need for staging laparotomy. Cancer 58: 883-885
Carde P, Mackintosh FR, Rosenberg SA et al. (1983) A dose and time response analysis of the treatment of Hodgkin's disease with MOPP chemotherapy. J Clin Oncol 1: 146-153
Carde P, Henry-Amar M, Burgers JMV et al. (1988) Clinical stages I and II Hodgkin's disease: a specifically tailored therapy according to prognostic factors. J Clin Oncol 6: 239-252
Carrasquillo JA, Bunn PA, Keenan AM et al. (1986) Radioimmunodetection of cutaneous T-cell lymphoma with ^{111}In-labeled T101 monoclonal antibody. N Engl J Med 315: 673-680
Carrasquillo JA, Mulshine JL, Bunn PA et al. (1987) Indium-111 T101 monoclonal antibody is superior to iodine-131 T101 in imaging of cutaneous T-cell lymphoma. J Nucl Med 28: 281-287
Castellino RA, Hoppe RT, Blank N et al. (1984) Computerized tomography, lymphography and staging laparotomy: correlations in initial staging of Hodgkin's disease. AJR 143: 37-41
Cosset JM, Henr-Amar M, Carde P et al. (1984) The prognostic significance of large mediastinal masses in the treatment of Hodgkin's disease. The experience of the Institut Gustave-Roussy. Hematol Oncol 2: 33-43
Diehl V, Kirchner HH, Burrichter H, Stein H, Fonatsch C, Gerdes J, Schaadt M, Heidt W, Uchanska-Ziegler B, Heintz F, Sueno K (1982) Characteristics of Hodgkin derived cell lines. Cancer Treat Rep 66: 615-632
Diehl V, Pfreundschuh M, Fonatsch C, Stein H, Falk M, Burrichter H, Schaadt M (1985) Phenotypic and genotypic analysis of Hodgkin's disease derived cell lines: histopathological and clinical implications. Cancer Surv 4: 399-419
Dillman RO, Beauregard J, Ryan KP et al. (1987) Radioimmunodetection of cancer with the use of indium-111-labeled monoclonal antibodies. NCI Monogr 3: 33-36
Endo K, Sakahara H, Nakashima T et al. (1987) Preparation and properties of antitumor monoclonal antibodies labeled with metallic radionuclides ^{111}indium, ^{67}gallium and ^{99m}technetium. NCI Monogr 3: 135-140
Ford PV, Linn AB, Abrams D et al. (1987) Use of total body gallium-67 scintigraphy in a patient with AIDS related complex. J Nucl Med 28: 402-404 (letter)
Gallez-Marechal D, Fayolle M, Henry-Amar M et al. (1984) Radiation injuries of the gastrointestinal tract in Hodgkin's disease: the role of exploratory laparotomy and fractionation. Radiother Oncol 2: 93-99
Ghose T (1987) Book review: immunoscintigraphy. J Nucl Med 28: 258-259
Gomez GA, Reese PA, Nave H et al. (1983) Staging laparotomy and splenectomy in early Hodgkin's disease: no therapeutic benefit. Am J Med 77: 205-210
Hibi S, Todo S, Mashuky S (1987) Thymic localization of gallium-67 in pediatric patients with lymphoid and non lymphoid tumors. J Nucl Med 28: 293-297
Kadin ME, Glatstein EJ, Dorfman RE (1971) Clinicopathologic studies in 177 untreated patients subjected to laparotomy for the staging of Hodgkin's disease. Cancer 27: 1277-1294
Kaplan HS (1980) Hodgkin's disease, 2nd ed. Harvard University Press, Cambridge, MA
Keenan AM, Weinstein JN, Mulshine JL et al. (1987) Immunolymphoscintigraphy in patients with lymphoma after subcutaneous injection of ^{111}indium labeled T101 monoclonal antibody. J Nucl Med 28: 42-46
Klein JL, Sandoz JW, Kopher KA et al. (1986) Detection of specific anti-antibodies in patients treated with radiolabeled antibody. Int J Radiol Oncol Biol Phys 12: 939-943
Lacher MJ (1983) Routine staging laparotomy for patients with Hodgkin's disease is no longer necessary. Cancer Invest 1: 93-99

Lenhard RE, Order SE, Spunberg JJ et al. (1985) Isotopic immunoglobulin: a new systemic therapy for advanced Hodgkin's disease. J Clin Oncol 3: 1296-1300

Linde R, Basso L (1987) Hodgkin's disease with hypercalcemia detected by thallium-201 scintigraphy. J Nucl Med 28: 112-115

Longo DL, Young RC, De Vita VT Jr (1982) Chemotherapy for Hodgkin's disease: the remaining challenges. Cancer Treat Rep 66: 925-936

Mach JP, Buchegger F, Forni M, Ritschard J, Berche C, Lumbroso JD, Schreyer M, Girardet C, Accola RS, Carrel S (1981) Use of radiolabeled monoclonal anti-CEA antibodies for the detection of human carcinomas by external photoscanning and tomoscintigraphy. Immunol Today 2: 239-249

Manil L, Motte PH, Pernas P, Troalen F, Bohuon C, Bellet D (1986) Evaluation of protocols for purification of mouse monoclonal antibodies: yield and purity in two-dimensional gel electrophoresis. J Immunol Methods 90: 25-37

Order SE, Klein JL, Ettinger D et al. (1980) Use of isotopic immunoglobulin in therapy. Cancer Res 40: 3001-3007

Paul R (1987) Comparison of fluorine-18-2-fluorodeoxyglucose and gallium-67 citrate imaging for detection of lymphoma. J Nucl Med 28: 288-292

Pfreundschuh M, Mommertz E, Meissner M (1988) Hodgkin and Reed-Sternberg cell associated monoclonal antibodies HRS-1 and HRS-2 react with activated cells of lymphoid and monocytoid origin. Anticancer Res 8: 217-224

Pressman D, Day ED, Blau M (1967) The use of paired labeling in the determination of tumor localizing antibodies. Cancer Res 17: 845-850

Rosenberg SA, Kaplan HS (1985) The evolution and summary results of the Stanford randomized clinical trials of the management of Hodgkin's disease: 1962-1984. Int J Radiat Oncol Biol Phys 11: 5-32

Saccavini JC, Bruneau J, Grzyb J (1985) Radiolabeling of monoclonal antibodies for in vivo diagnosis. In: Donato L, Britton K (eds) Immunoscintigraphy. Gordon and Breach Science, New York, p 23

Skarin AT (1985) Non-Hodgkin's lymphoma. In: Wiernik PH, Canellos GP, Kyle RA, Schiffer CA (eds) Neoplastic diseases of the blood, vol 2. Churchill Livingstone, Edinburgh, p 841

Somers R, Henry-Amar M (1987) MOPP versus alternating MOPP/ABVD in advanced Hodgkin's disease. Proc 3rd Intern Conf Malign Lymphoma, Lugano, p 32 (abst 23)

Stein H, Mason DY, Gerdes J, O'Connor N, Wainscoat J, Palleses G, Falini B, Selsoi G, Lemke H, Schwarting R, Lennert K (1985) The expression of the Hodgkin's disease associated antigen Ki-1 in reactive and neoplastic lymphoid tissue: evidence that Reed-Sternberg cells and histiocytic malignancies are derived from activated lymphoid cells. Blood 66: 848-858

Tsan M (1985) Mechanism of gallium-67 accumulation in inflammatory lesions. J Nucl Med 26: 88-92

Tubiana M, Henry-Amar M, van der Werf Messing B et al. (1985) A multivariate analysis of prognostic factors in early stage Hodgkin's disease. Int J Radiat Oncol Biol Phys 11: 23-30

Vaughan ATM, Yankuba SCS, Anderson P (1987) Antibodies labeled with metallic radionuclides: influence of nuclide chemistry on dose distribution. NCI Monogr 3: 141-144

Winzelberg GG, Medala GA, Hydovitz JD (1986) False-positive thallium-201 parathyroid scan of the mediastinum in Hodgkin's lymphoma. Am J Roentgenol 147: 819-821

Diagnostic Strategies and Staging Procedures for Hodgkin's Lymphoma: Bone Marrow Scintigraphy and Magnetic Resonance Imaging

H. Schicha, M. Franke, J. Smolorz, A. Linden, W. Waters, and V. Diehl

Institut für klinische und experimentelle Nuklearmedizin, Medizinische Klinik I, Universität Köln, Joseph-Stelzmann-Straße 9, 5000 Köln 41, FRG

Introduction

In Hodgkin's lymphoma the bone marrow is infrequently involved. Clinical diagnosis is usually made by bone marrow biopsy. Because biopsy is blind, false-negative results often occur. For imaging of the bone marrow, two methods are available: scintigraphy and magnetic resonance imaging.

Bone marrow scintigraphy (BMSc) is performed after injection of ^{99m}Tc microcolloid. About 5%-10% is phagocytosed by the reticuloendothelial System (RES) of the bone marrow [4]. In healthy adults the central skeleton and the proximal third of the humoral and femoral diaphyses are visualized (Fig. 1). Originally BMSc was used to demonstrate increased bone marrow activity, including peripheral expansion in polycythemia [4]. Recently it has been employed to detect localized lesions in malignant diseases [8]. The advantage of BMSc is its high sensitivity for estimating functional activity and expansion of the RES. Its disadvantage is its limited resolving power.

In contrast, magnetic resonance imaging (MRI) permits high resolution of bone marrow morphology [2, 5, 9]. Using the conventional T1-weighted spin-echo mode, bone marrow yields high signal intensity because of its high fat content (Fig. 2). Small differences between the active hematopoietic tissue and the fatty bone marrow can be observed. The relative intensities vary with age [3]. The advantages of MRI are its high resolution and the possibility of obtaining images from any orientation. A disadvantage is the long examination time, especially if the total body bone marrow is to be investigated. Furthermore, with MRI one cannot assess the functional activity and peripheral expansion of bone marrow [5, 9, 10].

In Hodgkin's disease patients we have frequently observed a pathological peripheral expansion of bone marrow on BMSc (Fig. 3). This could be a primary consequence of the disease, a secondary consequence of compensatory mechanisms, or an effect of therapy. Using MRI in Hodgkin's disease we have often observed a localized or a diffuse decrease of signal intensity in bone marrow (Fig. 4). This could also be due to different mechanisms: primary malignant infiltration or secondary displacement of the fatty bone marrow by active marrow. In order to

Recent Results in Cancer Research. Vol. 117

clarify which of these mechanisms are in operation we performed BMSc and MRI in a nonselected group of patients with Hodgkin's lymphoma. The results are presented here.

Material and Methods

Forty-three investigations were performed in 39 patients with Hodgkin's lymphoma (18 female, 21 male; age 14-72 years, mean 38 years).

The first examination was always BMSc. It was performed 1 h after intravenous injection of 15 mCi (555 MBq) ^{99m}Tc Nanocoll (Solco-Nuclear) using a large-field gamma camera (Nuclear Ohio) with computer (Philips). Ten to twelve images of the whole body were obtained with an acquisition time of 5 min each.

MRI was performed using a 1.5-T MR tomograph (Gyroscan S 15 m, Philips) in T1-weighted spin-echo mode (TR 450 ms, TE 30 ms). The standard examination included coronal slices of 8 mm thickness of the pelvic and lumbosacral regions, lumbar spine, both femurs and knees, and various other regions according to the pathological findings on BMSc.

The interpretation of both BMSc and MRI often turned out to be very difficult because of interobserver differences: results were classified as normal or pathological, or as showing differing degrees of pathology. In our latest attempt at classification, with the lowest possible interobserver variability, the degree and the extension of pathological findings were judged.

BMSc classification

0 Normal findings (Fig. 1)
I Pathological findings of intermediate degree, i.e., bone marrow expansion involving more than half of the femur but not affecting the tibia, no cold or hot lesions (Fig. 3)
II Pathological findings of severe degree, i.e., bone marrow expansion, more intensive in the femur than in the pelvis, extension to the tibia, and/or clear cold or hot lesions (Fig. 5)

MRI classification

0 Normal findings (Fig. 2)
I Pathological findings of intermediate degree, i.e., minor diffuse decrease of signal intensity and/or small, poorly defined focal lesions (Fig. 4)
II Pathological findings of severe degree, i.e., high-grade diffuse decrease of signal intensity with large extension to the distal femur or to the tibia and/or clear large or multiple localized defects (Fig. 6)

Histological classification

0 Normal findings,
I Reactive changes
II Tumor involvement

For BMSc, MRI, and histology, boderline cases were assigned to the lower class.

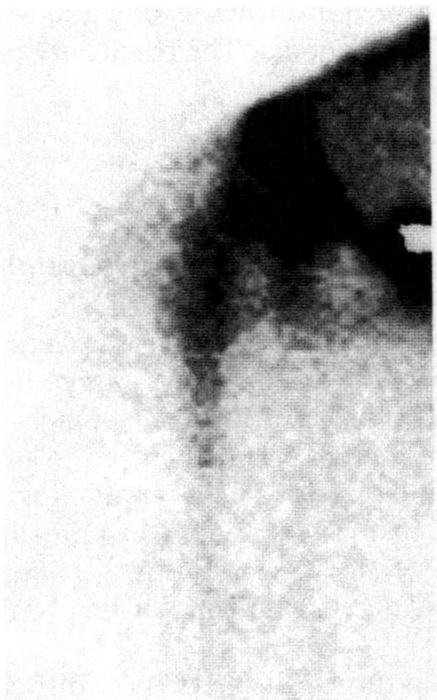
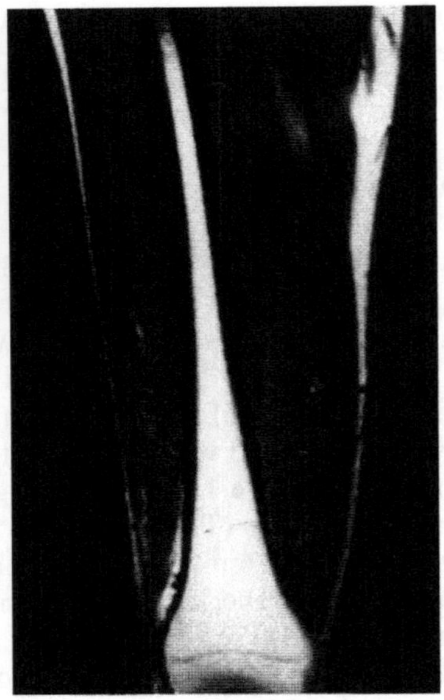

Fig. 1 *(left).* Bone marrow scintigraphy: normal findings (class 0)

Fig. 2 *(right).* Magnetic resonance imaging: normal findings (class 0)

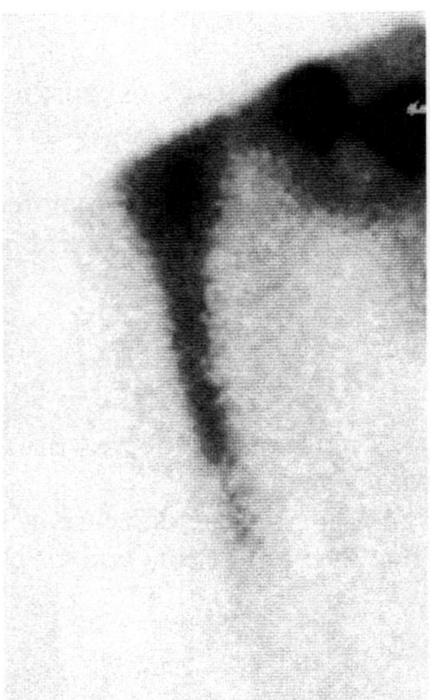
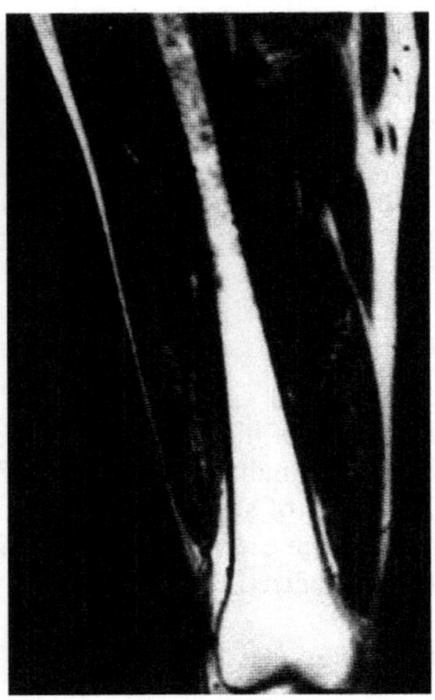

Fig. 3 *(left).* Bone marrow scintigraphy: intermediate pathological findings (class I)

Fig. 4 *(right).* Magnetic resonance imaging: intermediate pathological findings (class I)

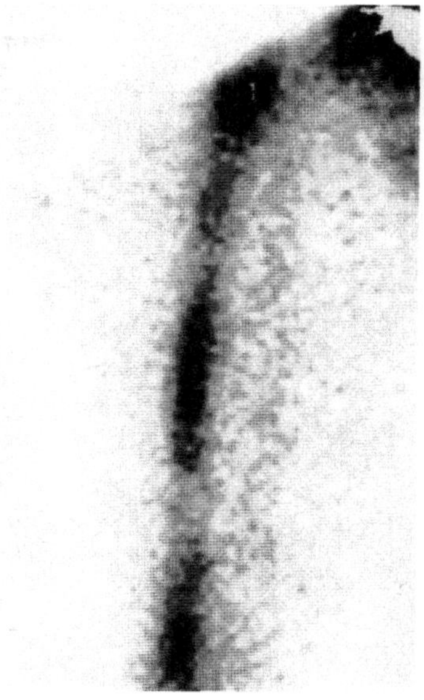

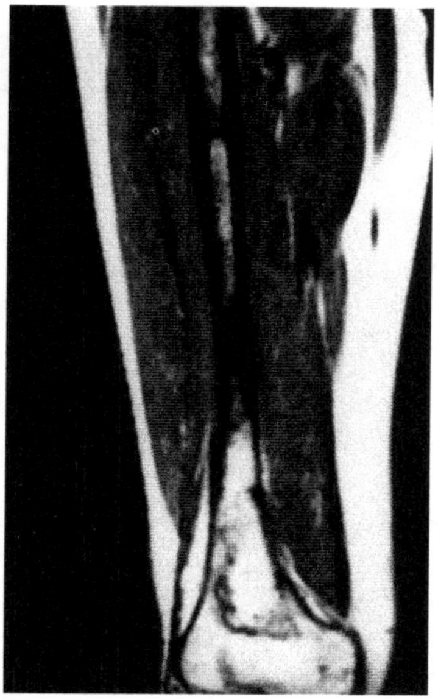

Fig. 5 *(left).* Bone marrow scintigraphy: severe pathological findings (class II)

Fig. 6 *(right).* Magnetic resonance imagning: severe pathological findings (class II)

Results

The findings of BMSc and MRI are shown in Fig. 7. Scintigraphically, the findings were normal in 23% of cases and 28% of patients had severe pathology. The remainder showed only a peripheral expansion of intermediate degree without focal defects. On MRI, 23% of patients had normal or nearly normal findings, 35% demonstrated intermediate pathology, and 42% had severe pathology. Only in 29 of 43 investigations (67%) was there complete agreement between BMSc and MRI. There were no differences between treated and untreated patients in the results of either examination (Fig. 8), neither was there a clear correlation between clinical stage and either BMSc or MRI findings (Fig. 9). In Figs. 10 and 11 the findings of BNSc and MRI are compared with those of posterior iliac crest bone marrow biopsy. The biopsies were obtained within ±3 weeks of the other examinations. In class O (BMSc and MRI), no patient showed tumor involvement of the bone marrow by histology. In contrast, in class II, biopsy showed reactive changes or tumor involvement in 10/11 cases for BMSc and 14/15 cases for MRI.

Discussion

The fact that in one-third of all investigations BMSc and MRI showed discrepant results indicates that there are differences between functional and morphological changes. In an attempt to improve our methods, we performed quantitative T1

MRI \ BMSc	0	I	II	Totals
0	7	3	0	10
I	3	11	1	15
II	0	7	11	18
Totals	10	21	12	43

Fig. 7. Results of BMSc versus MRI of bone marrow

a

MRI \ BMSc	0	I	II	Totals
0	5	1	0	6
I	0	3	0	3
II	0	2	3	5
Totals	5	6	3	14

b

MRI \ BMSc	0	I	II	Totals
0	2	2	0	4
I	3	8	1	12
II	0	5	8	13
Totals	5	15	9	29

Fig. 8. a Results of BMSc versus MRI of bone marrow in patients without treatment. **b** Results of BMSc versus MRI of bone marrow in patients with chemotherapy or radiotherapy

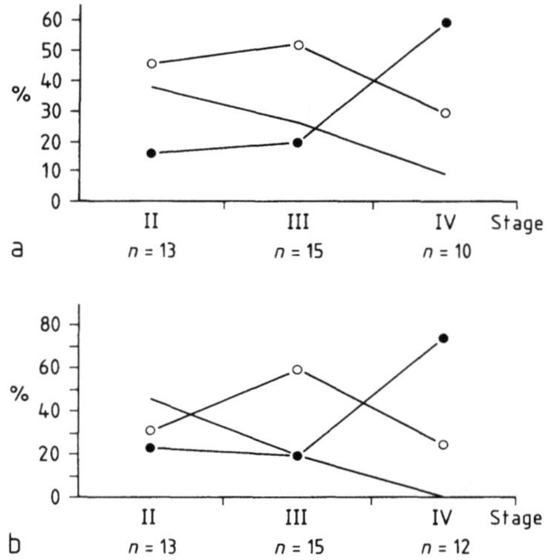

Fig. 9. a Results of BMSc versus clinical staging (Ann Arbor classification). There was only one patient with stage I. **b** Results of MRI of bone marrow versus clinical staging (Ann Arbor classification). There was only one patient with stage I. —, BMSc/MRJ class O; —○—, class I; —●—, class II

and T2 measurements in 20 patients, but this did not improve the results significantly [7]. The fast gradient-echo technique offers some promise, but further investigations are necessary.

The comparison of treated and untreated patients (Fig. 8) suggests that chemotherapy has no direct effect upon scintigraphic and MR imaging. In a small number of patients examinations were performed before and after therapy. In one patient with complete remission after chemotherapy (confirmed by biopsy) we observed a marked increase in signal intensity on MRI and a diminished expansion of the bone marrow on BMSc. The results, which must be confirmed in a larger patient population, suggest that follow-up studies may be useful in documenting therapeutic responses. Unfortunately there is no real "golden standard" to validate the findings of BMSc and MRI compared to clinical staging. Our findings (Fig. 9) may indicate that our parameters of classification are not valid, that the observed functional and morphological changes are not strongly correlated to the severity of the disease as measured by clinical staging, or that the clinical staging is inadequate or insufficient.

Even histology (Figs. 10, 11) does not provide sufficiently reliable confirmation of the findings of BMSc and MRI, since a negative result does not exclude tumor involvement in a nonbiopsied region. While biopsy showed absence of infiltration of the bone marrow in all cases of normal findings (class 0; BMSc $n=9$, MRI $n=10$), the positive predictive values of BMSc and MRI class I and II findings are lower. If only class II BMSc or MRI is judged as pathological, i.e., if class I changes are considered to represent nonspecific reactions, and if the non-golden standard of biopsy is used for comparison, then the following predictive values (pV) are obtained:

		Bone marrow biopsy 0	I	II	Totals
BMSc	0	4	5	0	9
	I	1	15	2	18
	II	1	8	2	11
	Totals	6	28	4	38

Fig. 10. Results of bone marrow biopsy of the posterior iliac crest versus BMSc

		Bone marrow biopsy 0	I	II	Totals
MRI	0	4	6	0	10
	I	1	14	0	15
	II	1	10	4	15
	Totals	6	30	4	40

Fig. 11. Results of bone marrow biopsy of the posterior iliac crest versus MRI of bone marrow

- BMSc class 0 or I: pV of a negative biopsy: 93%
- MRI class 0 orI: pV of a negative biopsy: 100%
- BMSc class II: pV of a positive biopsy: 18%
- MRI class II: pV of a positive biopsy: 27%

The positive predictive value of class II BMSc and MRI findings is depressingly low. This does not appear to be due only to the imaging methods. In our hands the positive predictive value of class II results in non-Hodgkin's lymphoma is 70%–75%. False-negative bone marrow biopsy results are a particular problem in Hodgkin's disease. Biopsies are routinely taken from the posterior iliac crest, a site accepted as representative for the whole bone marrow. However, bilateral posterior iliac crest biopsies have shown discrepant findings in over 40% of cases [1]. In six lymphoma patients (two Hodgkin's patients) with abnormal MRI findings of severe degree and normal or reactive histology, repeated biopsy or autopsy confirmed tumor involvement. Thus, neither the imaging methods nor biopsy appear to be an adequate golden standard. The imaging methods could usefully be performed first in order to identify regions likely to yield positive histological findings. It is obvious from our results that more detailed analysis of the data and further investigations are necessary:

1. Biopsies in the locations indicated by BMSc or MRI, to elucidate the pathological-anatomical correlation. There are limitations to this, because lesions identified by imaging are frequently located in regions unfavorable for biopsy, e.g., in the femoral shaft.
2. Examinations of patients before and after treatment in order to obtain a better understanding of the mechanisms of functional and morphological bone marrow changes [6].
3. Better classification of the results of BMSc and MRI [2, 4, 7, 9]. This is difficult, because there is no golden standard.
4. Quantification of all or some of the results in order to obtain more objective data with low interobserver variability. Computer-assisted evaluation is possible for BMSc [4, 8], but the time required is high. RES function and hematopoietic activity in the bone marrow may be different. Thus, evaluation of hematopoietic activity using ^{111}In-transferrin [4] would be desirable, but is expensive, time consuming, and labor intensive.

References

1. Brunnig RD, Bloomfield CD, McKenna RW, Peterson L (1975) Bilateral trephine bone marrow biopsies in lymphoma and other neoplastic diseases. Ann Intern Med 82: 365-366
2. Daffner RH, Lupetin AR, Dash N, Deeb ZL, Sefczek RJ, Schapiro RL (1986) MRI in the detection of malignant infiltration of bone marrow. Am J Radiol 146: 353-358
3. Dooms GC, Fisher MR, Hricak H, Richardson M, Crooks LE, Genant HK (1985) Bone marrow imaging: magnetic resonance studies related to age and sex. Radiology 155: 429-432
4. Hotze A, Mahlstedt J, Wolf F (1984) Knochenmarkszintigraphie. Methode, Indikationen, Ergebnisse. Giebeler, Darmstadt
5. Kangarloo H, Dietrich RB, Taira RT, Gold RH, Lenarsky C, Boechat MI, Feig SA, Salusky I (1986) MR imaging of bone marrow in children. J Comput Assist Tomogr 10: 205-209
6. McKinstry CS, Steiner RE, Young AT, Jones L, Swirsky D, Aber V (1987) Bone marrow in leukemia and aplastic anaemia: MR imaging before, during, and after treatment. Radiology 162: 701-707
7. Moore SG, Gooding CA, Brasch RC, Ehman RL, Ringertz HG, Ablin AR, Matthay KK, Zoger S (1986) Bone marrow in children with acute lymphocytic leukemia: MR relaxation times. Radiology 162: 237-240
8. Munz DL (1984) Knochenmarkszintigraphie: Grundlagen und klinische Ergebnisse. Nuklearmediziner 7. 251-268
9. Olson DO, Shields AF, Scheurich CJ, Porter BA, Moss AA (1986) Magnetic resonance imaging of the bone marrow in patients with leukemia, aplastic anemia, and lymphoma. Invest Radiol 21: 540-546
10. Widding A, Smolorz J, Franke M, Waters W, Diehl V, Schicha H (1987) Bone marrow investigation with Tc-99m microcolloid and magnetic resonance imaging in malignant lymphomas. In: Schmidt HAE, Csernay L (eds) Nuklearmedizin. New trends and possibilities in nuclear medicine. Schattauer, Stuttgart, pp 126-130

Therapeutic Strategies

Overviews

Radiotherapy Trials in Hodgkin's Disease

H. Sack

Strahlenklinik, Universitätsklinikum Essen, Hufelandstraße 55, 4300 Essen, FRG

History of Radiotherapy in the Treatment of Hodgkin's Lymphoma

Radiotherapy of Hodgkin's disease has a long tradition dating back to the beginning of the century. In 1920 Gilbert pioneered the irradiation of adjacent regions at risk in addition to clinically involved areas. Vera Peters (Toronto) introduced a systematic staging procedure for the determination of the anatomic extent of the disease and proved the value of higher radiation doses. She was able to improve the 5-year survival rate of all irradiated patients to 30%. High cure rates in stages I–III became possible with the advent of megavoltage radiotherapy and the introduction of large treatment portals by Henry Kaplan. For the past two decades Kaplan's group at Stanford has published numerous clinical papers on trials with sophisticated radiotherapy techniques and on the specific indication for either primary irradiation or combined modalities for patients in stages I–IIIA. Today's survival rates achieved by several groups are listed in Table 1. It becomes evident from this table that a disease which was uniformly fatal 20 years ago can now be successfully treated in the majority of patients.

Table 1. Survival of patients following total nodal irradiation or extended field irradiation in pathological stages I and II

Reference		Number of patients	Survival		Time (years)
			Total (%)	NED (%)	
Hoppe et al.	1982	109	84	77	10
Rosenberg and Kaplan	1985	35	80	80	17
Hellman and Mauch	1982	209	95	80	10
Hanks et al.	1983	235	88	77	4
Kapp et al.	1982	336	95	78	10

Recent Results in Cancer Research. Vol. 117

Features of Modern Radiotherapy

Major features of modern radiation therapy of Hodgkin's disease include the following:

Linear Accelerators. Linear accelerators allow for the treatment of large volumes in a single irradiation field; with this technique patients can be treated within a short time and with a more homogeneous dosage than would be possible with a cobalt unit.

High-Dose Irradiation. Irradiation of the tumor volume with high-dose irradiation improves long-term control; Kaplan (1966) first showed that the probability of local tumor recurrence was dependent on the dose of irradiation given to the respective area. Local recurrences are rare when doses of 40 Gy or more are given.

Homogeneous Dose Distribution. Large-field technique overcame major problems of radiation therapy because it avoids underdosage and overdosage in the margins of adjacent fields. Incorrect dosage in this area used to be a frequent reason for local recurrences and actinic damage to normal tissues.

Individualization of Radiation Portals. Radiation portals custom-tailored to the patient's anatomy in the treatment position are achieved by interposing precisely shaped blocks into the treatment portal. The blocks protect radiation-sensitive organs like the lung, liver, and gonads.

Special Issues of Modern Radiotherapy

Aggressivity of the Staging Procedure. One of the main questions in the management of Hodgkin's disease is that of how much surgical staging is necessary. In the EORTC H-2 trial infradiaphragmatic manifestation of Hodgkin's disease was detected by diagnostic laparotomy and splenectomy in 18%–36% of cases depending on the clinical stage (Tubiana et al. 1981; Table 2). When radiotherapy alone is planned, the information gained by a staging laparotomy is important for the delineation of the radiotherapy portals. In surgically defined stage III patients total nodal irradiation is necessary instead of extended fields; similarly, when splenic involvement is found patients will benefit from radiation treatment to the liver through thin blocks. The progress made by such an approach is demonstrated by the excellent results obtained by Zagars and Rubin (1985) for surgically staged patients in their new series (Table 3).

Irradiation of the Lung. When hilar adenopathy or parenchymal extension into the lung is present, results of radiotherapy can be improved by thin lung blocks. The ipsilateral lung can be irradiated to approximately 16 Gy over 4 weeks, a dose close to the radiation tolerance of the lung. When combined therapy is planned, the possible benefit of lung irradiation must be carefully weighed against the risks of damage to functional lung tissue and the bone marrow in the ribs.

Table 2. EORTC study H 2: frequency of abdominal manifestation of Hodgkin's disease. (Tubiana et al. 1981)

Clinical stage	*n*	%
IA	9/51	18
II_2A	5/25	
$I+II_2B$	15/43	36
II_3	9/23	

Table 3. Ten-year survival in Hodgkin's disease stages IA and IIA: Gains achieved by modern therapy. (Zagars and Rubin 1985)

	Old series		New series	
	All	NED	All	NED
Stage IA	39%	34%	92%	89%
Stage IIA	45%	42%	74%	47%

Old series: mantle radiotherapy only without laparotomy staging and without subdiaphragmatic treatment (49 patients).
New series: staging laparotomy, total or subtotal nodal irradiation, a small number with chemotherapy (98 patients).

Irradiation of the Liver. In patients with presumptive Hodgkin's disease of the liver a simple and safe technique for irradiating the liver through a thin liver block can be applied. Here, too, the small size of each fraction delivered to the liver is within the tolerance of the normal tissue and yet effective against the tumor. This does not compromise subsequent chemotherapy, as additive chemotherapy given after this sort of liver irradiation is usually well tolerated.

Risk Factors That Recommend Additional Chemotherapy

Currently, the generally recommended therapy for Hodgkin's disease stages I and II is radiotherapy alone. But when looking at the survival rates of different centers (Table 1), there are differences in total survival and freedom from relapse (survival NED) indicating patients at risk for relapse if they are treated by radiotherapy alone. A German multicenter study is now investigating whether the patients in the stages I-IIIA with the risk factors large mediastinal mass, extranodal disease, or massive splenic involvement definitely profit from additional chemotherapy (Diehl et al. 1986).

Several investigators have concluded that radiation therapy alone is a risk in patients with *massive mediastinal disease and extranodal involvement.* This is demonstrated in Tables 4 and 5. When the bulk exceeds one-third of the largest thoracic diameter the disease-free survival rate drops substantially. This decrease in disease-free survival can be avoided by additional chemotherapy within a combined modality approach.

Table 4. Recurrence and survival rates in patients with large mediastinal masses in stages I and II (Ann Arbor)

Reference	Patients	Survival (%)		Time (years)
		All	NED	
Prosnitz et al. 1980	$n=169$			
	$n=$ 69, no mediastinal mass	94	81	10
	$n=$ 38, <0.3	100	72	10
	$n=$ 24, >0.3	90	55	10
	$n=$ 38, >0.5	97	69	10
Mauch et al. 1982	$n=$ 27 >0.3	89	48	8
Fuller et al. 1982	$n=$ 62 >0.3	84	57	10

Table 5. Actuarial 5-year survival and disease-free survival versus mediastinal status. (Willett et al. 1987)

	No. of patients	Five-year survival (%)	
		All	NED
No mediastinal mass	52	89	85
Mediastinal disease			
<0.33	35	80	77
>0.33 without MOPP	28	96	45
>0.33 with MOPP	7	100	100
Mediastinal disease and extranodal involvement	7	100	100

Carmel and Kaplan (1976) demonstrated that the recurrence rate is also dependent on the extent of the disease and that it falls below 50% when radiotherapy alone is given to stage II patients with *seven or more involved lymph node regions* (Table 6). Similarly, in *stage IIIA* (Table 7) the relapse-free survival is not satisfactory when radiotherapy alone is applied. The results are better if additional chemotherapy is given. The EORTC study H5, similar to two other EORTC trials, revealed additional risk factors such as *sedimentation rate, B symptoms, or extranodal growth* (Tubiana et al. 1986).

Risks of a Combined Modality Approach

Though many physicians, well-known centers, and publications recommend the addition of chemotherapy to radiation treatment in patients with limited disease when risk factors are present, one must keep in mind that combination of chemo- and radiotherapy raises the risk of second cancers. Secondary leukemias most evidently appear in the first 8 years after treatment (Balaney et al. 1987) and are predominantly seen in patients treated with chemotherapy with or without irradiation, with only a few cases occurring in patients treated with irradiation alone

Table 6. Recurrence rate and number of involved regions in 220 patients with pathological stage IIA or IIB Hodgkin's disease: radiotherapy only, extended field or total nodal irradiation. (Carmel and Kaplan 1976)

Regions involved	Recurrence rate (%)
2	8
4	25
6	35
7-9	50

Table 7. Survival and recurrence-free survival in pathological stage IIIA

Reference	Patients	Survival (%)		Time (years)
		All	NED	
Hoppe et al. 1982	201 patients, stage IIIA, TNI	71	57	10
	TNI+CT	83	79	10
	Spleen+TNI	62	32	10
	TNI+CT	88	74	10
Mauch et al. 1983	130 patients, stage IIIA+B			
	III 1, TNI	73	47	10
	RT+CT	96	91	10

TNI, total nodal irradiation.

Table 8. Development of second cancers in patients treated with radiotherapy only: actuarial risk. (Rubin et al. 1986)

Institution	Year	Author	Actuarial risk of	
			Tumors	Leukemia
Italy, Multicenter	1980	Baccarani	0%	0%
Stanford	1982	Coleman	6.1%	0%
SWOG	1982	Coltman	1.5%	0%
Milan	1982	Valagussa	9.9%	0%
NCI	1984	Tester	7.0%	0%
Rochester	1985	Rubin	2.3%	0%

Table 9. Development of second cancers in patients treated by radiotherapy only: observed vs expected. (Rubin et al. 1986)

Institution	Year	Author	OER
NCI	1972	Arseneau	3.4
NCI	1975	Canellos	3.8
Memorial	1977	Brody	3.7
McGill/Harvard	1981	Boivin	0
Rochester	1981	Nelson	4.2
Rochester	1985	Rubin	3.7

OER, observed to expected ratio.

Table 10. Development of second cancers in patients receiving combined treatment: actuarial risk. (Rubin et al. 1986)

Institution	Year	Author	Actuarial risk of	
			Tumors	Leukemia
Italy (multicenter)	1980	Baccarani	2.3%	18.0%
Stanford	1982	Coleman	3.6%	17.4%
SWOG	1982	Coltman	2.9%	14.1%
Milan	1982	Valagussa	11.7%	9.7%
NCI	1984	Tester	7.0%	12.0%

Table 11. Development of second cancers in patients receiving combined treatment: observed vs expected. (Rubin et al. 1986)

Institution	Year	Author	OER
NCI	1972	Arseneau	3.3
NCI	1972	Arseneau	29.0
NCI	1975	Canellos	14.5
Memorial	1977	Brody	2.6
McGill/Harvard	1981	Boivin	270.0
McGill/Harvard	1984	Boivin	3.3
McGill/Harvard	1984	Boivin	28.8
Rochester	1981	Nelson	7.6
Rochester	1985	Rubin	7.3

OER, observed to expected ratio.

(Tables 8-11). This risk may be higher than 10%, in two studies 17% and 18%, and may reach 20% if solid tumors are included. The secondary leukemias are difficult to treat with low rates and short durations of complete remissions (Coleman et al. 1982; Rubin et al. 1986; Tester et al. 1984).

Reassessment of Prognostic Factors Following Modification of Radiotherapy

Taking the rate of secondary leukemias and of sterility after combined modality into consideration, primary radiotherapy without additional chemotherapy should be applied whenever justified. It has to be considered carefully how dangerous a relapse would be or how successful a failure after radiation therapy could be treated. Clear evidence that combined modality therapy in stages I-IIIA improves survival is still lacking (Rosenberg and Kaplan 1985), since the salvage rates that can be achieved with chemotherapy are significant in patients who relapse after radiotherapy only (Table 12).

When all possibilities of modern radiotherapy technique are utilized, significant improvements of relapse-free survival are possible. These possibilities include the above-mentioned irradiation of the lung and liver and a highly sophisticated treatment technique. Levitt et al. (1984) reported their results in patients with large me-

Table 12. Results of therapy for relapses following radiotherapy

Cooper et al. 1984	$n=137$	Complete remission	76%
		8-year survival	75% < 34 years
			30% > 34 years
Carde et al. 1983	$n=94$	Complete remission	74%
		12-year survival	39%
Timothy et al. 1979	$n=27$	Complete remission	85%
		5-year survival	80%

Table 13. Mediastinal masses in Hodgkin's disease: influence of radiotherapy technique. (Levitt et al. 1984)

Series 1970-1974		Series 1975-1980	
Mantle field without whole lung irradiation		Mantle field with whole lung irradiation	
$n=20$		$n=20$	
Complete remission	19	Complete remission	19
Recurrences	15	Recurrences	3
	=79%		=16%
Survival NED 5 years: 20%		Survival NED 5 years: 85%	

Table 14. Five-year survival for various risk groups with standard and modified radiotherapy. (Lee et al. 1987)

	Survival NED (%)		Total survival (%)	
	Treatment 1	Treatment 2	Treatment 1	Treatment 2
Symptoms				
A	61	81*	89	91
B	33	78*	67	70
Mediastinal mass				
None	71	82	95	91
Small	69	78	85	89
Large	24	85*	78	87
Number of sites				
3	50	76	93	83
4	32	70*	77	78
Stage				
IIB	25	67	63	57
IIIA	31	72*	81	89

* $P<0.01$ (Breslow's test).

diastinal lymphoma in two series (Table 13). They used thin lung blocks in a modified technique and were able to increase the relapse-free survival rate from 20% to 85%, thus reaching the same level as investigators with combined modality (Table 5). Lee and coworkers (Lee et al. 1985) also compared two radiotherapy techniques in 175 patients with stages IA, IB, IIA, IIB, or IIIA. Their modified protocols consisted of low-dose lung irradiation in patients with large mediastinal masses and/or hilar disease, and low-dose liver irradiation for stage IIIAS+ patients (treatment group 2, $N = 110$). Recurrence-free survival rates improved significantly for various risk groups (Table 14). Most of the risk factors seen following treatment 1 lost their significance as predictors of recurrence for patients receiving treatment 2.

With this short overview on the results of radiotherapy trials of recent years I want to show that a more precise and improved radiotherapy treatment technique is able to improve results considerably in patients who are at risk of relapse if they are treated by less sophisticated radiotherapy techniques. Even though there are no randomized trials, the results of radiotherapy employing the described new techniques seem to be as good as those obtained with combined modality treatment in patients with localized disease and risk factors. If this impression can be proven in randomized trials, it would be possible to spare chemotherapy for a considerable proportion of patients. This would enable us to lower the rate of secondary leukemias, the most distressing complication of combined modality treatment.

References

Blayney DW, Longo DL, Young RC et al. (1987) Decreasing risk of leukemia with prolonged follow-up after chemotherapy and radiotherapy for Hodgkin's disease. N Engl J Med 316: 710-714

Carde P, Mackintosh FR, Rosenberg SA (1983) A dose and time response analysis of the treatment of Hodgkin's disease with MOPP chemotherapy. J Clin Oncol 1: 146-153

Carmel RJ, Kaplan HS (1976) Mantle irradiation in Hodgkin's disease. Cancer 37: 813-825

Coleman CN, Kaplan HS, Cox R et al. (1982) Leukemia, non-Hodgkin's lymphomas and solid tumors in patients treated for Hodgkin's disease. Cancer Surv 1: 733

Cooper RM, Pajak TF, Gottlieb AJ, Glicksman AS et al. (1984) The effects of prior radiation therapy and age on the frequency and duration of complete remission among various four-drug treatments for advanced Hodgkin's disease. J Clin Oncol 2: 748-755

Diehl V, Pfreundschuh M, Hauser FE et al. (1986) Zwischenergebnisse der Therapiestudien HD1, HD2 und HD3 der deutschen Hodgkin-Studiengruppe. Med Klin 81: 1-6

Fuller LM, Hutchinson GB (1982) Collaborative trial for stage I and II Hodgkin's disease. Cancer Treat Rep 66: 775-787

Glicksman AS, Pajak TF, Gottlieb A et al. (1982) Second malignant neoplasms in patients successfully treated for Hodgkin's disease. Cancer Treat Rep 66: 1035-1044

Hanks GE, Kinzie JJ, White RL et al. (1983) Patterns of care outcome studies. Results of the national practice in Hodgkin's disease. Cancer 51: 569-573

Hellman S, Mauch P (1982) Role of radiation therapy in the treatment of Hodgkin's disease. Cancer Treat Rep 66: 915-923

Hoppe RT, Coleman CN, Cox RS, Rosenberg SA, Kaplan HS (1982) The management of stage I-II Hodgkin's disease with irradiation alone and combined modality therapy: the Stanford experience. Blood 59: 455-465

Kaplan HS (1966) Evidence for a tumoricidal dose level in radiotherapy of Hodgkin's disease. Cancer Res 26: 1221

Kapp DS, Prosnitz LR, Farber LR et al. (1982) Improved results in the management of Hodgkin's disease. The Yale experience. ASCO abstracts 1982: C-649

Lee CKK, Aeppli DM, Bloomfield CD, Levitt SH (1987) Hodgkin's disease: a reassessment of prognostic factors following modification of radiotherapy. Int J Radiat Oncol Biol Phys 13: 991-993

Levitt SH, Lee CKK, Bloomfield CD (1984) Radical radiation therapy in the treatment of laparotomy staged Hodgkin's disease patients. Int J Radiat Oncol Biol Phys 10: 265-274

Mauch P, Gorsheim D, Cunningham J, Hellman S (1982) Influence of mediastinal adenopathy on site and frequency of relapse in patients with Hodgkin's disease. Cancer Treat Rep 66: 809-817

Mauch P, Rosenthal DS, Canellos GP, Hellman S (1983) Improved survival for stage IIIA and IIIB Hodgkin's disease patients treated with combined radiation therapy and chemotherapy. Int J Radiat Oncol Biol Phys (Suppl 1) 9: 129

Prosnitz LR, Curtis AM, Knowlton AH et al. (1980) Supradiaphragmatic Hodgkin's disease. Significance of large mediastinal masses. Int J Radiat Oncol Biol Phys 6: 809-813

Rosenberg SA, Kaplan HS (1985) The evolution and summary results of Stanford randomised clinical trials of the management of Hodgkin's disease: 1962-1984. Int J Radiat Oncol Biol Phys 11: 5-22

Rubin P, Zagars G, Chuang C et al. (1986) Hodgkin's disease: is there a price for successful treatment? Int J Radiat Oncol Biol Phys 12: 153

Tester WJ, Kinsella TJ, Waller B et al. (1984) Second malignant neoplasms complicating Hodgkin's disease: The National Cancer Intitute experience. J Clin Oncol 2: 762-769

Timothy AR, Sutcliffe SBJ, Wrigley PEM, Jones AE (1979) Hodgkin's disease: chemotherapy for relapse following radical radiotherapy. Int J Radiat Oncol Biol Phys 5: 165-169

Tubiana M, Hayat M, Henri-Amar M et al. (1981) Five-year results of the EORTC randomised study of splenectomy and spleen irradiation in clinical stages I and II of Hodgkin's disease. Eur J Cancer 17: 355-363

Tubiana M, Henry-Amar M, van der Werf-Messing B et al. (1985) A multivariate analysis of prognostic factors in early stage Hodgkin's disease. Int J Radiat Oncol Biol Phys 11: 23-30

Willett CG, Linggood RM, Meyer J (1987) Results of treatment of stage IA and IIA Hodgkin's disease. Cancer 59: 1107-1111

Zagars G, Rubin P (1985) Hodgkin's disease stages IA and IIA, A long term follow-up study on the gains achieved by modern therapy. Cancer 56: 1905-1912

Current Status of Chemotherapy for Hodgkin's Disease

P. H. Wiernik

Department of Oncology, Montefiore Medical Center, Division of Medical Oncology, Albert Einstein College of Medicine, Bronx, New York, NY, USA

Combination chemotherapy is curative of advanced Hodgkin's disease in a large minority of patients. That fact first became evident during long-term follow-up studies of patients treated with MOPP (nitrogen mustard, vincristine, procarbazine, and prednisone) at the National Cancer Institute, United States (De Vita et al. 1980). The remarkable results obtained with chemotherapy in advanced disease led to a pilot comparative study of MOPP compared with combined modality treatment in early-stage disease (O'Dwyer et al. 1985), and a larger study in which early-stage patients were randomized to receive MOPP or radiotherapy (Longo et al. 1987). The results to date indicate that MOPP is highly effective as initial treatment for early-stage Hodgkin's disease also.

It is the purpose of this review to explore the current status of combination chemotherapy for Hodgkin's disease of all stages by selectively drawing upon some of the major literature on the subject, and our own work.

Advanced Disease

Patients with advanced disease confined to lymph nodes (stages III_2A, III_1B, III_2B) or extranodal disease not arising by direct extension from a nodal mass (stages IVA or B) are generally candidates for combination chemotherapy. The same prognostic factors influence outcome with all chemotherapeutic regimens, which suggests that all such treatments are essentially similar therapeutically. Although histologic subtype of disease is less important than initially thought, lymphocyte-depleted histology carries a poorer prognosis than other subtypes and often is associated with an aggressive Richter's syndrome - like rapidly progressive course, especially in elderly patients. In addition, patients with nodular sclerosing histology have achieved complete remission less frequently than patients with other histologies in some studies, or have had shorter remission duration. Bulky disease, as in most other malignancies, has a poor prognosis for disease-free and overall survival when treated with one modality. In Hodgkin's disease, bulky disease is most likely to be found in the mediastinum, where it is usually defined as a mass whose largest transverse diameter on a PA chest X-ray is one-third or more

Recent Results in Cancer Research. Vol. 117

of that of the chest at the same level. Combined modality therapy (Wiernik and Slawson 1982; Specht et al. 1985) is usually required for such patients. Excessive mass of disease without localized bulky disease also confers a relatively poor prognosis, as evidenced by the diminished success rate with patients who have multiple sites of extranodal involvement, or more than five splenic nodules of disease.

Patients with stages III or IV who have systemic manifestations of disease such as fever, night sweats, or weight loss (B signs) have a much poorer prognosis than those patients without symptoms. It has also recently been determined that those patients with nodular sclerosing histology who have severe pruritis have a poorer prognosis than those with the same histology who have mild or no pruritis (Gobbi et al. 1985). Whether this is due simply to a direct relationship of severity of pruritis to bulk of disease, or to factors independent of extent of disease is not clear at present.

In most series, patients with advanced age have a poorer prognosis than adolescent or young adults, and males tend to have a somewhat poorer outcome than females.

Dose intensity (Pillai et al. 1985) and adherence to regimen schedule are the major treatment factors that influence outcome. It is imperative that the maximum fraction of a planned dose be given on schedule if best results are to be obtained. Dosage reductions or violations of schedule for trivial or treatable toxicities should not be made. This is especially true for patients with B-signs or other poor prognostic factors. Such patients should be treated as aggressively as acute leukemia patients, with the same quality and quantity of supportive care. It should also be remembered that, while the majority of successfully treated patients require 6 months or less of treatment with chemotherapy, many patients require more prolonged therapy. Progressively responding patients should continue therapy until complete remission is achieved, and never be taken off therapy to which they are responding simply because a certain predetermined number of courses have been administered. Patients should receive one to two courses of therapy after complete remission, and complete remission should be confirmed histologically whenever possible.

Treatment regimens for patients with advanced disease fall into three major categories: (1) single chemotherapeutic regimens, (2) alternating non-cross-resistant regimens, and (3) combined modality treatment with chemotherapy and radiotherapy.

Treatment with a Single Chemotherapeutic Regimen

MOPP (De Vita et al. 1970) is the standard regimen to which all other treatments for advanced Hodgkin's disease much be compared. Complete response rates in previously untreated patients vary from 40% (Colonna et al. 1986) to more than 80% (De Vita et al. 1970), depending upon the prevalence of prognostic factors within the treatment groups, and whether or not the philosophy of treatment outlined above was employed. In most series more than 65% of complete responders have survived disease free at least 8-10 years (De Vita et al. 1980). Many minor and major variations of this regimen have been reported over the years with little evidence that improved efficacy has been achieved. Reduction of the rates of leukemogenesis and infertility has been claimed for the ABVD regimen (doxorubicin, bleomycin, vinblastine, and dacarbazine) compared with those of MOPP (Santoro et al. 1986; Santoro et al. 1987), but the rate of those two significant sideeffects has varied considerably among MOPP studies. In our own studies (Konits et al. 1981; O'Dwyer et al. 1985; Dutcher and Wiernik 1985) acute leukemia and infertility have not been observed more frequently with MOPP or MOPP and radiotherapy than in studies of ABVD or ABVD and radiotherapy reported by others.

We have also studied a regimen of streptozotocin, lomustine, doxorubicin, and bleomycin (SCAB) in previously untreated patients with advanced Hodgkin's disease, stages IIIB, IVA, and IVB (Diggs et al. 1980). Eighty percent of patients achieved a complete remission, and a recent update indicates that more than 90% of the complete responders continue alive and well in first remission 11 years posttreatment (Wiernik and Schiffer 1988). Although these data are provocative, whether this regimen is or is not superior to MOPP or its variants remains to be determined by a prospective, comparative trial.

A large number of multiagent regimens have been studied in patients who have relapsed from radiotherapy- or chemotherapy-induced remissions. The ABVD regimen has been significantly more effective in salvaging treatment failures than has MOPP in the hands of the Milan group (Santoro et al. 1986), but others have had less success with that regimen (Case et al. 1977; Krikorian et al. 1978; Sutcliffe et al. 1979; Harker et al. 1984). In these latter studies, subtle differences in patient selection and the delivery of the treatment make strict comparisons with the Milan study difficult, however (Canellos et al. 1983).

Cancer and Leukemia Group B (CALGB) (Schulman et al. 1987) studied a regimen of etoposide, cyclophosphamide, prednisone, moderate-dose methotrexate with leucovorin rescue, cytosine arabinoside, and vincristine (MOPLACE) in 30 previously treated patients, 70% of whom had B symptoms. Four to eight cycles of the regimen were given. Only five patients (17%) had achieved a complete remission at the time of the initial report, but several partial responders were still candidates for a better response. The regimen was toxic, especially to patients who had previously received bleomycin. These results appear to be inferior to those of an Eastern Cooperative Oncology Group (ECOG) study of Bleo-CCVPP (bleomycin, lomustine, cyclophosphamide, vincristine, procarbazine, and prednisone), although only radiotherapy failures were entered into the latter study whereas all MOPLACE patients previously failed chemotherapy alone, or combined modality therapy. A complete response rate of 76% was achieved in the ECOG Bleo-CCVPP study and 74% of those patients survived at least 3 years.

ECOG is currently evaluating a regimen consisting of lomustine, mitoxantrone, and vinblastine in relapsed patients with advanced-stage disease. The regimen was designed to incorporate the most active agents of the ABVD and SCAB regimens, and to reduce doxorubicin toxicity without loss of efficacy by substituting mitoxantrone for doxorubicin. The study has just begun to accrue patients and the first three stage IV patients have all had major objective disease regression after one cycle of treatment.

A new and promising approach to relapsed or refractory advanced disease is autologous bone marrow transplantation (ABMT) support for high-dose chemotherapy. A complete response rate of 61.5% was obtained with this technique in one small study (Carella et al. 1985). In that study nonfrozen marrow was used and patients were treated with high-dose carmustine alone, or in combination with large doses of cyclophosphamide and etoposide. Most patients had previously failed standard chemotherapy and irradiation therapy, including three patients who remained in complete remission for more than 1–3 years after ABMT. Three additional patients were disease free for 2–8 months post-ABMT at the time the data were reported. These results are excellent in this very poor risk group of patients, and give impetus to the further study of this approach.

Treatment with Alternating, Non-Cross-Resistant Regimens

The Milan group has pioneered the concept of alternating non-cross-resistant combination chemotherapy for advanced Hodgkin's disease. In a stud of 88 patients with stage IV disease, MOPP alternating with ABVD resulted in a complete remission rate of 88.9%, which was superior to the 74.4% rate obtained with MOPP alone in the same study (Bonadonna et al. 1986). None of the patients in either group had received prior chemotherapy but 25 patients (equally distributed between the groups) had relapsed after radiotherapy given for early-stage disease. The eight-drug treatment resulted in a significantly better 8-year disease-free survival (72.6%) compared with treatment with MOPP alone (45.1%). The major advantage for MOPP/ABVD with respect to complete remission rate was among patients who were previously untreated.

A similar approach in which all of the drugs utilized in the MOPP/ABVD program except dacarbazine were employed has been devised by Connors and Klimo (1987). That MOPP/ABV program is based on the same Goldie-Coldman hypothesis (Goldie et al. 1982) as is the Milan regimen, but total treatment time is condensed in the seven-drug treatment. A complete response was obtained in 96% of 74 patients treated with MOPP/ABV, and 90% of the complete responders are projected to continue disease-free for at least 5 years. These results appear to be superior to those obtained with MOPP/ABVD, but the median follow-up for the MOPP/ABV patients is only 3 years (Connors and Klimo 1987) compared with almost 8 years for the MOPP/ABVD patients (Bonadonna et al. 1986).

The higly encouraging results obtained with both treatments based on the Goldie-Coldman hypothesis have stimulated a number of multicenter trials designed to confirm or reject them.

An NCI study conducted in Bethesda and Baltimore randomized stage III–IV patients without bulky mediastinal disease to treatment with MOPP alone, or MOPP alternating with SCAB (Diggs et al. 1980). In an analysis of the first 79 patients (Young et al. 1985) the two treatments each resulted in a complete response rate of approximately 85%, and actuarial survivals in both groups exceeded 80% at 4 years. No significant differences had yet emerged between the treatment options.

The ECOG compared MOPP/ABVD with BCVPP (carmustine, cyclophosphamide, vinblastine, procarbazine, and prednisone) alone, and with BCVPP plus low-dose irradiation to involved sites in patients with advanced Hodgkin's disease. The results among the treatment options were similar, with 68%-76% of 294 evaluable patients obtaining a complete remission. The projected disease-free survival at 3 years for all complete responders is 66%, with no differences among the groups.

A joint study of CALGB and ECOG has recently been initiated in which patients with advanced disease are randomly assigned to treatment with six cycles of MOPP or MOPP/ABV. Complete responders to MOPP subsequently receive three courses of ABVD, and MOPP/ABV complete responders receive an additional two courses of MOPP/ABV. The study has not yet undergone an interim evaluation.

Combined Modality Treatment

The Yale group recently reported on 102 previously untreated stage IIIB and IV patients and 82 patients who relapsed with advanced disease after radiotherapy (Prosnitz et al. 1987). Initially, patients were treated with nitrogen mustard, vincristine, vinblastine, procarbazine, and prednisone (MVVPP) for three cycles over 6 months, which was followed by low-dose irradiation to inital sites of disease. Later in the study the chemotherapy was changed to MOPP/ABVD, and ultimately to a randomization between MOPP and MVVPP before radiotherapy. The three regimens were found to be equivalent. The overall complete response rate was 82% and the 15-year actuarial survival of all treated patients was 54% (71% if only deaths from Hodgkin's disease were considered).

The Memorial group reported no difference in outcome between patients treated with low-dose radiotherapy to involved sites combined with MOPP/ABVD or CAD (lomustine, phenylalanine mustard, and vindesine)/MOPP/ABVD and concluded that disease-free and overall survival in this study were similar to those reported for MOPP alone (Straus et al. 1987), although earlier results were more encouraging (Straus et al. 1984).

The Milan group recently reported 7-year results of a study in which 232 patients with stages IIB, IIIA, and IIIB were randomly assigned to treatment with six cycles of MOPP or ABVD, with total nodal radiotherapy given between the third and fourth courses of both drug regimens. The complete response rate was significantly higher in patients who received ABVD (92.4% vs. 80.7%) as were relapse-free survivals (87.7% vs. 77.2%) and overall survivals (77.4% vs. 67.9%) at 7 years (Santoro et al. 1987). Serious toxicity was also more frequent in the MOPP group. Although this was an extremely carefully planned and conducted study, Rosenberg (1987) points out that imbalances between the two treatment groups exist. The planned postirradiation three chemotherapy cycles were delivered to 80% of the ABVD group but to only half as many patients in the MOPP group. In addition, the ABVD group as a whole received a larger radiation dose in a shorter period than did the MOPP group. Thus, although one may conclude that patients who receive three cycles of ABVD tolerate additional radiotherapy and chemotherapy better than those who receive three cycles of MOPP, it may not be appropriate to conclude that the two treatment designs employed in this study have different therapeutic activity.

Early-Stage Disease

Combined Modality Therapy

Most studies of combination chemotherapy in early-stage disease employed the chemotherapy as an adjuvant to radiotherapy. We have reported results at 14 years of a study in which 87 previously untreated patients with pathologic stages IA, IB, IIA, IIB, or IIIA were randomly allocated to receive extended-field radiotherapy alone, or involved-field radiotherapy followed by six cycles of MOPP (Dutcher and Wiernik 1985). Disease-free survival was significantly better for all stages with combined modality therapy. Overall survival was also greater for all stages with combined modality therapy, but the difference between the two treatments was only significant for patients with stage IIIA. Further evaluation of the data revealed that only patients with large mediastinal masses, stage III_2A, or both had a significant improvement in disease-free and overall survival with combined modality therapy (Levi et al. 1977). The overall results of this small study were excellent. There was a 93% actuarial disease-free survival at 14 years for patients with stages I-II treated with combined modality therapy, compared with 64% for patients treated with radiotherapy alone. The 14-year actuarial survival for stage IIIA patients was 94% with combined modality treatment and 42% with radiotherapy alone. These data have not changed with an additional 2 years of observation and are consistent with many other reports of studies of similar design.

No regimen has been demonstrated to be superior to MOPP as adjuvant to radiotherapy for early-stage disease, and results appear to be independent of the sequence in which the modalities are given (Andrieu et al. 1985). The Memorial group (Koziner et al. 1987) compared MOPP with thio-tepa, vinblastine, and bleomycin (TVP), both given with involved-field radiotherapy, in 102 patients with stages IA and IIA. They found no difference in response rate or disease-free or overall survival between the two groups. In the Milan study in which ABVD and MOPP were compared as adjuvants to sandwiched radiotherapy (Santoro et al. 1987), 162 of the 232 patients had stage IIB or IIIA disease. Complete response rates for those early-stage patients were similar with both regimens. Relapse-free survival for stage IIB patients was not influenced by treatment, but stage IIIA patients had an actuarial disease-free survival at 7 years of 67% with MOPP and 95% with ABVD. The results in stage IIIA patients with MOPP and radiotherapy appear to be inferior to the results obtained by others (Dutcher and Wiernik 1985) with radiotherapy plus MOPP, or a MOPP variant alone (Lister et al. 1983), and this makes interpretation of the difference noted in the Milan study (Santoro et al. 1987) difficult.

Chemotherapy Alone

My colleagues and I began a pilot study in 1976 in which patients with pathologic stages IB to IIIA were randomized to treatment with extended-field radiotherapy followed by six cycles of MOPP, or MOPP alone (O'Dwyer et al. 1985). Complete remission was achieved in 94% of patients with combined modality therapy and 80% of those treated with MOPP alone. Two of three MOPP failures subsequently achieved a complete response with radiotherapy. To date, there is no significant difference in outcome between the two groups with respect to disease-free or overall survival, or freedom from relapse. Of the 17 patients in the combined modality group, one who failed to respond died of progressive Hodgkin's disease. Another patient relapsed and died 4 years later. Three additional patients died without recurrence of Hodgkin's disease: one of unknown causes at 44 months, an elderly patient died of lung cancer at 57 months, and another patient died of sepsis while undergoing treatment for chronic active hepatitis which was present before her diagnosis of Hodgkin's disease, at 28 months. The other 13 patients remain alive and well at 10 years. Of the 15 patients treated with MOPP alone, one nonresponder died of progressive Hodgkin's disease at 57 months, and one partial responder salvaged with radiotherapy died at 78 months of acute leukemia. The remaining 13 patients continue in first remission at 10 years, although one of those patients developed squamous cell carcinoma of the lung, but was apparently cured with surgery.

The NCI conducted a study in which 86 patients with stages IA-IIB were randomly allocated to treatment with six cycles of MOPP or subtotal radiation therapy. In an interim evaluation with a median observational period of 40 months (Longo et al. 1987), all 44 MOPP-treated patients had achieved complete remission, and only 4 had relapsed. The complete response rate was 95% for the radiotherapy group, but one-third had relapsed. Death had occurred in 9% of the MOPP patients and 22% of the radiotherapy group. There was a significant difference in disease-free survival in favor of the MOPP group, but overall survival was similar at the time of the analysis.

Conclusion

Previously untreated patients with advanced disease are candidates for combination chemotherapy alone, or for chemotherapy in addition to irradiation given to initial sites of bulky disease. MOPP has been the standard chemotherapy regimen and, although modifications may result in a more acceptable range of short-term toxicities, there are few data to support the notion that minor variations on the theme have resulted in more effective cancer treatment. Major alterations of MOPP, primarily those containing doxorubicin, a nitrosourea, or both have been reported to be more efficacious, but direct comparisons with MOPP are few. In some studies superiority of a given regimen over MOPP appears to result more from an imbalance of treatment intensity between the study groups, rather than from an intrinsic superiority of one treatment regimen over another. The concept of alternating non-cross-resistant regimens is intriguing but, unfortunately, it has been difficult to confirm the contention that such therapy is superior to standard treatment given at full dose and on schedule.

Relapsed or refractory patients with advanced disease represent a special challenge. Although many such patients are initially "salvaged" with a multidrug regimen, most ultimately die of Hodgkin's disease or acute leukemia. Such patients are usually candidates for entirely new approaches, such as bone marrow transplantation.

Most early-stage patients are treated with extended-field radiotherapy. A large minority of such patients with bulky mediastinal disease and/or stage III_2A fare significantly better with combined modality therapy. It has recently been demonstrated by the NCI, United States, that MOPP therapy alone is at least equivalent in efficacy to radiotherapy alone or combined modality therapy for early-stage patients without bulky disease. Long-term observation will be necessary to appreciate fully the relative merits and drawbacks of MOPP and radiotherapy for patients with early-stage Hodgkin's disease. However, at present it is clear that appropriately administered chemotherapy is highly effective therapy for early disease. This observation should have been anticipated because of the curability of combination chemotherapy in a sizeable proportion of patients with extensive disease that has been demonstrated over the past 20 years.

References

Andrieu J-M, Dana M, Desprez-Curely J-P, Jacquillat C, Weil M (1985) MOPP chemotherapy plus irradiation for Hodgkin's disease, stages IA to IIIB. Long-term results of the prospective trial H72. Hematol Oncol 3: 219-231

Bonadonna G, Valagussa P, Santoro A (1986) Alternating non-cross-resistant combination chemotherapy or MOPP in stage IV Hodgkin's disease. Ann Intern Med 104: 739-746

Canellos GP, Come SE, Skarin AT (1983) Chemotherapy in the treatment of Hodgkin's disease. Semin Hematol 20: 1-24

Carella AM, Santini G, Santoro A, Coser P, Frassoni F, Martinengo M, Nati S, Congiu A, Giordano D, Cerri R, Damasio E, Lercari G, Occhini D, Marmont AM (1985) Massive chemotherapy with non-frozen autologous bone marrow transplantation in 13 cases of refractory Hodgkin's disease. Eur J Cancer Clin Oncol 21: 607-613

Case DC Jr, Young CW, Lee BJ (1977) Combination chemotherapy of MOPP-resistant Hodgkin's disease with adriamycin, bleomycin, dacarbazine and vinblastine (ABVD). Cancer 39: 1382-1386

Colonna P, Andrieu JM, Kubisz P, Malou M, Zouaoui Z, Abad MT, Touhami M, Faid L (1986) Traitement de la maladie de Hodgkin par chimiotherapie MOPP seule. Presse Med 15: 599-602

Connors JM, Klimo P (1987) MOPP/ABV hybrid chemotherapy for advanced Hodgkin's disease. Semin Hematol 24: 35-40

De Vita VT Jr, Serpick AA, Carbone PP (1970) Combination chemotherapy in the treatment of advanced Hodgkin's disease. Ann Intern Med 73: 881-895

De Vita VT Jr, Simon RM, Hubbard SM, Young RC, Berard C, Moxley JH III, Frei E III, Carbone PP, Canellos GP (1980) Curability of advanced Hodgkin's disease with chemotherapy. Ann Intern Med 92: 587-595

Diggs CH, Wiernik PH, Sutherland JC (1980) Treatment of advanced Hodgkin's disease with SCAB - an alternative to MOPP. Cancer 47: 224-228

Dutcher JP, Wiernik PH (1985) Combined modality treatment of Hodgkin's disease confined to lymph nodes. Results 14 years later. In: Cavalli F, Bonadonna G, Rozencweig M (eds) Malignant lymphomas and Hodgkin's disease: experimental and therapeutic advances. Martinus Nijhoff, Boston, pp 317-328

Gobbi PG, Cavalli C, Gendarini A, Crema A, Ricevuti G, Frederico M, Di Prisco U, Ascari E (1985) Reevaluation of prognostic significance of symptoms in Hodgkin's disease. Cancer 56: 2874-2880

Goldie JH, Coldman AJ, Gudauskas GA (1982) Rationale for the use of alternating non-cross-resistant chemotherapy. Cancer Treat Rep 66: 439-449

Harker WG, Kushlan P, Rosenberg SA (1984) Combination chemotherapy for advanced Hodgkin's disease after failure of MOPP: ABVD and B-CAVe. Ann Intern Med 101: 440-446

Konits PH, Wiernik PH, Aisner J, Rulenz CA (1981) Pregnancy outcome in patients treated for Hodgkin's disease. Proc Am Soc Clin Oncol 22: 381

Koziner B, Myers J, Kirsch J, Redman J, Bajorunas L, Nisce L, Straus DJ, Lee BJ, Fuks Z (1987) Randomized comparison of two different chemotherapy programs (MOPP vs TVB) with segmental radiotherapy (RT) in clinical stages IA and IIA Hodgkin's disease. Proc Am Soc Clin Oncol 6: 201

Krikorian JG, Portlock CS, Rosenberg SA (1978) Treatment of advanced Hodgkin's disease with adriamycin, bleomycin, vinblastine and imidazole carboxamide (ABVD) after failure of MOPP therapy. Cancer 41: 2107-2111

Levi JA, Wiernik PH, O'Connell MJ (1977) Patterns of relapse in stages I, II and IIIA Hodgkin's disease: influence of initial therapy and implications for the future. Int J Radiat Oncol Biol Phys 2: 853-862

Lister TA, Dorreen MS, Faux M, Jones AE, Wrigley PFM (1983) The treatment of stage IIIA Hodgkin's disease. J Clin Oncol 1: 745-749

Longo D, Glatstein E, Young R, Duffey P, Hubbard S, De Vita V Jr (1987) Randomized trial of MOPP chemotherapy vs. subtotal nodal radiation therapy in patients with laparotomy-documented early stage Hodgkin's disease. Proc Am Soc Clin Oncol 6: 206

O'Dwyer PJ, Wiernik PH, Stewart MB, Slawson RG (1985) Treatment of early stage Hodgkin's disease: a randomized trial of radiotherapy plus chemotherapy versus chemotherapy alone. In: Cavalli F, Bonadonna G, Rozencweig M (eds) Malignant lymphomas and Hodgkin's disease: experimental and therapeutic advances. Martinus Nijhoff, Boston, pp 329-336

Pillai GN, Hagemeister FB, Velasquez WS, Sullivan JA, Johnston DA, Butler JJ, Schullenberger CC (1985) Prognostic factors for stage IV Hodgkin's disease treated with MOPP, with or without bleomycin. Cancer 55: 691-697

Prosnitz LR, Farber LR, Scott J, Bertino JR (1987) Fifteen years experience with combined modality therapy for advanced Hodgkin's disease. Proc Am Soc Clin Oncol 6: 193

Rosenberg SA (1987) ABVD versus MOPP: which is better? J Clin Oncol 5: 7

Santoro A, Viviani S, Villarreal CJR, Bonfante V, Delfino A, Valagussa P, Bonadonna G (1986) Salvage chemotherapy in Hodgkin's disease irradiation failures: superiority of doxorubicin-containing regimens over MOPP. Cancer Treat Rep 70: 343-348

Santoro A, Bonadonna G, Valagussa P, Zucali R, Viviani S, Villani F, Pagnoni AM, Bonfante V, Musumeci R, Crippa F, Tess JDT, Banfi A (1987) Long-term results of combined chemotherapy-radiotherapy approach in Hodgkin's disease: superiority of ABVD plus radiotherapy versus MOPP plus radiotherapy. J Clin Oncol 5: 27-37

Schulman P, Propert K, Cooper MR, Norton L, Gottlieb AJ (1987) Phase II study of MOPLACE in previously treated Hodgkin's disease. Proc Am Soc Clin Oncol 6: 188

Specht L, Nissen NI, Walbom-Jorgensen S (1985) Therapeutic implications of mediastinal involvement in advanced Hodgkin's disease. Scand J Haematol 35: 166-173

Straus DJ, Myers J, Lee BJ, Nisce LZ, Koziner B, McCormick B, Kempin S, Mertelsmann R, Arlin Z, Gee T, Poussin-Rosillo H, Hansen H, Clarkson BD (1984) Treatment of advanced Hodgkin's disease with chemotherapy and irradiation. Am J Med 76: 270-278

Straus DJ, Myers J, Gaynor J, Caravelli JF, Nisce LZ, Koziner B, Lee BJ, Clarkson BD (1987) Long-term results of treatment of advanced Hodgkin's disease with potentially non-cross-resistant drug combinations and moderate-dose radiation therapy. Proc Am Soc Clin Oncol 6: 202

Sutcliffe SB, Wrigley PFM, Stansfeld AG et al. (1979) Adriamycin, bleomycin, vinblastine and imidazole carboxamide (ABVD) therapy for advanced Hodgkin's disease resistant to mustine, vinblastine, procarbazine and prednisone (MVPP). Cancer Chemother Pharmacol 2: 209-213

Wiernik PH, Schiffer CA (1988) Long-term follow-up of advanced Hodgkin's disease patients treated with a combination of streptozotocin, lomustine (CCNU), doxorubicin and bleomycin (SCAB). J Cancer Res Clin Oncol 114: 105-107

Wiernik PH, Slawson RG (1982) Hodgkin's disease with direct extension into pulmonary parenchyma from a mediastinal mass: a presentation requiring special therapeutic considerations. Cancer Treat Rep 66: 711-716

Young RC, Longo DL, Glatstein E, Duffey PL, Winkler CF, Wiernik PH, De Vita VT Jr (1985) The current status of NCI trials in Hodgkin's disease. In: Cavalli F, Bonadonna G, Rozencweig M (eds) Malignant lymphomas and Hodgkin's disease: experimental and therapeutic advances. Nijhoff, Boston, pp 293-298

Report on Recent and Current Clinical Trials

*Risk Factor Adapted Treatment of Hodgkin's Lymphoma: Strategies and Perspectives**

M. Loeffler, M. Pfreundschuh, U. Rühl, E. Hiller, H. Gerhartz, R. Roloff, M. Adler, W. Schoppe, U. Hagen-Aukamp, G. Schmitt, J. Teichmann, V. Brust, R. Fuchs, B. Makoski, H. Bartels, J. Entzian, G. Kühn, A. Hellriegel, H. Kirchner, H. Emminger, G. Dölken, W. Oertel, W. Wilhelmy, B. Krüger, S. Petsch, H. Pflüger, R. Hinterberger, P. Worst, W. Gaßmann, K. Smith, D. Hasenclever, and V. Diehl (chairman)

Medizinische Universitätsklinik I, Joseph-Stelzmann-Straße 9, 5000 Köln 41, FRG

Introduction

Although modern therapy strategies have considerably improved the prognosis of Hodgkin's lymphoma, in general the results for stages I, II, and IIIA with mediastinal tumor, spleen, or extranodal involvement and those for advanced stages IIIB and IV are not yet satisfactory. In a number of different trials large mediastinal tumor (Cosset et al. 1984; Lee et al. 1980; Mauch et al. 1982), E-stages (Kaplan 1980; Musshoff 1970; Pillai et al. 1985; Prosnitz et al. 1981), and massive splenic involvement (Desser et al. 1977) have been reported to be unfavorable prognostic indicators if found in localized or IIIA stages. In most trials involving advanced stages complete remission rates range between 50% and 90% with a 5-year survival rate of less than 60% (Bonadonna et al. 1975; Cosset et al. 1984; Hancock 1986; Longo et al. 1986). With respect to the possibility and necessity of therapy intensification for selected groups of patients, it is useful to examine likewise treated patients for heterogeneity of response. We therefore performed a prognostic risk factor analysis based on the data of patients registered in our trials.

In 1982, nationwide multicenter therapy trials were started in the Federal Republic of Germany with the following main objectives: (1) standardization of diagnostic and therapeutic approaches in the FRG; (2) comparison of two combined modality treatments in stages I, II, and IIIA with MT, E, or S presentation (HD1 protocol); (3) comparison of radiotherapy versus chemotherapy in advanced stages IIIB/IV as consolidation for COPP+ABVD-induced complete remissions and evaluation of salvage regimens (HD3 protocol).

Recruitment for the trials started in July 1983 and ended in spring 1988. In order to allow for later rigorous statistical testing we decided to explore possible prognostic factors on the first half of the patients expected to enter the trial, thus

* Report of the German Hodgkin Study Group.

Recent Results in Cancer Research. Vol. 117

generating test hypotheses for the second half presently not yet evaluable. As the trial is ongoing no information about differences in the randomized treatment arms will be presented.

The choice of the end point is a particular problem in evaluating the results of clinical trials with long treatment periods, intermediate CR rates, and relatively long survival. With survival as an end point even major differences in the efficacy of the primary treatments may remain undetected. Relapse-free survival excludes all cases that do not achieve CR by the standard treatment. We use the parameter of Freedom From Treatment Failure (FFTF), the definition of which differs slightly from the parameter "freedom from progression" defined by others (e.g., Carde et al. 1988; Rosenberg 1985) in that it also takes other failures than progression into account (any status that is not proven CR).

Methods

Patients

Untreated patients between 15 and 60 years of age with histologically proven Hodgkin's lymphoma were eligible. Alcoholics, drug addicts, or nonresidents were not excluded from registration. Written consent was required.

Participating Centers

In 1982 three prospective multicenter trials were started for treatment of Hodgkin's lymphoma in the Federal Republic of Germany (HD1 protocol for localized stages with risk factors, HD2 protocol for stage IIIA,N, HD3 protocol for advanced stages). The trials started with 25 participating hospitals. According to the aim of standardizing diagnostic and therapeutic measures throughout the Federal Republic of Germany, new participating centers were accepted during the recruitment phase if radiotherapists and chemotherapists agreed to obey the study protocol. By October 1987 the number of centers participating in all studies had increased to 52, 20 of which are university hospitals, 26 regional hospitals, and 6 oncologists in private practice. One center was excluded due to systematic protocol violation. The annual registration rate for all studies has increased from about 100 to 220, an estimated fifth of the incidence in the Federal Republic of Germany. Twenty-two institutions contributed patients for the subsequent analysis of data gathered in the HD1 protocol and 29 centers for the HD3 protocol (see "Appendix").

Staging Procedure

The staging procedure included the following examinations: size of all enlarged lymph nodes, size of liver and spleen, presence and duration of symptoms; chest X-ray, abdominal CT scan and ultrasound, bone marrow biopsy, scanning of the skeletal system and laboratory tests (ESR, AP, WBC, Hb, thrombocytes, differential blood counts, liver enzymes) as obligatory examinations; thoracic CT scan, liver and spleen scintiscan, and X-ray of the skeleton were optional investigations. Laparotomy was routinely performed in stages CSI-IIIA if no contraindications were present for the invasive procedure. If laparotomy could not be performed, liver biopsy, bone marrow biopsy, and bipedal lymphangiography were obligatory to exclude infradiaphragmal disease. Laparotomy was performed according to international standards (Paglia et al. 1973). Staging results were classified according to Ann Arbor. Histological diagnosis was made by regional pathologists and copies of the reports were sent to the study center.

These diagnoses are presently being reviewed by a reference panel of pathologists. Among 394 cases reviewed so far only 2 have been identified as having been non-Hodgkin's lymphomas (no solid tumor). They were eliminated from the data base.

Therapy Allocation

HD1 Protocol. PS I, II, IIIA with one or more of the following presentations: extranodal disease, massive spleen involvement (diffuse or more than five involved nodules), large mediastinal tumor with more than one-third of the maximum transverse thoracic diameter above the diaphragm as determined by chest X-ray.

HD3-Protocol. CS and PS IIIB, IV.

Therapy Protocol of the HD1 Trial

Patients received two double cycles of alternating COPP+ABVD and were then randomized to receive additional radiotherapy 20 Gy EF (40 Gy on primary bulk disease) versus 40 Gy EF. In stage IIIA TNI was given instead of EF. The design is summarized in Fig. 1.

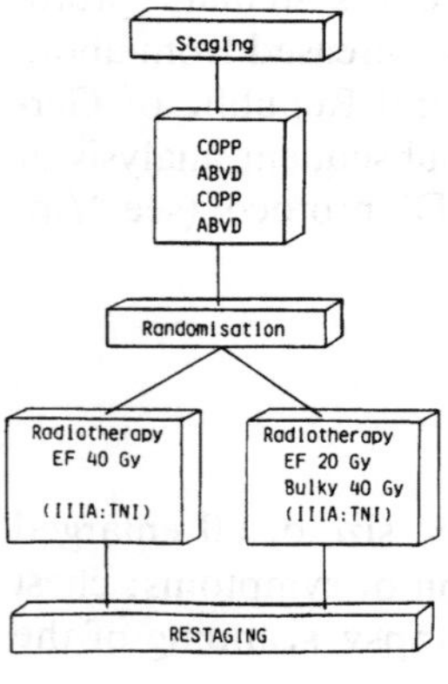

Fig. 1. Design of the HD1 trial for patients in stage I, II, IIIA with large mediastinal tumor or E-stage disease or massive spleen involvement. Patients received 2 × (COPP+ABVD) and were then randomized to either 20 Gy EF (40 Gy bulk) or 40 Gy EF (stage III: TNI instead of EF). As the trial is still ongoing no comparisons of these groups are given

Therapy Protocol of the HD3-Trial

Patients received three double cycles of alternating COPP+ABVD followed by complete restaging. Patients in complete remission were randomized to receive consolidation therapy by radiotherapy (involved field radiotherapy with 20 Gy) or chemotherapy (one double cycle COPP+ABVD). Patients not in complete remission received salvage therapy. This consisted of radiotherapy in the case of persisting nodal involvement (IF 20 Gy, persisting tumor 40 Gy). In the case of persisting disseminated or organ involvement the salvage therapy consisted of four cycles of CEVD (see below). The study design is summarized in Fig. 2.

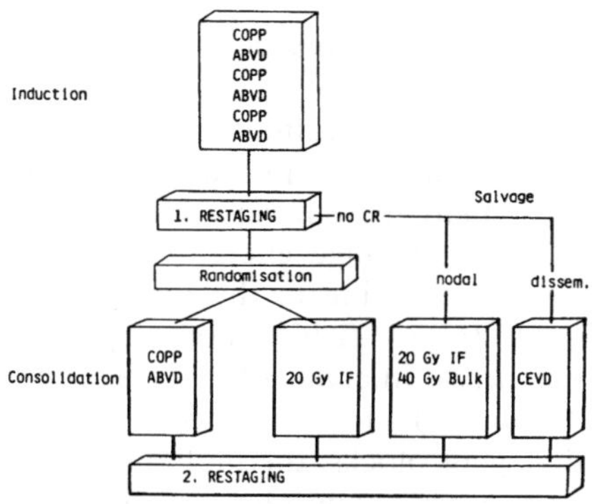

Fig. 2. Design of the HD3 trial for patients in advanced stage Hodgkin's disease CS/PS IIIB, IV. Patients received 3× (COPP+ABVD) as induction therapy. If CR was achieved they were randomized for comparison of different consolidation therapies. If CR was not achieved by induction treatment a salvage therapy followed with 4× CEVD for disseminated disease or 40 Gy radiotherapy in the case of nodal disease. As the trial is still ongoing no comparisons of these groups are given

Chemotherapy Protocols

As a modification of the MOPP scheme of DeVita et al. (1970) COPP was given, with mustargen being substituted by cyclophosphamide (Morgenfeld et al. 1975).

Cyclophosphamide	650 mg/m^2	i.v.	day 1, 8
Vincristine	1.4 mg/m^2	i.v.	day 1, 8
Procarbazine	100 mg/m^2	p.o.	day 1-14
Prednisone	40 mg/m^2	p.o.	day 1-14
Recycle			day 29

ABVD was given according to the Milan protocol (Bonadonna et al. 1975):

Doxorubicin	25 mg/m^2	i.v.	day 1, 15
Bleomycin	10 mg/m^2	i.v.	day 1, 15
Vinblastine	6 mg/m^2	i.v.	day 1, 15
DTIC	375 mg/m^2	i.v.	day 1, 15
Recycle			day 29

CEVD was given as described by our group (Pfreundschuh et al. 1987):

CCNU	80 mg/m^2	p.o.	day 1, 15
Etoposide	120 mg/m^2	p.o.	day 1-5, 22-26
Vindesine	3 mg/m^2	i.v.	day 1, 22
Dexamethasone	3 mg/m^2	p.o.	day 1-8
Dexamethasone	1.5 mg/m^2	p.o.	day 9-26
Recycle			day 43

The study protocol provided detailed instructions on the degree of dose reduction if myelopoietic toxicity exceeded given thresholds.

Radiotherapy Protocol

Radiotherapy treatment plans were designed by the study center according to the documentation of primary disease or involvement persisting after chemotherapy. Wide-field radiotherapy was delivered on a 4- to 10-MeV linear accelerator or ^{60}Co-megavoltage energy source. The radiation dose to each region was 20 or 40 Gy according to the respective protocol, given at five 1.8-2.0-Gy fractions/week. If wide-field radiotherapy was delivered to both supra- and infradiaphragmal areas, intervals of 3-4 weeks between each treatment course were permitted. Technical factors employed included use of treatment simulation, individualized divergent blocks, thermoluminescent dosimetric monitoring, and equal treatment from anterior and posterior fields. Adherence to the protocol was controlled by a review of the radiotherapy protocols and in the case of relapse by reevaluation of field control roentgenograms.

Evaluation of Therapy

The effect of therapy was evaluated by a restaging 4-8 weeks after therapy. It consisted of the control of all Hodgkin's disease manifestations by adequate clinical and histological methods. Complete remission (CR) was defined as the disappearence of all disease manifestations for at least 2 months. Patients were considered evaluable if they had started and completed therapy according to the protocols. Patients who did not complete the treatment were also considered evaluable. In this case the reasons for incomplete treatment were classified according to the following categories: progression (PRO) under therapy, excessive toxicity, violation of protocol, intercurrent death, or refusal of further therapy by the patient. In all these cases the remission status was also classified (CR, PR, PRO, or unclear). The decisions were made by a review committee. Freedom From Treatment Failure (FFTF) was defined as the time from the start of therapy (including laparotomy) to the first of the following events: death, progressive disease, non-CR status (partial remission or unclear status) at the end of (incomplete or complete) therapy, or relapse (Loeffler et al. 1988). For survival (SV) all deaths were included.

Table 1. Characteristics of patients in stages I, II, IIIA with large MT, E, or massive S involvement (HD1 trial)

Parameters	Overall results	Subgroups (not mutually exclusive)		
		Extranodal disease	Large mediastinal tumor	Massive spleen involvement
Patients in study	145			
Patients evaluable for response	89	16	55	32
Sex (% male)	54%	38%	40%**	75%**
Age, (years) median[a]	28	27	29	28
Staging				
Laparotomies	54%	13%	31%	100%
Stage IA[b]	1	0	1	0
Stage IIA	29	8[c]	27[c]	0
Stage IIB	21	5[d]	20[d]	1
Stage IIIA	38	3	7[e]	31[e]
Lung involvement	12	5	8	3
Only E or MT or spleen involvement	75	3	43	29
Laboratory				
ESR (mm/h)	39	35	52	20
(median, range)	(2-163)	(4-150)	(3-163)	(2-64)
AP (IU/ml)	155	150	158	143
(median, range)	(10-654)	(73-226)	(10-654)	(89-283)
Histology[f]				
LD	6%	0	2%	13%
NS	63%	69%	80%*	34%*
MC	26%	25%	16%*	44%*
LP	1%	0	0	3%
EP	2%	0	0	6%
NC	2%	6%	2%	0
CR-rate	74 (83%)	13 (81%)	47 (85%)	25 (78%)
FFTF events	18 (20%)	4 (25%)	9 (18%)	9 (28%)
During treatment				
Progressive disease	8	3	6	2
Intercurrent death	1	0	0	1
Excessive toxicity	0	0	0	0
Protocol violation	2	0	0	2
Refusal of therapy	2	0	0	2
PR at end of treatment	2	0	2	0
Relapses after treatment	3	1	1	2
Deaths	7 (8%)	1 (6%)	3 (6%)	4 (12%)

* Different at a significance level $P \leq 0.05$; ** different at a significance level $P \leq 0.01$.

[a] All ranges are (15-56).

[b] No patient with stage IB.

[c] Six patients had E stage and large MT.

[d] Five patients had E stage and large MT.

[e] One patient had large MT and spleen involvement; two patients had E stage and large MT.

[f] Subgrouping according to primary histology.

Biostatistics

The endpoint parameters freedom from treatment failure (FFTF) and survival (SV) were evaluated according to the Kaplan-Meier method. Univariate tests were performed using the Cox-Mantel test. For multivariate analysis a proportional hazard model was used with stepwise Cox regression. All variables used were dichotomized. Consequently relative risk values could be given.

Results

Results for Stages I, II, IIIA with Large MT, E-Stage, or Massive Spleen Involvement

Patients with stage I, II, IIIA and large MT, E, or S involvement were entered into the HD1 protocol, whose design is summarized in Fig. 1. Patients were randomized into two combined modality treatments. By October 1987 145 patients had started therapy according to this protocol, of which 89 had finished therapy. The results are summarized in Table 1.

Although males and females were about equally represented in the trial, large mediastinal tumor (MT) was more often found in females ($P < 0.01$) and massive spleen involvement more often in males ($P < 0.01$). Laparotomy was performed in 54% of all patients. Presentation of E disease (mostly massive lung involvement) and large MT with the associated respiratory problems were frequently contraindications for the invasive procedure. Only one patient in stage I was recorded. Apparently stage I in general is rarely associated with MT or E disease. Most patients were stage II (50/89). Systemic symptoms were observed in 42% of these patients. Stage IIIA was present in 38 patients. Only one case had primary infradiaphragmal disease with massive spleen involvement. Large MT was primarily found in stage II. E stage was rare and frequently associated with large MT. In 75 patients only one factor, i.e., either E disease or MT or S involvement was present while 14 patients had two such factors. ESR and AP did not differ between the subgroups. Nodular sclerosis histology was correlated with large MT ($P < 0.05$) and mixed cellularity with massive spleen involvement ($P < 0.05$).

After combined modality treatment 74 patients (83%) achieved CR with similar results in all subgroups. Eight patients suffered from progressive disease. One pa-

Fig. 3. a Survival *(SV)* and freedom from treatment failure *(FFTF)* of 89 CS/PS I, II, IIIA ▷ patients with large MT, E-stage, or massive spleen involvement evaluable for response in the HD1 trial presented as Kaplan-Meier plots. The median time of observation is 30 months for both SV and FFTF. FFTF is defined as time from entry into the study until the first of the following events: death, progressive disease, non-CR at the end of therapy, relapse. **b** FFTF curves according to stage and constitutional symptoms. (30 patients in stage IA and IIA, median observation time 28 months; 21 patients in stage IIB, 32 months, 38 patients in stage IIIA, 30 months). **c** FFTF curves for patients with only one site of massive tumor burden with either a large mediastinal tumor (*MT*, 43 patients) or massive spleen involvement (*S*, 29 patients) or with two sites of massive tumors (11 *MT & E*, 1 *MT & S*, 2 *E & S*). The median times of observation are 31, 25, and 32 months respectively

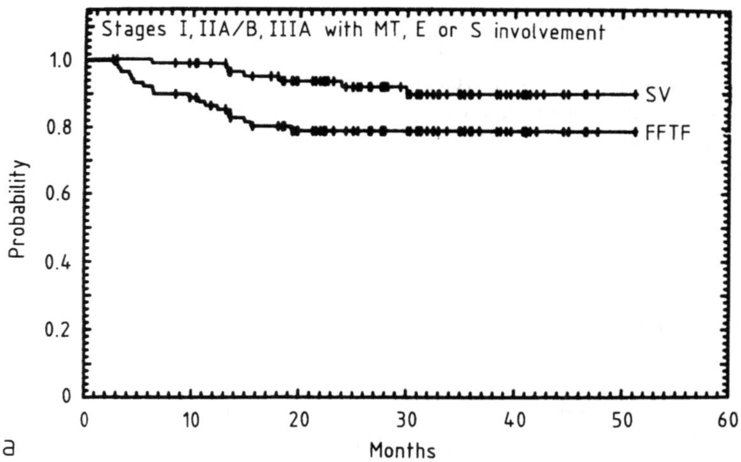
Stages I, IIA/B, IIIA with MT, E or S involvement
SV
FFTF
Probability
Months

a

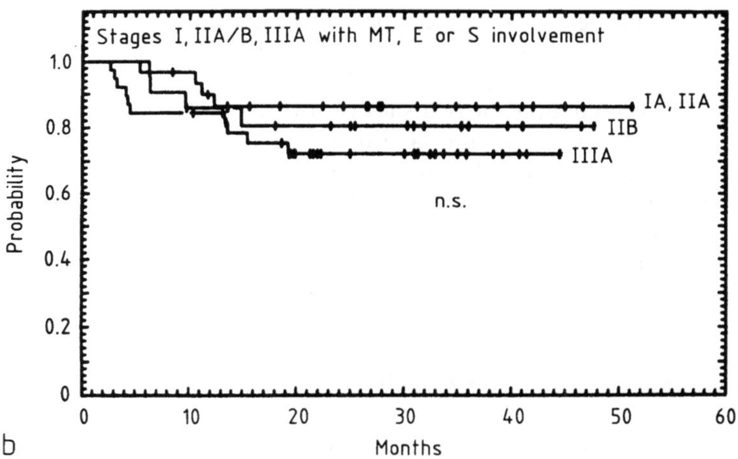
Stages I, IIA/B, IIIA with MT, E or S involvement
IA, IIA
IIB
IIIA
n.s.
Probability
Months

b

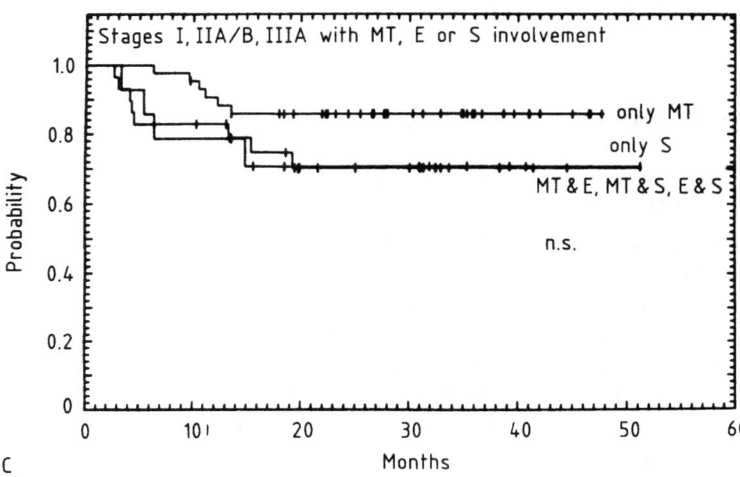
Stages I, IIA/B, IIIA with MT, E or S involvement
only MT
only S
MT & E, MT & S, E & S
n.s.
Probability
Months

c

tient died. In two cases therapy was not given according to the protocol. Two patients deliberately terminated treatment without a proven CR. Two patients had a PR at the end of the complete treatment. Only three relapses have been observed so far. The 3-year overall survival is 92% (±6%; 95% confidence interval) and FFTF is 80% (±8%; 95% confidence interval), both demonstrating no differences between the subgroups.

Figure 3a gives the probability of survival (SV) and freedom from treatment failure (FFTF) of all 89 patients.

Prognostic Factor Analysis

In a univariate explorative data analysis none of the following staging parameters had a prognostic significance: stage, sex, age, laparotomy, systemic symptoms, stage, MT, E stage, S involvement, histology, ESR, AP, leukocytes, lymphocytes. Figure 3b gives the FFTF curves for stages IA and IIA, IIB, and IIIA. They are statistically not different. Likewise the presentation of only one site of massive tumor burden (MT or E or S) does not differ from that of combinations. Figure 3c gives the FFTF curves if only a large MT or a massive spleen involvement was present versus combinations of two sites of involvements. It is furthermore noteworthy that FFTF curves for the laparotomized and not laparotomized groups match almost precisely (not shown). Multivariate analysis using a proportional hazard model showed that inclusion of the factors mentioned above could not significantly improve a hazard model in which none of the factors was considered. Thus, at least at the present stage of analysis, the group of patients that entered the HD1 trial behave fairly homogeneously with respect to prognosis.

Results for Advanced Disease (CS/PS IIIB, IV)

By October 1987, 230 patients in stages CS/PS IIIB/IV had started therapy according to the HD3 protocol and 137 patients had finished therapy. The results are summarized in Table 2.

Patient Characteristics and Overall Results

The majority of patients were male. They were about equally distributed in stage IIIB and IV. Only a minority (8%) did not exhibit systemic symptoms. Median erythrocyte sedimentation rate (ESR) was 62 mm/h, median serum alkaline phosphatase (AP) 181 IU/ml.

After induction therapy, 87 patients (63%) achieved complete remission. Another 17 patients achieved complete remission by salvage therapy. The overall complete remission rate after the end of therapy was 76%. Eleven patients suffered from progressive disease. In 18 patients treatment was prematurely terminated without a (proven) CR. The reasons were therapy-related death (2), intercurrent death (1), protocol violation (4), and refusal of further therapy (11). Among the lat-

Table 2. Patient Characteristics and overall results

Parameters	HD3
Patients in study	230
Patients evaluable for response	137
Sex (% male)	68%
Age (median, range)	33 (15-60)
Staging	
Laparotomies	35 (26%)
CS/PS IIIB	77 (56%)
CS/PS IVA	11 (8%)
CS/PS IVB	49 (36%)
Organ involvement[a]	
Liver	15 (11%)
Bone	8 (8%)[b]
Bone marrow	25 (19%)[c]
Lung	18 (13%)[d]
Laboratory findings	
ESR (mm/h) (median, range)	62 (2-167)[e]
AP (IU/ml) (median, range)	181 (58-1122)[f]
Histology[g]	
LD	8 (6%)
NS	65 (47%)
MC	43 (31%)
LP	10 (7%)
EP	6 (4%)
NC	5 (4%)
Results of treatment	
CR after induction	87 (63%)
CR through salvage	17
CR at end of therapy	104 (76%)
FFTF events (overall)	54 (39%)
During treatment phase	
Progressive disease	11
Premature termination of treatment in non-CR status	
Therapy-related death	2
Intercurrent death	1
Excessive toxicity	0
Protocol violation	4
Refusal of further therapy	11
PR at end of treatment	4
Relapse after treatment	21
Deaths (overall)	17

FFTF, freedom from treatment failure.
[a] Not mutually exclusive.
[b] Thirty-seven values missing.
[c] Five values missing.
[d] Two values missing.
[e] Two values missing.
[f] Two values missing.
[g] Subgrouping according to primary histology.

ter patients are two alcoholics and two nonresident persons who left the country before treatment was finished. All these cases were counted as treatment failures because their status was unclear. Four patients had a PR at the end of the regular treatment. We observed 21 relapses among the 104 patients in CR. In total 17 deaths have been observed so far. Figure 4a gives the probability of survival (SV) (84% ± 8%; 95% confidence interval) and freedom from treatment failure (FFTF) (56% ± 10%; 95% confidence interval) of all 137 patients. It is apparent that FFTF declines more rapidly than SV, which makes it a more sensitive end point for prognostic factor analysis.

Univariate Analysis of Prognostic Factors

Sex, age (above versus below 40 years), laparotomy, bone marrow, liver, and bone involvement do not appear as univariate prognostic factors with respect to FFTF or SV ($P > 0.25$). Stage as the classical Ann Arbor factor plays a minor role and is not significant ($P > 0.08$) (see Fig. 4b). In contrast, pretreatment values of ESR and AP seem to discriminate subgroups of patients with respect to FFTF (see Fig. 4c, d). As discriminators we selected threshold values of 80 mm/h for ESR and 230 IU/ml for AP. The univariate difference between the 97 patients with low ESR and the 38 cases with high ESR is significant (two cases missing) (see Fig. 4c). Likewise the univariate difference between the 91 patients with low AP and the 44 cases with high pretreatment AP was significant (two cases missing) (see Fig. 4d).

Multivariate Analysis of Prognostic Factors

To assess the independent contribution of the various prognostic factors, a multivariate analysis was performed using a proportional hazard model with dichotomous variables. Cox regression analysis was used to test whether given hypothetical prognostic models could significantly be improved by taking further factors into account.

Table 3 summarizes the results of the four most important scenarios. If stage is the only prognostic factor included in the model (model I), the test statistics show that this model can be significantly improved if ESR ($P = 0.0021$) or AP

Fig. 4. **a** Overall survival *(SV)* and overall freedom from treatment failure *(FFTF)* of ▷ 137 CS/PS IIIB, IV patients evaluable for response in the HD3 trial presented as Kaplan-Meier plots. The median time of observation is 25.5 months for SV and 27.5 months for FFTF. Univariate prognostic factor analysis using freedom from treatment failure (FFTF) shows: **b** Patients in stage IIIB do not perform significantly better than patients in stage IV ($P > 0.08$), (77 patients in stage IIIB, 60 patients in stage IV). **c** Patients with a pretreatment ESR below 80 mm/h have a significantly better prognosis than those with a higher ESR (97 patients with ESR ≤ 80, 38 patients with ESR > 80). **d** Patients with a pretreatment AP below 230 IU/ml do significantly better than those with increased values (91 patients with AP ≤ 230, 44 patients with AP > 230)

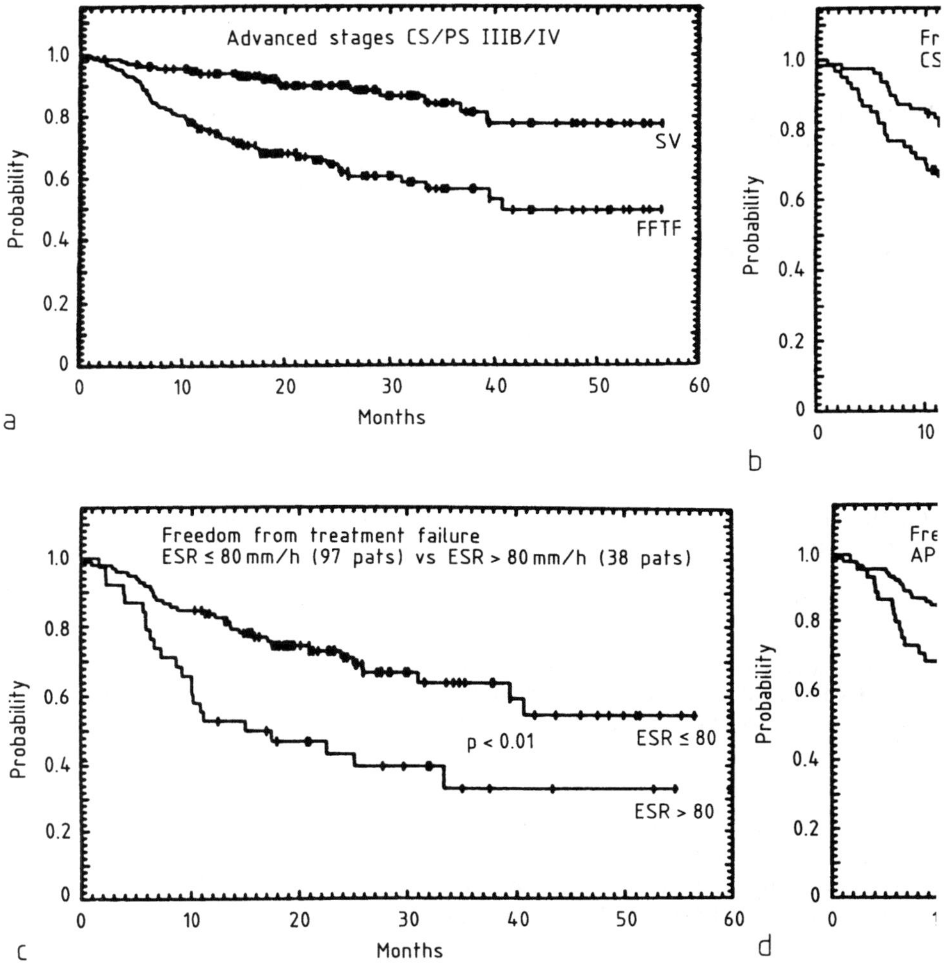
Advanced stages CS/PS IIIB/IV
SV
FFTF
Probability
Months
a
Fr
CS
b
Freedom from treatment failure
ESR ≤ 80 mm/h (97 pats) vs ESR > 80 mm/h (38 pats)
p < 0.01
ESR ≤ 80
ESR > 80
c
Fre
AP
d

Table 3. Summary of Cox-regression analysis ($N=134$)[a]

Models considered (factors included)		Significance of factors for improvement of the model[b] (χ^2-value to enter or remove from the model, P value)		
		ESR	AP	Stage
I. Stage ($0\equiv$ III, $1\equiv$ IV)	χ^2	9.45	7.15	2.95
	P	0.0021	0.0075	0.086
(RR = 1.61)				
II. ESR ($\equiv >80$; $0\equiv \leq 80$)	χ^2	8.61	3.52	3.80
	P	0.0033	0.061	0.051
(RR = 2.31)				
III. AP ($1\equiv >230$; $0\equiv \leq 230$)	χ^2	3.74	8.39	1.71
	P	0.053	0.0038	0.19
(RR = 2.25)				
IV. ESR and/or AP	χ^2	0.18	0.01	1.76
($0\equiv$ both low, $1\equiv$ else)	P	0.68	0.96	0.18
(RR = 2.80)				

RR, relative risk.

[a] Three cases excluded because of incomplete data.

[b] Besides ESR, AP, stage the contribution of the following factors is not shown because significance was lacking: sex, age, bone marrow, bone, liver, laparotomy (χ^2-value ≤ 1.25, $P > 0.25$).

($P=0.0075$) are taken into account. Consideration of stage alone does not seem to exploit the maximum information. This confirms the impression from the above univariate analysis. If in contrast a prognostic factor model is designed which is entirely based on ESR (model II), the improvement of the model by additional consideration of AP ($P=0.061$) and/or stage ($P=0.051$) is less dramatic than in model I. Apparently ESR already carries a lot but not all of the information of the parameters AP and stage. If AP is considered as the only prognostic factor (model III), the model can be improved more by ESR ($P=0.053$) than by stage ($P=0.19$). Models II and III show that ESR as well as AP alone provide better prognostic predictors than stage. They seem to have about equal strength but do not discriminate the same subgroups. ESR seems to be less dependent on stage than AP. All the other prognostic factors like sex, age, laparotomy, bone marrow, liver, and bone involvement do not play a role.

It is therefore tempting to combine ESR and AP into a new dichotomous prognostic factor (model IV in Table 3). One group is characterized by ESR below 80 mm/h and AP below 230 IU/ml while the other group includes cases for which one or both parameters exceed the thresholds. The regression analysis shows that this model cannot be improved significantly by taking further factors (e.g., stage) into account. The relative risk between the two groups differs by a factor of 2.8, which is higher than in the previous models I–III (1.6, 2.3). Other models combining ESR with stage or AP with stage did not result in an increased relative risk (not shown).

The subgroups just examined appear to be comparable. With respect to sex, age, stage, frequency of laparotomy, organ involvement, and histology no significant difference could be found between the model II and III subgroups [i.e., of low ESR vs. high ESR (Fig. 4c), low AP vs. high AP (Fig. 4d)]. Only with respect to laboratory findings could a weak correlation between pretreatment ESR and AP values be found ($r=0.4$).

Discussion

The interim analysis of the HD1 protocol demonstrates several interesting results about patients with localized and stage IIIA disease exhibiting MT, S, or E involvement:

Although the group of patients recruited is fairly heterogeneous with respect to stage (II and III) as well as sites of tumor burden (thoracical, extranodal, splenic), the fate of these patients under our combined modality treatment seems to be very similar as is shown by the failure to identify prognostic factors within this group. The prognostic homogeneity in our eyes corroborates the inclusion criteria of the HD1 trial. However, this conclusion may change if more patients and a longer follow-up can be evaluated. In view of the particular selection of the patients a 3-year FFTF of 80% and a 3-year survival of 92% can be considered as encouraging as they come close to the values observed for patients with limited disease. The EORTC has, for example, obtained similar results in the H5F trial in favorable PSI, II patients (age ≤ 40 years, ESR ≤ 70 mm/h, LP or NS subtype) receiving only mantle or mantle plus paraaortic 40 Gy irradiation (Carde et al. 1988). For patients with unfavorable PSI, II receiving TNI the results were also similar and only combined modality treatment generated better results than those found in our group of more severely afflicted patients. In particular, our patients with large mediastinal tumor showed a very promising result with only 1 relapse out of 47 patients.

The interim evaluation of the HD3 protocol for advanced Hodgkin's disease revealed several interesting results:

The first observation is that in patients with CS/PS IIIB/IV the complete remission rate achieved with the alternating COPP+ABVD regimen is better than in a pilot study with COPP. We now observe 76% compared with 55% previously.

The remission rates obtained by our group with COPP+ABVD are similar to the results obtained by an ongoing multicenter study of the EORTC for MOPP+ABVD (Somers et al. 1988) and appear to be slightly better than those reported by the British Lymphoma Group for MOPP or LOPP (Hancock 1986). However, our rates are inferior to those obtained by the Milan group (Bonadonna et al. 1986).

The second interesting finding is that patients in stages CS/PS IIIB/IV with a pretreatment ESR > 80 mm/h have a significantly shorter duration of freedom from treatment failure ($P<0.01$) than those with a lower ESR. Likewise patients with a pretreatment AP > 230 IU/ml have a significantly shorter duration of FFTF than those with a lower AP. In comparison with these two parameters, all other factors tested (including stage) were of minor importance. ESR and AP had about

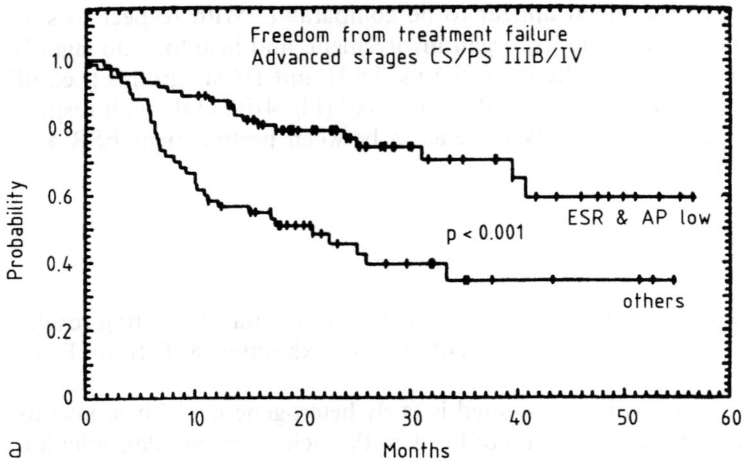

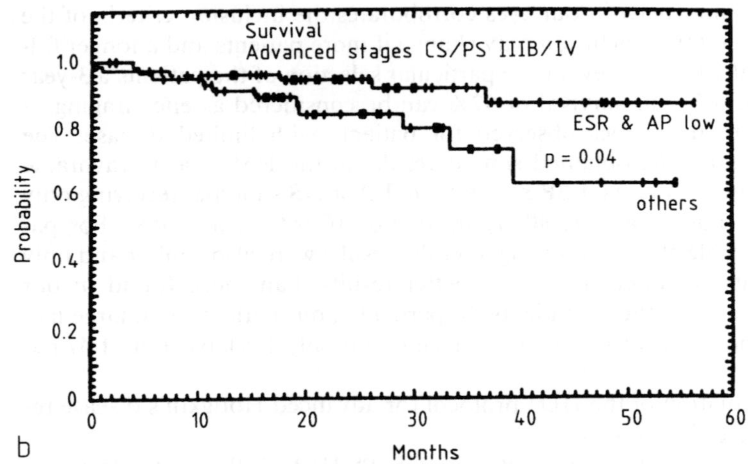

Fig. 5 a, b. Comparison of patients with low ESR and low AP with all other CS/PS IIIB, IV patients. **a** Freedom from treatment failure *(FFTF):* (74 patients with low ESR and AP, median observation time 26 months, 60 other patients, median observation time 30 months). **b** Survival (SV) (74 patients with low ESR and AP, median observation time 26 months; 60 other patients, median observation time 23 months)

the same discriminative ability if taken individually. If one of them was favored (i.e., included as a factor in a Cox model), the relevance of the second diminished considerably (see Table 3) but did not completely disappear. For this reason the parameters ESR and AP do not appear as two fully independent risk factors. A combination of ESR and AP (model IV, Table 3) gives the best discrimination of two prognostic subpopulations making the contribution of all other risk factors tested insignificant ($P > 0.18$).

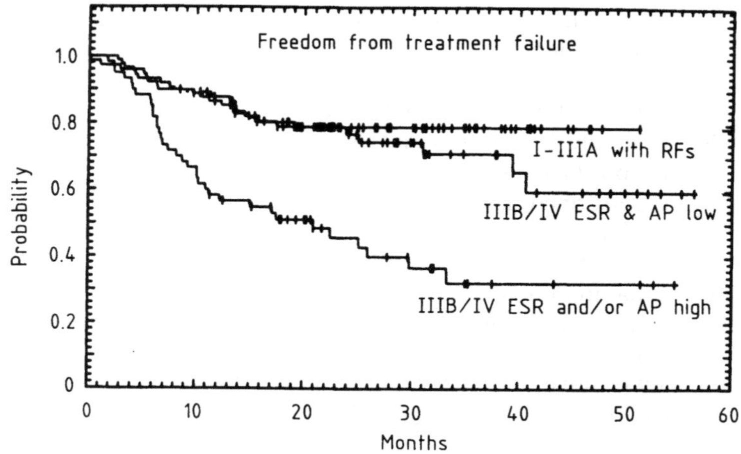

Fig. 6. FFTF- Summary. According to the previous analysis there are three different groups of patients with a 3-year FFTF of about 30% (high risk), of about 60% (intermediate risk), and of about 80% (favorable). The curves are redrawn from Figs. 3a and 5b

One may speculate that factors such as high ESR or AP reflect a biologically more aggressive form of Hodgkin's lymphoma or a more extensive disease which is not detectable by our current diagnostic procedures. As the freedom from treatment failure of this group is below 40% at 40 months, the question is raised whether a more aggressive first-line therapy such as autologous bone marrow transplantation (Carella et al. 1985) is indicated in this group.

Taken together, the results of the HD1 and HD3 trials and the prognostic factor analysis suggest that there may exist four groups of patients which differ considerably in prognosis (Fig. 5): Advanced stage IIIB and IV with high ESR or AP (3-year FFTF of about 30%), advanced stage IIIB and IV with low ESR and low AP (3-year FFTF of about 60%), stages I, II, IIIA with MT, S or E involvement (3-year FFTF about 80%). This is again summarized in Fig. 6. In addition a fourth even more favorable group may include stages I, II with small tumor loads (not studied by us). For this group FFTF may not be the appropriate end point because relapses after radiotherapy can well be salvaged by chemotherapy (Carde et al. 1988).

A subdivision into four groups of patients as suggested by our analysis implies that the role of the classical Ann Arbor stages as prognostic indicators becomes questionable. Their relevance has also been challenged recently by other groups (Gobbi et al. 1988; Haybittle et al. 1985; Tubiana et al. 1985). Based on retrospective analyses of largely differently treated groups of patients they concluded that other factors like ESR, age, and histological subtype (NS1 vs. NS2) may be of equal or even higher relevance for survival than a topological staging or the presence of systemic symptoms. This may also hold true for FFTF as an endpoint as suggested by our analysis. In the HD3 trial a homogeneous group of patients all having advanced stage could be subdivided into more favorable and less favorable groups by laboratory parameters. In the HD1 trial patients with quite a topological heterogeneity of involved sites were found to have a fairly similar prognosis. The failure to identify ESR or AP as prognostic factors in the HD1 trial does not contradict a more general role of these parameters for Hodgkin's disease in general as suggested by others (Gobbi et al. 1988; Haybittle et al. 1985; Tubiana et al. 1985). It should be kept in mind that the HD1 patients are already a highly selected group on their own and a comparison with unselected PS I, II, IIIA patients was not performed here.

The major setback of prognostic factor analyses performed so far by us and others (e.g., Haybittle et al. 1985; Tubiana et al. 1985; Gobbi et al. 1988) is their retrospective nature. Each study group extensively explores its data material for possible parameters which leads into the known problems of retrospective data analysis associated with statistical overtesting, uncontrolled biases, and a tendency to report only positive results. As a consequence the results can only be used to derive statistical hypotheses which should properly be tested again on an independent set of data.

The analysis presented here was such an exploratory data analysis. Although the material shown appears to be consistent we interpret the "statistical significance" with caution and use the results only as a working hypothesis. Whether ESR and/or AP really are prognostic factors in advanced stages will be tested by examining the next cohort of patients becoming evaluable for response in our trial.

But even if our hypothesis is confirmed, the results may only hold true for our specific diagnostic and therapeutic strategy. The question of whether they have a more general meaning for the prognosis of advanced Hodgkin's disease under intensive chemotherapy in general can only be assessed if other study centers examine their data in a similar way.

Besides retrospectivity the prognostic factor analyses nowadays often suffer from a second problem whose relevance is difficult to estimate: Interactions between treatment and prognostic factors are rarely taken into account. Typically patients from many trials performed in one study center over the past 2 decades are pooled and analyzed by a proportional hazard model (Haybittle et al. 1985; Tubinana et al. 1985; Gobbi et al. 1988). This procedure neither takes into account the diversity of treatments (of which many were designed as "risk factor adapted") nor the allocation strategies to these treatments. This attitude neglects observations that one can successfully design treatments which eliminate previously existing prognostic risk factors (e.g., mediastinal tumor; Lee et al. 1980), while new ones (e.g., ESR and AP) may appear. The question at present is perhaps not so much which risk factors emerge from the history of individual study centers (where old and abandoned treatment strategies will necessarily be overrepresented). Of more relevance should be the question which new risk factors now emerge under modern treatment strategies (e.g., serum parameters, tumor volume), which ones disappear (e.g., systemic symptoms, histology), and how we have to change the allocation procedures in order to achieve an improved risk factor adapted treatment. Having this is mind we believe that a prognostic risk factor analysis should be seen in relation to the specific treatment strategy under which it was obtained.

In order to assess the general validity of the prognostic risk factors under discussion and their dependency on treatment strategies, we advocate a systematic international collaborative assessment of prognostic factors in which similar evaluations should be undertaken on data from different groups. Over 15 years after Ann Arbor the scientific community has gathered a large amount of data on Hodgkin's disease patients with reasonably well documented and comparable staging. These data are available in electronic data bases and could be brought together into one reference data base. It would be very challenging to evaluate these data in a common effort. The task is, however, not trivial. It requires a careful comparison of the staging and restaging data. The interaction of treatment strategies with risk factors may require new statistical procedures. But the chance to clarify a number of debates may well validate the effort.

Summary

In a national multicenter trial in the Federal Republic of Germany, patients with Hodgkin's lymphoma in stages I, II and IIIA presenting with large mediastinal tumor (MT), extranodal (E), or massive spleen (S) involvement received a combined modality treatment with 2 × (COPP + ABVD) followed by 20 or 40 Gy EF radiation (HD1 protocol). By October 1987, 89 patients aged 15-60 years had finished therapy and were evaluable for response. Of these 74 (83%) achieved complete remission (CR). After 3 years freedom from treatment failure (FFTF) is 80% (±8%, 95% confidence interval) and survival (SV) 92% (±6%, 95% confidence interval). In a univariate and multivariate analysis using FFTF as endpoint we could not identify any particularly prominent prognostic risk factor among the following examined: stage, constitutional symptoms, MT, E stage, S involvement, age, sex, histology, laparotomy, erythrocyte sedimentation rate (ESR), leukocytes, lymphocytes, and alkaline phosphatase (AP). These data suggest that the inclusion criteria have selected a fairly homogeneous group of patients with respect to prognosis.

In a separate trial (HD3 protocol) patients in stages IIIB/IV received induction chemotherapy with 3 × (COPP + ABVD). Patients in complete remission (CR) received consolidation therapy by either radiotherapy (20 Gy IF) or further chemotherapy (COPP + ABVD). Patients not in CR received salvage therapy (40Gy in the case of persisting nodal disease, or else 4 × CEVD chemotherapy). By October 1987, 137 patients had finished therapy and were evaluable for response. Of these 86 (63%) achieved CR after induction chemotherapy. Including salvage therapy a total of 104 patients (76%) achieved CR. After 3 years FFTF is 56% (±10%, 95% confidence interval) and SV 84% (±8%, 95% confidence interval). Univariate and multivariate prognostic risk factor analyses were performed using FFTF as endpoint. Sex, age, stage, splenectomy, bone marrow, and liver and bone involvement had no prognostic impact. In contrast, a pretreatment erythrocyte sedimentation rate (ESR) above 80 mm/h and a serum alkaline phosphatase (AP) above 230 IU/ml each appeared as significant prognostic factors ($P < 0.01$; relative risk, 2.3). The two parameters can be combined to separate two groups (A: ESR and AP both low; B: ESR and/or AP high) which differ significantly for FFTF ($P < 0.001$) and survival ($P < 0.04$). The decision for risk-adapted treatment requires identification of groups of patients in the frame of specified diagnostic and therapeutic strategies. In our case we have indications for three groups of Hodgkin's patients, one with poor prognosis (3-year FFTF ≈ 30%), one intermediate (3-year FFTF ≈ 60%), and one favorable (3-year FFTF ≈ 80%). A fourth group with limited disease may be added but has not been examined by us.

Acknowledgments. The German Hodgkin Study Group is supported by the Federal Minister of Science and Technology (BMFT) [Therapiestudien bei bösartigen Neubildungen: Morbus Hodgkin, Förderungskennzeichen (01ZP550A)]. We thank K. Siewert and H. Nisters-Backes for help with the manuscript. We are grateful to Prof. Schaadt (Köln) for valuable suggestions.

Appendix: Participating Institutions

Hospitals and Practitioners (Listed According to Recruitment). München Großhadern (E. Hiller, H. Gerhartz, R. Rohloff); Köln Med. Universitätsklinik I (V. Diehl, M. Pfreundschuh, P. Müller, M. Adler); Düsseldorf Universitätsklinik (W. Schoppe, H. Kürten, U. Hagen-Aukamp); Berlin Steglitz (J. Teichmann, H. Ernst); Duisburg St. Johannes-Hospital (M. Westerhausen, R. Fuchs, B. Makoski); Lübeck Städt. Krankenhaus-Süd (H. Bartels, J. Entzian); Berlin Moabit (U. Rühl, G. Kühn, H. Hellriegel); Hannover Med. Hochschule (H. Kirchner, H. J. Schmoll, H. Emminger); Freiburg Universitätsklinik (G. Dölken, T. Hecht, H. Hinkelbein); Berlin Charlottenburg (W. Oertel); Berlin Neukölln (W. Wilhelmy); Mainz Universitätsklinik (B. Krüger, K. Kutzner); Erlangen Universitätsklinik (J. König, S. Petsch, R. Sauer, B. Grabenbauer); Marburg Universitätsklinik (H. Pflüger, R. Pfab); Mainz Praxis Schniepp/Hinterberger; Mannheim Klinikum (P. Worst, R. Lütgemeier); Kiel Städtisches Krankenhaus (W. Gaßmann, T. Brix); Frankfurt Universitätsklinik (K. Schalk); Wiesbaden Praxis Dr. Hildesheim; Städt. Krankenhaus (J. Preiß, W. Gärtner); Bremen Zentralkrankenhaus links der Weser (T. Luska); Ravensburg St. Elisabethen-Krankenhaus (W. Mende); Hannover Praxis Dr. Wysk; Karlsruhe St. Vincentius Krankenhäuser (S. Theml, R. Staiger); Osterholz-Scharmbeck Städt. Krankenhaus (A. Behboudi); Berlin Praxis Dr. Weißenfels; Neuss Lukas-Krankenhaus (P. Czygan); Oldenburg Evangelisches Krankenhaus (F. Hinrichs, A. Temmesfeld); Köln Krankenhaus Merheim (E. Renner, M. Cohen); Köln Med. Universitätsklinik I (B. Mödder); Lübeck Medizinische Hochschule (T. Wagner); Dormagen Kreiskrankenhaus (H. Solbach); Trier Mutterhaus der Borromäerinnen und der Barmherzigen Brüder (H. Hennekeuser, H. Siebner, K. H. van de Weyer, D. Dornhoff); Bonn Universitätsklinik (U. Loos, I. Boldt).

Radiotherapy Reference Centre. U. Rühl (Berlin)
Reference Pathology. A. Georgii (Hannover), R. Fischer (Köln), K. Hübner (Frankfurt), E. W. Schwarze (Dortmund)
Data Management. H. Nisters-Backes (Köln)
Statistics. M. Loeffler, K. Smith, D. Hasenclever (Köln)
Study Coordinators. M. Pfreundschuh, M. Loeffler (Köln)
Chairman. Volker Diehl (Köln)

References

Bonadonna G, Zucali R, Monfardini S, De Lena M, Uslenghi C (1975) Combination chemotherapy of Hodgkin's disease with adriamycin, bleomycin, vinblastin and imidazole carboxamide versus MOPP. Cancer 36: 252-259

Bonadonna G, Valagussa P, Santoro A (1986) Alternating non-cross-resistant combination chemotherapy or MOPP in stage IV Hodgkin's disease. Ann Intern Med 104: 739-746

Carde P, Burgers JMV, Henry-Amar M, Hayat M, Sizoo W, Van der Schueren E, Monconduit M, Noordijk EM, Lustman-Marechal J, Tanguy A, de Pauw B, Casset JM, Cattan A, Schneider M, Thomas J, Meerwaldt JH, Somers R, Tubiana M (1988) Clinical stages I and II Hodgkin's disease: a specifically tailored therapy according to prognostic factors. J Clin Oncol 2: 239-252

Carella AM, Santini G, Santoro A, Coser P, Frassoni F, Martinengo M, Nati S, Congiu A, Giordano D, Cerri R, Damasio E, Lercari G, Ochini D, Marmont AM (1985) Massive chemotherapy with non-frozen autologous bone marrow transplantation in 13 cases of refractory Hodgkin's disease. Eur J Cancer Clin Oncol 21: 607-613

Cosset JM, Henry-Amar M, Carde P, Clarke D, LeBourgeois JP, Tubiana M (1984) The prognostic significance of large mediastinal masses in the treatment of Hodgkin's disease. The experience of the Institut Gustave Roussy. Hematological Oncology 2, 33-43

Desser RK, Golomb HM, Ultman JE, Ferguson DJ, Moran EM, Griem ML, Vardiman J, Miller B, Oetzel N, Sweet D, Lester EP, Kinzie J, Blough R (1977) Prognostic classification of Hodgkin's disease in pathologic stage III, based on anatomic considerations. Blood 49, 883-893

DeVita VT, Serpick AA, Carbone PP (1970) Combination chemotherapy in the treatment of advanced Hodgkin's disease. Ann Intern Med 73, 881-895

Diehl V, Pfreundschuh M, Hauser FE, Loeffler M, Rühl U, Brücher H, Georgii A (1986) Zwischenergebnisse der Therapiestudien HD1, HD2 und HD3 der deutschen Hodgkin-Studiengruppe. Med Klinik 81, 1-6

Gobbi PG, Cavalli C, Frederico M, Bertoloni D, Di Prisco UA, Rossi A, Silingardi V, Mauri C, Ascardi E (1988) Hodgkin's disease prognosis: a directly predictive equation. Lancet 3: 675-678

Hancock BW (1986) Randomised study of MOPP (musine, oncovin, procarbazine, prednisone) against LOPP (leukeran substituted for mustine) in advaned Hodgkin's disease. Radiother Oncol 7, 215-221

Haybittle JL, Easterling MJ, Bennet MH, Hudson BV, Hayhoe FG, Jelliffe AM, Hudson GV, MacLennan KA (1985) Review of British National Lymphoma Investigation studies of Hodgkin's disease and development of prognostic index. Lancet 1: 967-972

Kaplan HS (1980) Hodgkin's Disease, 2nd edn. Harvard University Press, Cambridge MA

Lee CGK, Bloomfield CD, Goldman AI, Levitte SH (1980) Prognostic significance of mediastinal involvement in Hodgkin's disease treated with curative radiotherapy. Cancer 46: 2403-2409

Lee CK, Aeppli DM, Bloomfield C, Levitt SH (1987) Hodgkin's disease: a reassessment of prognostic factors following modification of radiotherapy. Int J Radiat Oncol Biol Phys 13: 983-991

Loeffler M, Hasenclever D (1988) Probleme der Konzeption und Auswertung von Therapiestudien bei langer Behandlungsdauer, mittlerer CR-Rate und langer Überlebensdauer. In: Selbmann HK, Dietz (eds) Medizinische Informationsverarbeitung und Epidemiologie im Dienste der Gesundheit. 32. Jahrestagung der GMDS Tübingen, October 1987 Proceedings, Springer, Berlin Heidelberg New York Tokyo (1988)

Longo DL, Young RC, Wesley M, Hubbard SM, Duffey PL, Jaffe ES, DeVita VT Jr (1986) Twenty years of MOPP therapy for Hodgkin's disease. J Clin Oncol 4: 1295-1306

Mauch P, Gorsheim D, Cunningham J, Hellman S (1982) Influence of mediastinal adenopathy on site and frequency of relapse in patients with Hodgkin's disease. Cancer Treat Rep 66: 809-817

Morgenfeld M, Pavlovsky A, Surez A, Pavlovsky S, Palan M, Barros CA (1975) Combined cyclophosphamide, vincristine, procarbazine, and prednisone (COPP) therapy of malignant lymphoma. Evaluation of 190 patients. Cancer 36: 1341-1347

Musshoff K (1970) Therapy and prognosis of two different forms of organ involvement in cases of malignant lymphoma (Hodgkin's disease, reticulum cell sarcoma, lymphosarcoma) as well as a report about stage division in these diseases. Klin Wochschr 48: 673-678

Paglia MA, Lachner MJ, Hertz RE, Geller W, Watson RC, Lewis JL Jr., Nisce LJ, Liebermann PH (1973) Surgical aspects and results of laparotomy and splenectomy in Hodgkin's disease. Am J Roentgenol 117, 1: 12-18

Pfreundschuh M, Schoppe WD, Fuchs R, Pflüger KH, Loeffler M, Diehl V (1987) CCNU, etoposide, vindesine and dexamethasone (CEVD) in Hodgkin's lymphoma refractory to COPP and ABVD. Cancer Treat Rep 71, 12: 1203-1207

Pillai GN, Hagemeister FB, Velasquez WS, Sullivan JA, Johnston DA, Butler JJ, Shullenberg CC (1985) Prognostic factors for stage IV Hodgkin's disease treated with MOPP, with or without bleomycin. Cancer 55: 691-697

Prosnitz LR, Farber LR, Kapp DS, Bertino JR, Nordlund M, Lawrence R (1981) Combined modality therapy for advanced Hodgkin's disease: long-term follow-up data. Cancer Treat Rep 66: 871-879

Rosenberg SA (1985) The current stages of the Stanford randomized clinical trials of the management of Hodgkin's disease. In: Cavalli F, Bonadonna G, Rozencweig M (eds) Malignant lymphomas and Hodgkin's disease: experimental and therapeutic advances. Martinus Nijhoff, Boston, pp 281-292

Somers R, Henry-Amar M, Carde P, Najman A (1988) MOPP versus 2 MOPP/2ABVD in advanced Hodgkin's disease (HD). Proc Am Soc Clin Oncol 7: 236

Tubiana M, Henry-Amar M, van der Werf-Messing B, Henry J, Abbatucci J, Burgess M, Hayat M, Somers R, Langier H, Carde P (1985) A multivariate analysis of prognostic factors in early stage Hodgkin's disease. Int J Radiat Oncol Biol Phys 11: 23-30

Reduced Combined Radiotherapy and Chemotherapy for Hodgkin's Disease – Risk-Adapted Treatment Approach

M. Herold, G. Anger, K. Keinert, D. Höche, R. Kästner, and K. Wutke

Klinik für Innere Medizin, Medizinische Akademie Erfurt, Nordhäuser Straße 74,
5010 Erfurt, GDR

Introduction

Over the past two decades major progress has been made in the treatment of Hodgkin's disease. It may be held as an example of successful tumor therapy. But still today certain subgroups of patients with Hodgkin's disease do not respond to treatment or relapse early after standard treatment programs. These are particularly patients with advanced disease (stages III B and IV), but also patients in limited stages of Hodgkin's disease (stages I–III A) with the presence of risk factors. Such risk factors which influence outcome in patients with limited stages of Hodgkin's disease adversely are:

- Mediastinal bulky disease (diameter greater than one-third of chest diameter) (Prosnitz 1983; Hoppe 1985; Specht and Nissen 1986)
- Extranodal (E) lesions (especially of the lung) (Prosnitz 1983; Rubin et al. 1986)
- Constitutional (B) symptoms (Prosnitz 1983; Rubin et al. 1986; Crnkovich et al. 1986)
- Extended abdominal disease (stage III_2) (Desser et al. 1977; Prosnitz 1983; Mauch et al. 1985; Diehl et al. 1986)
- An unfavorable histological subtype (LD, not classified) (Löffler 1982; Mauch et al. 1985)

Combined modality treatment proved to be very effective in the management of Hodgkin's disease. However, full-dose radio- and chemotherapy bears an unacceptable risk of long-term complications such as second malignancies [especially acute nonlymphocytic leukemia (ANLL)] and permanent sterility (Coltman and Dickson 1982; Rubin et al. 1986; Bookman and Longo 1983). So it cannot be regarded as the treatment of choice in Hodgkin's disease (Schmidt 1982).

To reduce these delayed complications of combined modality treatment while preserving its effectiveness, we designed a study of a reduced combined radio- and chemotherapy approach for patients with Hodgkin's disease in limited stages with risk factors and in advanced stages. Figure 1 demonstrates the strategy in diagnostic and therapeutic management of this study. Important points of this strategy are: staging laparotomy is restricted to patients in whom a treatment decision de-

Recent Results in Cancer Research. Vol. 117

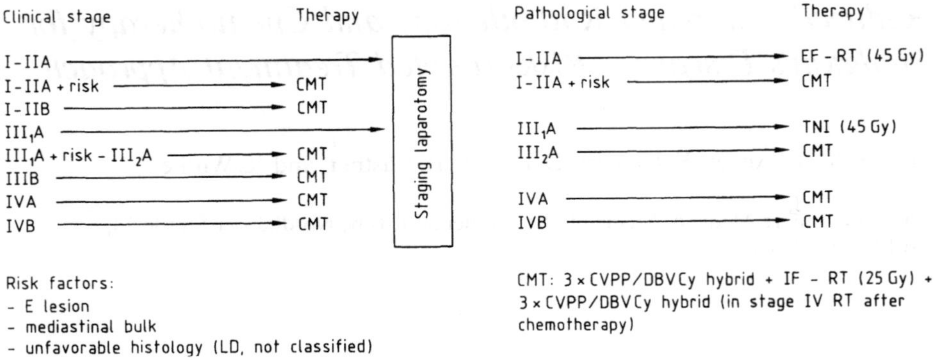

Fig. 1. Risk-adapted strategy in diagnostics and treatment of Hodgkin's disease. *CMT*, combined modality treatment; *EF*, extended field; *TNI*, total nodal irradiation; *LD*, lymphocytic depletion; *IF*, involved field. For other abbreviations see Table 1

Table 1. CVPP/DBVCy hybrid program

Cyclophosphamide	600 mg/m² i.v.	day 1
Vinblastine	6 mg/m² i.v.	day 1
Procarbazine	100 mg/m² p.o.	day 1-7
Prednisolone	40 mg/m² p.o.	day 1-14
Rubomycin	25 mg/m² i.v.	day 8
Bleomycin	10 mg/m² i.m.	day 8
Vincristine	1.4 mg/m² i.v.	day 8
Cytostasan	30 mg/m² i.v.	day 8-12

Day 15-28, no treatment.

pends on this procedure, radiation therapy is limited to involved fields with a reduced dosage of 25 Gy, and chemotherapy consists of six cycles of the cyclophosphamide, vinblastine, procarbazine, prednisolone/rubomycin, bleomycin, vincristine, cytostasan (CVPP/DBV Cy) hybrid program only (Table 1).

Patients and Methods

From May 1985 to September 1988 45 previously untreated patients were entered into the ongoing study. Staging was made according to the Ann Arbor criteria (Carbone et al. 1971), histological subclassification according to the Rye classification (Lukes et al. 1966). Diagnostic workup and treatment followed the guidelines given in Fig. 1. Up to October 1986 chemotherapy consisted of cyclic alternating CVPP/DBV Cy; since then the CVPP/DBV Cy hybrid program (Table 1) has been used. Low-dose (25 Gy) involved field radiotherapy is sandwiched between the six cycles of chemotherapy. In stage IV radiotherapy is delivered after completion of chemotherapy. There is no maintenance treatment. The patients' characteristics are given in Table 2. Evaluation of response was made 1 month after the end

Table 2. Patients' characteristics ($n = 35$) in combined reduced radio- and chemotherapy in Hodgkin's disease

Parameter	n
Male	22
Female	13
Age (years)	34,5 (18-65)
Stage	
IIB	8
III_2A	6
IIIB	16
IVA	1
IVB	4
Total A	7
Total B	28
E stage	3
Mediastinal bulky disease	6
Histology	
LP	6
NS	15
MC	10
LD	4

of the treatment program. Complete remission was stated to occur when all signs and symptoms of disease disappeared (clinical restaging, and bone marrow biopsy in cases of primary bone marrow innvolvement); partial remission was defined as disappearance of B-symptoms and reduction by greater than 50% of tumor masses and biochemical activity.

Results

On September 30 1988 35 patients were evaluable for response to treatment. The response rates are given in Table 3. Of the 35 patients, 28 (80%) achieved complete remission, 4 (11%) achieved partial remission, and 3 patients (9%) failed to respond to treatment. Of these patients, 31 were treated according to the schedule, while 4 stopped treatment after the fourth cycle of chemotherapy due to severe nausea and vomiting; however, all 4 are complete responders and still in complete remission. Survival data of the patients evaluable are shown in Table 4. The median follow-up is now 25 months, and 25 patients are still in first complete remission. The rate of relapse-free survival after 1 year is 100%, and after 2 years 94%; overall survival after 1 year is 97%, and after 2 years 94%.

So far two patients have died, both from the nonresponder group, 5 and 16 months after diagnosis. In the CR group 3 patients had a nodal relapse (14, 30, 31 months). The acute toxicity of the treatment in general was acceptable (Table 5). Dose reductions under chemotherapy were necessary in six patients; two received more than 90% of the dose planned, three 75%-90%, and one patient

Table 3. Results of reduced combined modality treatment (30 September 1988; $n=35$)

	n (%)
CR	28 (80)
PR	4 (11)
NR	3 (9)

Table 4. Survival data of reduced combined modality treatment

	n	%
In first CR	25/28	89
Median observation (months)	25 (13-38)	
Overall survival, 1 year	34/35	97
Relapse-free survival		
1 year	28/28	100
2 years	14/15	93

Table 5. Acute toxicity of reduced combined modality treatment ($n=35$)

Toxicity	Total	Chemotherapy	Radiotherapy
Bone marrow	9	8	6
Gastrointestinal tract	27	25	12
Neuropathy	9	9	–
Fever	2	2	–
Other	3	2[a]	1[b]
No side effects	1	2	17

[a] One case of cyclophosphamide-induced cystitis.
[b] Pneumonitis.

only 50%–75%. During radiation therapy in two patients myelotoxicity caused interruption of treatment for some days. One case of pneumonitis occurred after radiation to the mediastinum, and a hemorrhagic cystitis was caused by cyclophosphamide.

Discussion

The preliminary results of our study are promising. A complete remission rate of 80% (28/35 patients) and the survival data after 1 and 2 years indicate the value of this treatment approach. Of course the data are from only a limited number of patients and the time of observation is short, so they should not be overemphasized; however, they do show a trend.

The results are clearly superior to those of a study published recently comparing CVPP and alternating CVPP/DBVCy in advanced Hodgkin's disease (Herold et

Table 6. Results from studies of reduced combined modality treatment in Hodgkin's disease

Program (reference)	No. of patients	CR (%)	RFS (%)	(years)
MOPP+RT (Santoro et al. 1983)	109	81	65	(5)
ABVD+RT (Santoro et al. 1983)	107	91	81	(5)
BOPP+RT (Gomez et al. 1983)	17	71	-	
CAD/MOPP/ABV+RT (Straus et al. 1984)	34	82	90	(3)
MOPP/ABVD+RT (Straus et al. 1984)	37	78	75	(3)
Chemotherapy+RT (Koletsky et al. 1986)	183	82	68	(10)
COPP+RT (Diehl et al. 1986)	26	65	-	
COPP/ABVD+RT (Diehl et al. 1987)	93	71	88	(2)
ABVD+RT (De Lena et al. 1986)	22	100	100	(2)
CVPP/DBVCy+RT (present study)	35	80	93	(2)

MOPP, mustine, oncovin, procarbazine, prednisone; RT, radiotherapy; ABVD, Adriamycin, bleomycin, vinblastine, dacarbazine; BOPP, BCNU, vincristine, procarbazine, prednisolone; COPP, cyclophosphamide, vincristine, procarbazine, prednisolone; DBVCy, rubomycin, bleomycin, vincristine, cytostasan; CAD, cyclophosphamide, Adriamycin, doxirubicin.

al. 1987), giving CR rates of 60% and 57.5% respectively. Remission rates and survival data achieved are also comparable to the results which have been reported by other authors using reduced combined modality treatment previously (Table 6). Acute toxicity of reduced combined radio- and chemotherapy is acceptable and does not differ from the results of other investigators (Straus et al. 1984). Due to the short time of observation it is impossible to report about long-term side effects at present.

In conclusion we regard our treatment approach with a combination of reduced radiotherapy and limited chemotherapy as useful in patients with limited stages of Hodgkin's disease presenting with risk factors and in advanced Hodgkin's disease. The preliminary results of the pilot study allow us to continue the study as a multicenter trial.

References

Bookman MA, Longo DL (1983) Concomitant illness in patients treated for Hodgkin's disease. Cancer Treat Rev 13: 77-111

Carbone PP, Kaplan HS, Musshoff K, Smithers DW, Tubiana M (1971) Report of the committee on Hodgkin's disease staging classification. Cancer Res 31: 1860-1861

Coltman CA, Dixon DO (1982) Second malignancies complicating Hodgkin's disease: a Southwest Oncology Group 10-year follow up. Cancer Treat Rep 66: 1023-1033

Crnkovich MJ, Hoppe RT, Rosenberg SA (1986) Stage II B Hodgkin's disease: the Stanford experience. J Clin Oncol 4: 472-479

De Lena M, Altomare E, Colluci G, Marzullo F, Timurian A (1986) Combined ABVD (3 cycles) and radiotherapy (RT) in Hodgkin's disease. 14th International Cancer Congress, Budapest 1: 443

Desser RK, Golomb HM, Ultman JE, Ferguson DJ, Moran GM, Gian MC, Vardiman J, Miller B, Oetzel N, Sweet D, Lester EP, Kinzu JJ, Blough R (1977) Prognostic classification of Hodgkin's disease in pathological stage III, based on anatomic considerations. Blood 49: 883-893

Diehl V, Pfreundschuh M, Hauser FE, Löffler M, Rühl U, Brücher H, Georgii A, Miller E, Gerhartz H (1986) Zwischenergebnisse der Therapiestudien HD 1, HD 2 und HD 3 der deutschen Hodgkin-Studiengruppe. Med Klin 81: 1-6

Gomez GA, Panahon AM, Stutzman U, Han T, Ozer H, Henderson ES (1983) Simultaneous low dose radiation (RT) and low dose chemotherapy (CT) in the treatment of stage IV Hodgkin's disease. Proc Am Soc Clin Oncol 1: C 811

Herold M, Anger G, Höche D, Kästner R (1987) Vorläufige Ergebnisse einer zyklisch alternierenden Chemotherapie (CVPP/DBVCy) bei fortgeschrittenem Morbus Hodgkin. Med Klin 82: 345-349

Hoppe RT (1985) The management of stage II Hodgkin's disease with a large mediastinal mass: a prospective program emphasizing irradiation. Int J Radiat Oncol Biol Phys 11: 349-355

Koletsky AJ, Bertino JR, Farber LR, Prosnitz LR, Kapp DS, Fisher D, Portlock CS (1986) Second neoplasma in patients with Hodgkin's disease following combined modality therapy - the Yale experience. J Clin Oncol 4: 311-317

Löffler H (1982) Prognostische Faktoren als Indikator des Malignitätsgrades bei Leukämien und malignen Lymphomen. Verh Dtsch Ges Inn Med 88: 474-484

Lukes RJ, Craver LF, Hale TC, Rappaport H, Ruben P (1966) Report of the nomenclature committee. Cancer Res 26: 1311

Mauch P, Goffman T, Rosenthal DS, Canellos GP, Come SE, Hellman S (1985) Stage III Hodgkin's disease: improved survival with combined modality therapy as compared with radiation therapy alone. J Clin Oncol 3: 1166-1173

Prosnitz LR (1983) The role of combination chemotherapy alone or as an adjuvant to radiation therapy in limited stages of Hodgkin's disease. In: Bennet JM (ed) Controversies in the management of lymphomas. Martinus Nijhoff, Boston, pp 151-166

Rubin P, Constine L, Bennet JM (1986) Hodgkin's disease. IIB or not to be - using irradiation alone or in combination with chemotherapy? That is the question! J Clin Oncol 4: 455-457

Santoro A, Viviani S, Zucali R, Ragni G, Bonfante V, Valagussa P, Banfi A, Bonadonna G (1983) Comparative results and toxicity of MOPP vs. ABVD combined with radiotherapy in PS IIB, IIIA, B Hodgkin's disease. Proc Am Soc Clin Oncol 2: C 872

Schmidt CG (1982) Antwort zum Thema: Bei welchen Lymphomen und in welchen Stadien setzen Sie die kombinierte Chemo-Strahlentherapie unter kurativer Zielstellung in die primäre Behandlung ein? Med Welt 33: 196

Specht L, Nissen NJ (1986) Prognostic significance of tumor burden in Hodgkin's disease PS I and II. Scand J Haematol 36: 367-375

Straus DJ, Myers J, Lee BJ, Nisce CZ, Koziner B, McCormick B, Kempin S, Mertelsmann R, Arlin Z, Gee T, Poussin-Rosillo H, Hanson H, Clarkson BD (1984) Treatment of advanced Hodgkin's disease with chemotherapy and irradiation. Am J Med 76: 270-278

Hodgkin's Disease: The Milan Cancer Institute Experience with MOPP and ABVD

G. Bonadonna, P. Valagussa, A. Santoro, S. Viviani, V. Bonfante, and A. Banfi

Divisions of Medical Oncology and Radiotherapy, Istituto Nazionale Tumori,
Via Venezian, 1, Milan, Italy

This paper briefly reports the Milan Cancer Institute experience on Hodgkin's disease treated with an Adriamycin- and bleomycin-based combination (ABVD: Adriamycin, bleomycin, vinblastine, and dacarbazine). Following the initial publication on MOPP (mustine, oncovin, procarbazine, and prednisone) chemotherapy (De Vita et al. 1970), many investigators including ourselves have attempted to improve further the therapeutic results by modifying the original drug combination. Published findings have indicated that none of the modified MOPP regimens has indeed achieved superior results (Bonadonna and Santoro 1982). In 1974, we developed a four-drug combination (ABVD) whose components were totally different from those included in the MOPP combination. Our aim was to utilize ABVD in MOPP-resistant patients. Through successive randomized studies we have gained experience on the efficacy of ABVD alternated with MOPP and tested versus MOPP within a combined modality treatment.

MOPP Versus ABVD Within a Combined Modality Therapy

To reduce some of the delayed sequelae, namely sterility and acute leukemia, associated with MOPP, particularly when this regimen is combined with intensive irradiation, in 1974 we designed a prospective randomized study testing MOPP versus ABVD within a combined modality approach for the intermediate stages of Hodgkin's disease. The 7-year results were recently published (Santoro et al. 1987) and we will summarize the essential comparative findings in terms of treatment efficacy and iatrogenic morbidity.

The study included 232 patients with pathologic stages IIB-IIIA-IIIB. Both MOPP and ABVD were administered according to the classical NCI (De Vita et al. 1980) and Milan (Bonadonna et al. 1986) schedules. In the absence of tumor progression following the first three cycles of either MOPP or ABVD, treatment was continued with high-energy irradiation. The general plan for radiotherapy included 45 Gy to the involved lymphoid areas and 30 Gy to the uninvolved areas. Following completion of irradiation, in the absence of tumor progression, three additional cycles of either drug combination were planned. The interval before

Recent Results in Cancer Research. Vol. 117

Table 1. Main comparative treatment results following combined modality therapy

	MOPP group (114 cases) (%)	ABVD group (118 cases) (%)	*P* value
CR, total	81	92	<0.02
Stage IIB	73	88	
IIIA	97	97	
IIIB	78	94	
RFS, total	77	88	0.06
Stage IIB	86	86	
IIIA	67	95	
IIIB	78	85	
Survival, total	68	77	0.03

starting irradiation and the second phase of chemotherapy ranged from 4 to 6 weeks, except for patients showing prolonged myelosuppression. Treatment response was assessed at the end of treatment.

Table 1 summarizes the main comparative treatment results. With the exception of one case, all patients not achieving complete remission (CR) showed progressive lymphoma in one or more sites during the various phases of therapy. About 55% of all patients presented with bulky disease. ABVD plus radiotherapy induced CR in 88% of patients and the relapse-free survival (RFS) was 81%; the corresponding findings for MOPP plus radiotherapy were 80% and 71%, respectively. As far as stage IIIA is concerned, the impact of the number of involved lymph node sites on the relapse rate was analyzed. In the presence of ≤ 3 involved sites, all patients regardless of treatment attained CR, but relapse occurred in 2 of 10 patients who received MOPP and in one of 18 who received ABVD. In patients with >three involved sites all but one in each treatment group achieved CR. Of 19 complete responders following MOPP, 7 (37%) showed relapse, while all 14 patients treated with ABVD remained in CR.

Table 2 presents the comparative findings related to gonadal dysfunction and second neoplasms. It appears that in patients in whom tests were repeated irreversible gonadal failure occurred only in patients given MOPP. Second neoplasms were detected so far in both treatment groups but acute leukemia was documented only in patients treated with MOPP plus radiotherapy (Santoro et al. 1987).

Cardiac and pulmonary function tests were evaluated in 50 patients (MOPP 24, ABVD 26) who were in continuous CR for more than 5 years and geographically accessible. ABVD-treated patients failed to present evidence of dramatic clinical alterations in the laboratory tests used (Santoro et al. 1987). Between MOPP and ABVD, the only significant difference was observed in the mean left ventricular diameter in end diastole (LVEDd). Despite a more evident decrease of mean values following ABVD, neither vital capacity nor forced expiratory volume was significantly different between the two treatment groups. Although the incidence of parenchymal lung damage, as observed on chest roentgenograms, was more

Table 2. Gonadal dysfunction and second neoplasms

	MOPP group	ABVD group
Azoospermia[a]	13/13	9/25
Oligospermia[a]	-	5/25
Recovery of spermatogenesis	1/10	13/13
Amenorrhea[b] > 6 months	5/20	0/24
Second neoplasms	3	4
Acute leukemia	2	0
High-grade lymphomas	1	0
Soft-tissue sarcoma	0	1
Colon carcinoma	0	1
Non-oat-cell lung cancer	0	2

[a] Males less than 45 years old.
[b] Woman less than 40 years old.

frequently detected in the ABVD group (46%), compared with the MOPP group (13%), there was no clear radiographic evidence of bleomycin-related toxicity.

Alternating MOPP/ABVD

The knowledge that about 20% of patients treated with MOPP do not have CR, and that about 40% of initial complete responders relapse within the first 5 years from the end of treatment, led us to test the administration of MOPP and ABVD alternating monthly to overcome the problem of drug-resistant cells. Table 3 summarizes the essential 8-year results based on 88 consecutive patients with pathologic stage IV Hodgkin's disease (Bonadonna et al. 1986). All patients were previously untreated with chemotherapy and were prospectively randomized to receive either 12 monthly cycles of MOPP or 6 cycles of MOPP alternating monthly with 6 cycles of ABVD.

The most important aspect of our randomized trial was the superiority of alternating chemotherapy in the subsets known to be either prognostically unfavorable or less affected by MOPP chemotherapy. The subsets included in particular pa-

Table 3. Main comparative treatment results following MOPP versus MOPP/ABVD

	MOPP (43 cases) (%)	MOPP/ABVD (45 cases) (%)	*P* value
CR, total	74	89	
FFP,[a] total	36	65	<0.005
RFS, total	45	73	<0.01
Survival, total	62	76	
Tumor mortality	36	16	<0.06

[a] Freedom from progression.

tients aged over 40 years, systemic symptoms, nodular sclerosis, and bulky lymphoma. After treatment with MOPP/ABVD, all patients over 40 years old achieved complete remission and 66% remained disease free at 8 years. In patients with systemic B symptoms there was a clear superiority in incidence and duration of complete remission, as well as in total survival, after alternating chemotherapy compared with MOPP. In the subgroup with nodular sclerosis the observed difference achieved in the comparative frequency of complete remission was even more evident in the rates of freedom from relapse (MOPP, 58%, compared with MOPP/ABVD, 86%) and total survival (MOPP, 73%, compared with MOPP/ABVD, 95%).

Our current study attempts to meet more closely the stringent requirements of the Goldie and Coldman (1985) hypothesis. In July 1982, random testing was begun of the efficacy of cyclical MOPP/ABVD through two different sequences (Bonadonna 1982) in patients with stage IIA disease having bulky lymphoma, IIB, III (A and B), and IV Hodgkin's disease previously untreated with chemotherapy. The control arm consists of MOPP/ABVD (or MM/AA) as administered in the previous study, and in the experimental arm a half-cycle of MOPP is alternated within a 1-month period with a half-cycle of ABVD (MA/MA). Both treatments are administered until there is complete remission plus two additional consolidation cycles (minimum six cycles). Moderate doses of irradiation following maximal tumor shrinkage are administered only to the mediastinal or paraaortic area if it is bulky at the start of chemotherapy. Knowing the close relationship between tumor volume and primary drug resistance (Goldie and Coldman 1985), we hope through the addition of local radiotherapy to ensure maximum tumor control in these important nodal sites.

A total of 270 patients have so far been entered into the randomized study and 209 are evaluable. Table 4 presents the comparative CR in the various subsets. At

Table 4. Main comparative treatment results following MM/AA versus MA/MA

	MM/AA		MA/MA	
	No.	CR (%)	No.	CR (%)
Total	105	91	104	90
Stage I-IIA	12	92	11	91
IIB	33	88	31	94
IIIA	13	100	17	88
IIIB	14	93	14	79
IV	16	81	18	89
Failing on RT	17	100	13	100
Bulky Yes	48	85	46	89
No	57	96	58	91
Nodal sites ≤ 3	59	97	51	98
>3	46	85	53	83
Nodular sclerosis	78	91	69	90
Other histologies	27	93	35	91
Symptoms "A"	38	97	45	91
"B"	67	88	59	90

present there are no important differences between the two treatment groups. At a median observation time of 30 months from the start of chemotherapy, 76% and 79% of patients, respectively, remain in continuous CR.

Conclusions

The above-reported results indicate that an Adriamycin- and bleomycin-based combination, such as ABVD, is useful in the management of Hodgkin's disease. In particular, ABVD can be utilized to improve both incidence and durability of complete remission in given subsets when alternated with MOPP or combined with irradiation and to decrease certain types of iatrogenic morbidity such as sterility and acute leukemia.

The usefulness of alternating two non-cross-resistant combinations has been so far confirmed by Klimo and Connors (1985) with the hybrid regimen. Other investigators (Somers and Henry-Amar 1987; Gams et al. 1986; Vinciguerra et al. 1986) have not observed a significant improvement of alternating regimens versus MOPP. However, it should be pointed out that the latter treatments were not administered following the same principles and/or technique (i.e., two equally effective and non-cross-resistant regimens, monthly alternation) as done with MOPP/ABVD and hybrid. Our ongoing study is attempting to confirm previous findings also in the intermediate stages of Hodgkin's disease. The study also involves irradiation of the initially bulky nodal site to avoid, or at least decrease, the chance of recurrence in the anatomical area where the fraction of drug-resistant cells is highest (Goldie and Coldman 1985).

The proper role and extent of radiotherapy when combined with chemotherapy, particularly in patients with stages IIB and III, remains to be properly assessed. Although there are numerous randomized studies testing wide-field radiotherapy with or without chemotherapy, there is at present only one study concerning radiotherapy plus chemotherapy versus chemotherapy alone. The Southwest Oncology Group (Grozea et al. 1987) has recently reported the results of a clinical trial in which 92 patients with stage IIIA were randomized to receive 10 cycles of MOPP plus low-dose bleomycin versus 3 cycles of the same chemotherapy followed by total nodal irradiation. The comparative 8-year RFS (65% vs. 75%) and total survival (78% vs. 83%) were not significantly different between the two treatment groups. The majority of relapses on chemotherapy alone occurred in original sites of disease. The authors conclude that for the initial therapy of stage IIIA Hodgkin's disease chemotherapy or combined modality treatment may be equally effective. However, considering the pattern of relapse following chemotherapy (Young et al. 1978), particularly in patients with bulky disease, the addition of radiotherapy to complete or partial responders could be useful but remains to be more precisely assessed through a prospective randomized study.

A prolonged follow-up will also more precisely delineate the pattern of second solid tumors and whether it will be possible to identify, as for acute leukemia, a substantial difference between MOPP and ABVD.

Summary

The report includes the essential therapeutic and toxic results following MOPP vs. MOPP/ABVD and MOPP vs. ABVD within a combined modality setting. The findings indicate that in stage IV the alternating regimen is superior to MOPP while in stage IIB-III ABVD plus radiotherapy yielded superior results associated with no treatment-induced sterility and leukemia compared with MOPP plus radiotherapy.

References

Bonadonna G (1982) Chemotherapy strategies to improve the control of Hodgkin's disease: the Richard and Hinda Rosenthal Foundation Award Lecture. Cancer Res 42: 4309-4320

Bonadonna G, Santoro A (1982) Evolution in the treatment strategy of Hodgkin's disease. Adv Cancer Res 36: 257-293

Bonadonna G, Valagussa P, Santoro A (1986) Alternating non-cross-resistant combination chemotherapy or mechlorethamine, vincristine, procarbazine, and prednisone in stage IV Hodgkin's disease. A report of 8-year results. Ann Intern Med 104: 739-746

De Vita VT Jr, Serpick AA, Carbone PP (1970) Combination chemotherapy in the treatment of advanced Hodgkin's disease. Ann Intern Med 73: 881-895

De Vita VT, Simon RM, Hubbard SM et al. (1980) Curability of advanced Hodgkin's disease with chemotherapy. Long term follow-up of MOPP-treated patients at the National Cancer Institute. Ann Intern Med 92: 587-595

Gams RA, Omura GA, Velez-Garcia E et al. (1986) Alternating sequential combination chemotherapy in the management of advanced Hodgkin's disease. A Southeastern Cancer Study Group Trial. Cancer 58: 1963-1968

Goldie JH, Coldman AJ (1985) Genetic instability in the development of drug resistance. Semin Oncol 12: 222-230

Grozea PN, Depersio EJ, Coltman CA et al. (1987) Chemotherapy alone vs. combined modality treatment for stage IIIA Hodgkin's disease: an eight-year follow-up of a Southwest Oncology Group (SWOG) study No. 7518. Proc Third Intl Conference on Malignant Lymphoma, Lugano, Switzerland, p 91

Klimo P, Connors JM (1985) MOPP/ABV hybrid program: combination chemotherapy based on early introduction of seven effective drugs for advanced Hodgkin's disease. J Clin Oncol 3: 1174-1182

Santoro A, Bonadonna G, Valagussa P et al. (1987) Long-term results of combined chemotherapy-radiotherapy approach in Hodgkin's disease: superiority of ABVD plus radiotherapy versus MOPP plus radiotherapy. J Clin Oncol 5: 27-37

Somers R, Henry-Amar M (1987) MOPP vs alternating MOPP/ABVD in advanced Hodgkin's disease. Proc Third Intl Conference on Malignant Lymphoma, Lugano, Switzerland, p 32

Vinciguerra V, Propert KJ, Coleman N et al. (1986) Alternating cycles of combination chemotherapy for patients with recurrent Hodgkin's disease following radiotherapy. A prospectively randomized study by the Cancer and Leukemia Group B. J Clin Oncol 4: 838-846

Young RC, Canellos GP, Chabner BA et al. (1978) Patterns of relapse in advanced Hodgkin's disease treated with combination chemotherapy. Cancer 42: 1001-1007

EORTC Lymphoma Cooperative Group Studies in Clinical Stage I–II Hodgkin's Disease 1963–1987

R. Somers[1], M. Tubiana[2], and M. Henry-Amar[2]

[1] The Netherlands Cancer Institute, Plesmanlaan 121, 1066 CX Amsterdam, The Netherlands
[2] Institut Gustave Roussy, 94800 Villejuit, France

The European Organization for Research and Treatment of Cancer (EORTC) Lymphoma Cooperative Group was founded in 1963 as the Radiotherapy-Chemotherapy Group. It changed its name in 1986 because the main study objects of the group were Hodgkin's disease and non-Hodgkin's lymphoma. Its aim is to organize multicenter trials between European hospitals, mainly Dutch, French, and Belgian, but recently extended with groups from Switzerland.

In this presentation an update of the prospective randomized studies in clinical stage I and II Hodgkin's disease is presented. The studies have been published previously more extensively (Carde et al. 1988; Noordijk et al. 1987; Tubiana et al. 1984, 1985; Van der Werf-Messing 1973). The aim of these studies was to arrive at an adapted treatment based on prognostic factors.

H1 Trial

The first study was carried out between 1963 and 1971. Its protocol was relatively simple: all patients were treated with mantle-field irradiation, followed by randomization between no further treatment or treatment with vinblastine chemotherapy as weekly injections for 2 years.

This trial included 288 patients. Relapse-free survival (RFS) was higher in the vinblastine-treated group (61% vs. 39% at 12 years, $P<0.001$), but this was not reflected in a statistically clear survival advantage (70% vs. 61% at 12 years, $P>0.10$) due to the efficacy of rescue treatment after relapse (Table 1).

The beneficial effect of adjuvant CT was more pronounced in the group of patients with mixed cellularity histology. The incidence of relapses was high in non-irradiated areas, especially in the paraaortic lymph nodes.

H2 Trial

The second study was performed in the period 1972–1976. Since laparotomy with splenectomy had been introduced as part of the treatment in Hodgkin's disease in

Recent Results in Cancer Research. Vol. 117

that period, the aim of this study was to investigate the therapeutic usefulness of this procedure and to assess the prognostic significance of a positive laparotomy.

In the H2 trial all patients received mantle-field irradiation and irradiation of the paraaortic area. Patients were randomized between spleen irradiation and splenectomy. Based on the data of the H1 trial, the subset of patients with mixed cellularity and lymphocytic depletion histological types were randomized between either vinblastine, weekly injection for 2 years, or vinblastine alternated with procarbazine during the same period. The results of this study can be summarized as follows (Table 1): there was no difference in RFS or survival between the groups of patients treated with splenectomy or with radiotherapy (RT) of the spleen: RFS, 76% vs. 68% at 10 years ($P>0.10$) and survival 80% vs. 79% at 10 years ($P>0.10$). The total survival, however, was higher when compared with the H1 trial, showing that the extension of the treatment to the upper abdomen was of benefit in clinical stage (CS) I and II patients.

No difference was found between the two chemotherapy regimens.

The conclusions about the equality of splenectomy and RT of the spleen were valid for the patients without chemotherapy as well as for the patients who received chemotherapy. In this study it was demonstrated that the findings at laparotomy had a strong influence on the probability of relapse in nonirradiated lymphatic areas and extranodal relapse: if the spleen was not involved RFS was 83% at 10 years. In the group with positive laparotomy this was only 56% ($P<0.001$).

Table 1. Results of the EORTC trials

Cohort	Treatment group	*n*	% RFS (SD)		*P* value	% Survival (SD)		*P* value
H1 (1963-1971)	Mantle RT	152	39 (4)	12 years	<0.001	61 (4)	12 years	>0.10
	Mantle RT + vinblastine	136	61 (4)			70 (4)		
H2 (1972-1976)	Spleen RT (no lap)	156	68 (4)	10 years	>0.10	80 (3)	10 years	>0.10
	Splenectomy (lap)	144	76 (4)			79 (4)		
	Neg. laparotomy	107	83 (4)	10 years	<0.001	80 (4)	10 years	>0.10
	Pos. laparotomy	37	56 (8)			76 (7)		
H5 (1977-1981)	Mantle RT (lap. –)	100	68 (6)	8 years	>0.10	93 (4)	8 years	>0.10
	Mantle + paraaortic RT (lap. –)	98	73 (5)			90 (4)		
	TNI (no lap. or lap. +)	152	65 (5)	8 years	<0.001	71 (6)	8 years	0.07
	MOPP-mantle RT-MOPP (no lap. or lap. +)	144	82 (4)			88 (3)		
H6 (1982-1987)	Splenectomy (lap. – = RT lap. + = CT + RT)	101	95 (3)	3 years	<0.01	96 (2)	3 years	>0.10
	Spleen RT (no lap.)	99	72 (7)			95 (4)		
	MOPP-mantle RT-MOPP	114	79 (5)	3 years	0.06	91 (4)	3 years	>0.10
	ABVD-mantle RT-ABVD	106	90 (4)			95 (3)		

This was not reflected, however, in the overall survival, which was 80% in the laparotomy-negative and 76% in the laparatomy-positive group, again due to the good results of rescue treatment with combination chemotherapy after relapse.

H5 Trial

The H5 study took into account prognostic factors as the basis of management strategy. Laparotomy had its early and late complications. It was felt that groups of patients with a high likelihood of relapse might benefit from a combined treatment of CT and RT from the start and a laparotomy was no longer indicated. The use could be restricted to defined groups of patients which could be treated by radiotherapy alone. In this study two groups of patients were outlined. The favorable group consisted of patients, younger than 40 years of age with stage I or stage II HD with mediastinum involvement, and with lymphocytic predominance (LP) or nodular sclerosis (NS) as histology, with a low erythrocyte sedimentation rate (ESR). Patients had to fulfil all these criteria. In the favorable group a laparotomy was performed; if it was negative patients remained in the favorable group and were randomized between mantle-field irradiation and mantle-field with para-aortic field irradiation. Hence the aim of the study was to evaluate the hecessity of paraaortal RT in patients with good prognostic indications and a negative laparotomy.

The unfavorable group comprised patients with poor prognostic indicators or initially favorable patients with positive laparotomy. Patients were randomized between total node irradiation (TNI) and combined modality treatment: three courses of MOPP (mustine, oncovin, procarbazine, and prednisone) followed by mantle-field irradiation, followed by three courses of MOPP.

The H5 study included 494 patients: 198 were included in the favorable group. There was no difference between mantle-field or mantle-field with paraaortic field irradiation in the favorable group (68% vs. 73% RFS at 8 years).

In the unfavorable patients group the RFS was higher in the group treated with MOPP+RT compared with TNI (82% vs. 65% at 8 years, $P<0.001$); the difference in survival was borderline significant (88% vs. 71% at 8 years, $P=0.07$). The lower survival in the TNI group was mainly due to the poor rescue in older patients above 40 years who had relapsed after TNI and to an imbalance of unrelated deaths between the two groups.

Prognostic Factors

A Cox model analysis of prognostic factors was performed (Table 2) on all the patients included in the H1, H2, and H5 trials after adjustment for the treatment. This enabled the independent impact of each of the prognostic factors on RFS and S to be assessed. It was shown that the influence of the factors investigated such as age, sex, histology, number of involved lymph node areas, and mediastinum involvement is different for RFS or death because the probability of death is largely related to the efficacy of rescue treatment after relapse.

Table 2. Cox model analysis of Hodgkin's disease CS I–II: independent prognostic significance after adjustment for therapeutic group (1139 patients)

		RR for relapse	Statistical significance (*P*)	RR for death	Statistical significance (*P*)
Sex	M F	1 0.77	0.0393	1 0.67	0.0201
Age	<40 >40	1 1.51	0.0026	1 3.12	0.00001
A, B	<50 mm, or[a] <30 mm	1 2.31	0.00001	1 1.55	0.0043
LP+NS MC+LD		1 1.16	0.22 NS	1 1.57	0.0038
I		1		1	
II_2		1.20	0.21 NS	1.05	0.92 NS
II_3		1.59	0.01	0.85	0.54 NS
II_4		2.37	0.001	1.94	0.0208
Med. + Med. –		0.83 1	0.12 NS	1.10 1	0.61 NS

RR, relative risk; NS, not significant.
[a] vs A ≥ 50 mm, or B ≥ 30 mm.

For relapse, the most important indicators are the number of involved areas and the presence of systemic symptoms (SS) + elevated ESR; age has a smaller influence, and mediastinal involvement by itself has only a slight influence. For survival, age has by fart the greatest influence; this indicates a smaller effectiveness of rescue treatment in the older age group of Hodgkin's patients (Table 2).

H6 Trial

The management strategy which has been successful in the H5 trial was pursued in the H6 trial, in which patients were also subdivided into favorable and unfavorable groups. However, advantage was taken of the results of the analysis of the prognostic factors for a better delineation of the two groups. In the H6 trial, they were defined as follows:

Favorable patients were stage I or stage II_2 patients (<2 lymph node areas), with low ESR (defined as A < 50 mm, B < 30 mm), aged less than 40 years. The thorax-mediastinum ratio had to be below 0.5. Based on the Cox model predictions, these patients had a low risk for relapse or death. They were randomized between radiotherapy on mantle field and paraaortic field and spleen versus laparotomy. After laparotomy, treatment depended on the findings of the operation: in the case of negative laparotomy mantle-field radiation was given in a case of histological-type nodular sclerosis; mantle-field with paraaortic field radiation was given in mixed-cellularity type HD. In the case of positive laparotomy, patients were treated according to the protocol for unfavorable patients.

The aim of the study was not primarily to compare RFS, which ought to be higher in the laparotomy arm since the treatment was better adapted to the characteristics of the patients. It was to compare in a pragmatic way the long-term survival in the two arms. If this proves to be identical, the aim of the study will be to weigh the inconveniencies of the laparotomy in terms of side effects and sequelae against those of a relapse and salvage therapy in a large proportion of patients. Presently as expected, the RFS is superior in the laparotomy group with adapted treatment after surgery in comparison with the RT group (95% vs. 72% at 3 years; $P < 0.001$). However, survival is identical in the two arms but the follow-up is relatively short. The unfavorable group trial compared three courses of MOPP-mantle-field irradiation three courses of MOPP versus three courses of ABVD-mantle-field irradiation 3 courses of ABVD. Its aim was to compare the long-term toxicities of the two combinations of chemotherapy and radiotherapy; pulmonary and cardiac function as well gonadal toxicity in males and females and secondary tumors will be systematically evaluated. The study was completed in 1987 and it is too early fully to evaluate toxicity. With regard to RFS, at this point in time there is no significant difference between the two arms, although there is a trend in favor of ABVD.

Design of the H7 Study (Table 3)

Based on the prognostic factors mentioned above and a retrospective analysis of the results of the different treatment modalities in the previous studies, we are presently designing a new study: the H7 trial. A treatment strategy can be approached from at least two opposite points of view. First, the goal is to prevent relapses at any cost. Therefore one is led to give maximal treatment to all the patients. The risk of such a strategy is overtreatment and unnecessary hazards, such as sterility and secondary tumors, for some patients. The other strategy is to tailor the treatment to the individual characteristics of the disease of each patient. The aim is to avoid in a large proportion of patients an intensive and potentially toxic therapy without compromising the chance of long-term survival. In such a strategy one accepts the risk of an increased relapse rate as long as a rescue treatment initiated after the relapse can salvage the patient. Acutally we have shown that patients can be cured after a relapse, since in some of the studies (H1 and H5) despite the lower RFS in one of the two arms the long-term survival is identical. However, this is not true for all patients and we have shown for example that in older patients rescue treatment does not produce satisfactory results. Moreover relapse is associated with a serious psychological trauma, and too high a probability of relapse might be unacceptable for some subsets of patients. Based on these considerations, we outlined for the H7 study, which will start in the near future, three subgroups of patients with a different treatment approach (Table 3). It is felt that laparotomy is no longer indicated because we expect from our data that in the favorable groups in which laparotomy may provide useful information first-line treatment modalities and rescue treatment will guarantee long-term survival for nearly all patients.

Table 3. Design H7 study CS I-II Hodgkin's disease

Group	Selection	Results of old studies					Treatment selected
			M	M+PA			
VF (6%)	CSI female,[a] <40 years, A, ESR<50, LP-NS, MT<0.35	5 years RFS	74%	91%			M
		10 years S	85%	94%			
			M	STNI	TNI	M+MOPP	
F (54%)	<50 years CSI, II_{2-3} not VF or U	5 years RFS	58%	81%	81%	84%	R: STNI 36-40 Gy / 6 EBVP+IF 36 Gy
		7 years S			89%	85%	
		10 years S	67%	84%			
			M	TNI	M+MOPP		
U (40%)	≥50 years,[b] CS II_{4-5}, A, ESR≥50, B, ESR≥30, MT≥0.35	5 years RFS	50%	66%	87%		R: 6 EBVP +IF 36 Gy / 6 MOPP/ABV +IF 36 Gy
		7 years S		68%	87%		
		10 years S	71%				

VF, very favorable; F, favorable; U, unfavorable; M, mantle-field irradiation; PA, paraaortic (+spleen) irradiation; (S) TNI, (sub)total nodal; EBVP, EpiAdria, bleomycin, vinblastine, prednisone; RFS, relapse-free survival; S, survival; MT, mediastinum-thorax ratio.
[a] Patient has to fulfil all those criteria.
[b] Patient belongs to this category if he or she fulfils one of those criteria.

In the small subgroup called "very favorable," mantle-field irradiation alone is sufficient treatment. This subgroup comprises only 6% of the patients who have the following characteristics: women below 40 years of age, stage I, without SS and with NS or LP, mediastinum-thorax (M/T) ratio below 0.35. In this subgroup, mantle-field irradiation alone can achieve RFS equal to 85% at 10 years, and survival at 10 years may even be higher due to improved rescue chemotherapy. At the other extreme, there are the unfavorable patients: those above 50 years, patients with four or more involved areas (II_4 and II_5), patients with a high M/T ratio, or patients with A ESR ≥50, or B ESR ≥30 mm. When treated with RT alone, they have 68% survival, but when treated with CT+RT from the start the survival at 7 years is 87%.

It was decided to treat these patients with a combination of chemotherapy and radiotherapy. The question posed is how much chemotherapy and which drugs can be given with the aim of reducing as much as possible the incidence and severity of long-term side effects. We decided to give on one hand the best standard treatment (MOPP-ABV)+radiotherapy and to compare this with a less intensive nonleukemogenic chemotherapy consisting of epiadriamycin, bleomycin, vinblastine, and prenisone+radiotherapy as tested by the Pierre-Marie Curie group (Hoerni, personal communications).

For the intermediate group (so-called favorable), subtotal nodal irradiation is being considered as standard treatment (which produced a 10-year total survival equal to 84%). In this subset of patients, total survival is identical in patients treated by subtotal node irradiation (STNI), TNI, or combination of CT + RT. In the H7 trial, we plan to compare STNI with the nonleukemogenic therapy + iceberg radiotherapy, which will be used in the unfavorable arm. We hope further to demonstrate in this study that a treatment modulated according to risk categories can produce the highest survival with the least toxicity.

The chairmen of the EORTC cooperative group have been M. Tubiana (1964-1967), K. Breur (1967-1969), V. van der Werf-Messing (1969-1972), J. Henry (1973-1975), J. Abbatucci (1975-1977), M. Burgers (1977-1980), M. Hayat (1980-1982), R. Somers (1983), statistician M. Henry-Amar, scientific secretaries A. Laugier (1964-1972), M. Hayat (1972-1975), M. Urbajtel (1975-1977), E. van der Schueren (1977-1980), P. Carde (1980-1985), J. H. Meerwaldt (1985-).

Cooperating centers are: The Netherlands: Antoni van Leeuwenhoek Ziekenhuis, Amsterdam (M. Burgers, R. Somers), University Hospital, Leiden (E. M. Noordijk), St. Radboud Academic Hospital, Nijmegen (D. Wagener, B. de Pauw. C. Haanen), Rotterdams Radiotherapeutisch Instituut, Rotterdam (B. van der Werf-Messing, M. Quasim, W. Sizoo, J. H. Meerwaldt), Ziekenhuis Leyenburg, The Hague (H. Kerkhofs), University Hospital, Utrecht (H. van Peperzeel, J. G. Nyssen, L. Verdonck); Belgium: Institut Jules Bordet, Bruxelles (J. Henry, J. Lustmann-Marechal, B. L. Reinier), Centre René Goffin, La Louvière (J. C. Goffin), Hôpital de Bavière, Liège (R. Lemaire), Acad. Ziekenhuis St. Rafael, Leuven (E. Van der Schueren, J. Thomas); Italy: Instituto di Radiologia, Universita di Firenze (G. de Giuli, L. Cionini); France: Fondation Bergonié, Bordeaux (C. Lagarde), Centre Francois-Baclesse, Caen (J. S. Abbatucci, A. Tanguy), Centre Léon-Bérard, Lyon (J. Papillon, L. Revol), Centre Antoine-Lacassagne, Nice (C. Lalanne, M. Schneider), Hôtel-Dieu, Paris (R. Zittoun), Institut Jean-Godinot, Reims (A. Cattan), Centre Henri-Becquerel, Rouen (H. Piguet, M. Monconduit), Institut Gustave-Roussy, Villejuif (M. Tubiana, M. Hayat, P. Carde, J. M. Cosset).

References

Carde P, Henry-Amar M, Burgers JMV, Hayat M, Van der Schueren E, Noordijk EM, Tanguy A, de Pauw B, Cosset JM, Somers R, Tubiana M (1988) Clinical stages I and II Hodgkin disease; a specifically tailored therapy according to prognostic factors. J Clin Oncol 6: 239-252

Noordijk EM, Henry-Amar M, Cosset JM (1987) Can irradiation alone be useful in unfavourable stages I-II Hodgkin's disease. European Society Therapeutic Radiology and Oncology, Congress, 6th annual meeting, 25-28 May, Portugal (Lisbon), p 73

Tubiana M, Henry-Amar M, Hayat M, Burgers JMV, Quasim M, Somers R, Sizoo W, Van der Schueren E (1984) The EORTC treatment of early stages of Hodgkin's disease. The role of radiotherapy. Int J Radiat Oncol Biol Phys 10: 197-210

Tubiana M, Henry-Amar M, van der Werf-Messing B, Henry J, Abbatucci J, Burgers JMV, Hayat M, Somers R, Laugier A, Carde P (1985) A multivariate analysis of prognostic factors in early stage Hodgkin's disease. Int J Radiat Oncol Biol Phys 11: 23-30

Van der Werf-Messing B (1973) Morbus Hodgkin's disease, stage I and II trial of the EORTC. Hodgkin Disease Symposium, Stanford University 1972. N C I Monogr 36: 381-386

*Current Stanford Clinical Trials for Hodgkin's Disease**

R. T. Hoppe, S. J. Horning, S. L. Hancock, and S. A. Rosenberg

Department of Radiation Oncology and Department of Medicine,
Division of Medical Oncology, Stanford University, Stanford, CA 94305, USA

Prospective randomized clinical trials for Hodgkin's disease were initiated at Stanford University by Dr. Henry S. Kaplan and Dr. Saul A. Rosenberg in 1962. In the subsequent 25 years, nearly 1000 patients have been accrued to four generations of clinical trials. In addition, during the same period, a similar number of patients were treated with programs identical to those used on the clinical trials, but were not assigned by randomization (Table 1). Each generation of clinical trials has built strongly upon the experience obtained in the preceding trials and has also incorporated new concepts in Hodgkin's disease management developed by other investigators. The use of standardized treatment programs among the nonrandomized patients and the careful follow-up of all patients permit evaluation of the entire cohort of nearly 2000 patients for identification of prognostic factors, long-term complications of therapy, etc.

The L studies, which were initiated in 1962, included patients with clinical stage I-III disease, all of whom underwent lymphography as a component of their staging. Patients with stage IV disease were excluded. Patients with clinical stage I-II disease were randomized to treatment with either involved field or extended field irradiation to a dose of 44 Gy (L1 study). The long-term results of the

Table 1. Stanford University clinical trials

Years	Studies	Randomized patients	Nonrandomized patients	Total
1962-1967	L	155	93	248
1968-1973	H (K, R)	351	231	582
1974-1979	S	268	317	585
1980-1987	C	175	401	576
Total		949	1042	1991

* Supported in part by Grant CA 34233 from the National Cancer Institute, National Institutes of Health. Dr. Rosenberg is an American Cancer Society Clinical Professor of Medicine.

Recent Results in Cancer Research. Vol. 117

L1 study are inconclusive, largely because of failure to anticipate the presence of subdiaphragmatic disease in patients who had undergone clinical staging only. For patients with stage III disease, treatment was randomized between palliative therapy with low-dose (16.5 Gy) involved field irradiation versus high-dose (44 Gy) total lymphoid irradiation (L2 study). This experimental approach to curative therapy for these patients was one of the first to be proved to be effective in stage III disease. With follow-up as long as 22 years, 50% of patients with stage IIIA and 33% of patients with stage IIIB Hodgkin's disease are alive, without evidence of disease.

The H studies, which were initiated in 1967-1968, were the first to incorporate routine staging laparotomy and splenectomy as a component of the initial staging. These trials are often referred to as the "adjuvant MOPP trials" since the majority of the treatment protocols for stage I-III disease randomized patients between treatment with radiation therapy alone (usually total lymphoid irradiation, 44 Gy) versus similar radiation treatment followed by six cycles of adjuvant MOPP (mustine, oncovin, procarbazine, and prednisone) chemotherapy. These studies confirmed the utility of staging laparotomy by demonstrating the high likelihood of detection of occult disease in subdiaphragmatic sites, especially the spleen. They also demonstrated the feasibility of combining aggressive radiation therapy and chemotherapy programs. In general, these studies showed a superiority of combined modality therapy over radiation therapy alone for patients with stage III disease. In stage I-II, combined modality therapy decreased the relapse risk compared with radiation alone, but did not confer a survival benefit.

The S studies were initiated in 1974. They tested the ability of MOPP for the treatment of occult disease in PS I-II, confirming its equivalence to radiation therapy in that respect (S1 study). For intermediate and advanced stages of disease, refinements of combined modality therapy treatment programs were investigated, including the substitution for MOPP of a less toxic drug regimen, PAVe (procarbazine, l-phenylalanine mustard, and vinblastine), and an altered sequence of combined modality management in which chemotherapy and radiation therapy were alternated.

A detailed analysis of the concept, design, and results of the L studies, H studies, and S studies has been the subject of recent reports (Rosenberg and Kaplan 1985; Hoppe et al. 1985). In this manuscript, we will deal primarily with the design and preliminary results of the C studies, which were initiated in 1980. These studies incorporate an assessment of prognostic factors to assist in assignment of patients to treatment programs and include assessment of toxicity as well as outcome in all trials.

C1-3 Studies

Table 2 summarizes the eligibility criteria and treatment options on the C-1-3 studies. These studies were designed for patients with "favorable" disease characterized as PS I-IIA, PS I-IIB, or PS IIIA in the absence of bulky disease such as a large mediastinal mass, multiple extralymphatic (E) lesions, or extensive splenic involvement. All patients undergo laparotomy staging. The standard treat-

Table 2. C1-3 studies

Eligibility	Laparotomy staging
	C1 PS I-IIA
	C2 PS I-IIB
	C3 PS $IIIA_S-$, PS $IIIA_S+_{min}$
Treatments	A. STLI (C1-2) or TLI (C3)
	B. IF+VBM

STLI, subtotal lymphoid irradiation; TLI, total lymphoid irradiation; IF, involved field irradiation; VBM, velban, bleomycin, and methotrexate chemotherapy; S^{+}min, minimal splenic involvement (<5 nodules).

ment arm is with radiation alone. Patients with supradiaphragmatic stage I-II disease are treated with subtotal lymphoid irradiation which includes sequential treatment to the mantle (44 Gy) and spade (paraaortic, splenic pedicle, common iliac) (40 Gy) fields. Patients with subdiaphragmatic PS II and all patients with PS IIIA disease receive treatment to the pelvic lymph nodes as well. Prophylactic irradiation of the preauricular nodes (36 Gy) is included whenever there is adenopathy in the high cervical region. Low-dose (16.5 Gy) prophylactic irradiation of the ipsilateral lung is included in the presence of pulmonary hilar lymph node involvement, and low-dose (22 Gy) prophylactic irradiation of the liver is incorporated in the presence of splenic involvement. The detailed techniques of treatment are as described by Kaplan (1980).

Patients treated on the experimental arm of this trial receive high-dose irradiation (44 Gy) to the involved field followed by adjuvant chemotherapy with six cycles of vinblastine, bleomycin, and methotrexate (VBM) chemotherapy. The schema for VBM chemotherapy is outlined in Table 3. The cycle is repeated every 28 days until six cycles of adjuvant therapy have been completed. The VBM combination was developed because each of the three agents has single-agent activity in the management of Hodgkin's disease and none of the agents are known to be associated with the severe long-term complications of sterility or secondary leukemia.

Sixty-seven patients have been accrued to this trial. The maximum follow-up is 6 years and the median follow-up is 3 years. There have been 4 relapses among the 35 patients treated with irradiation alone and 1 relapse among the 32 patients treated with involved field irradiation followed by VBM chemotherapy. The actuarial survival and freedom relapse for these patients are shown in Fig. 1.

The results of three sequential trials which have utilized involved field irradiation with or without adjuvant chemotherapy for PS I-II Hodgkin's disease are summarized in Table 4. The groups of patients in each of these studies are not per-

Table 3. VBM chemotherapy

Vinblastine	6 mg/m^2	i.v. Days 1, 8
Bleomycin	10 mg/m^2	i.v. Days 1, 8
Methotrexate	30 mg/m^2	i.v. Days 1, 8

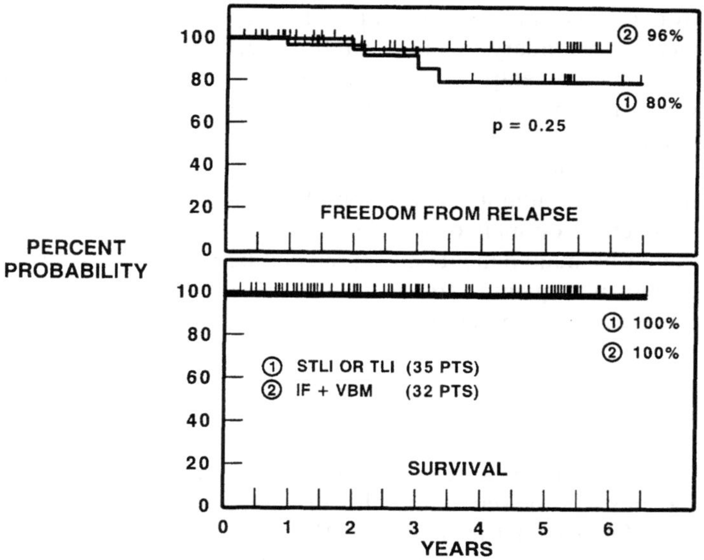

Fig. 1. Freedom from relapse and survival for the C1-3 studies for patients with "favorable" PS I-IIA, I-IIB, or IIIA Hodgkin's disease

Table 4. Involved field irradiation with/without adjuvant chemotherapy: sequential Stanford trials

Trial(s)	Stage	No. pts.	Adjuvant chemo-therapy	Survival (%)		Freedom from relapse (%)		
				3-year	5-year	3-year	5-year	
H1A	I-IIA	28	-	96	93	43	36	$P=0.0001$ (H1A vs S1B, R1B); $P<0.0001$ (H1A vs C1B, C2B)
S1B, R1B	I-IIA	45	MOPP	98	93	82	77	$P=0.04$ (S1B, R1B vs C1B, C2B)
C1B, C2B	I-IIA/B	26	VBM	100	100	100	100	

fectly comparable. In the C studies, in contrast to the earlier studies, patients with large mediastinal masses (mediastinal mass ratio exceeding 1/3) and patients with multiple extralymphatic (E) lesions were excluded. However, the current studies include symptomatic as well as asymptomatic stage I-II patients. Despite these differences, some reasonable comparisons can be made. The survival of patients in all trials is excellent, even when the relapse risk is high. This is attributable to the fact that patients who relapse after limited treatment, such as involved field irradiation alone, are usually effectively treated with curative intent by the use of MOPP chemotherapy with or without additional irradiation.

The relapse risk in the H1 study was quite high, with only 36% of patients disease free at 5 years. The addition of adjuvant MOPP in the S1B and R1B trials improved the 5-year freedom from relapse to 77% ($P=0.0001$). With involved field irradiation followed by VBM chemotherapy (the C1-2B trials) the 5-year freedom from relapse appears to be improved even further. Even if this additional improvement is secondary to more careful selection of patients, at least it appears that VBM chemotherapy is not inferior to MOPP chemotherapy when used as an adjuvant after involved field irradiation. These preliminary results support the continued use of VBM in clinical trials.

Patients on these studies are also undergoing prospective serial assessment of fertility and pulmonary function. As anticipated, VBM has had relatively little adverse effect on male or female fertility. Seven normal live births, four children fathered by male patients and three children born to our female patients, have occurred subsequent to treatment with VBM. Effects on pulmonary function are the subject of a more detailed report (see Horning et al. 1988).

C4-6 Studies

The C4-6 studies were designed to provide appropriately tailored management for patients who present with bulky disease, primarily by virtue of a large mediastinal mass [mediastinal mass ratio (MMR) > 1/3]. Previous analyses of data from our institution indicated that patients with a large mediastinal mass who had undergone routine staging with laparotomy and were treated with irradiation alone had a relapse risk of approximately 50% (Hoppe et al. 1982). This relapse risk could be decreased significantly by the addition of adjuvant chemotherapy. In an analysis of survival of these patients, it was shown that patients treated initially with irradiation alone who died usually died of progressive Hodgkin's disease while those who died after combined modality therapy generally died of complications of their treatment. The C4-6 studies were designed with the expectation of achieving similar results to previous combined modality therapy programs but to reduce the intensity of the staging and treatment programs in order to minimize long-term morbidity and mortality related to treatment.

The eligibility criteria and treatment programs for patients on these studies are summarized in Table 5. Since all patients are treated systemically, laparotomy and splenectomy are not incorporated as part of the routine staging. All patients are treated with combined modality therapy in a split course or alternating fashion in which the chemotherapy is the initial treatment component. The chemotherapy randomization is between PAVe and ABVD (Adriamycin, bleomycin, vinblastine, dacarbazine). These two drug combinations were chosen because of their proved efficacy in combined modality treatment programs (Bonadonna et al. 1977; Ro-

Table 5. C4-6 studies

Eligibility	C4 CS I-II A
	C5 CS I-II B
	C6 CS III A
	Bulky disease or multiple E lesions
	Clinical staging plus bone marrow biopsy
Treatments	A. PAVe/modified IF/PAVe
	B. ABVD/modified IF/ABVD

Table 6. PAVe chemotherapy

l-Phenylalamine mustard	7.5 mg/m^2 p.o.	Days 1-2, 8-9
Vinblastine	6.0 mg/m^2 i.v.	Days 1, 8
Procarbazine	100 mg/m^2 p.o.	Days 1-14

senberg and Kaplan 1985). The toxicities of these two chemotherapy programs, however, should be expected to be quite different. PAVe should be associated with significant effects on fertility and the potential for secondary leukemia, while ABVD has more acute toxicity such as nausea and vomiting, is associated with more severe epilation, and has potential long-term cardiac and pulmonary morbidity. In view of these different toxicities, patients undergo pre- and post-treatment evaluation of cardiac and pulmonary function as well as fertility.

The radiotherapy as defined in these studies is "modified involved field." Irradiation is to a dose of 44 Gy and limited to the mantle, paraaortic-spleen, or pelvic fields, whichever are initially involved by clinical staging studies. When a single region is involved, the radiotherapy is delivered in a single course between cycles 1-3 and 4-6 of chemotherapy. If more than one region is to be irradiated, the chemotherapy and irradiation are administered in an alternating fashion, similar to that utilized in advanced-stage Hodgkin's disease (Hoppe et al. 1979).

PAVe chemotherapy was developed as an alternative to MOPP chemotherapy and is administered on a schedule similar to MOPP (Table 6). L-Phenylalanine mustard is substituted for nitrogen mustard and is administered orally on days 1, 2 and 8, 9. Vinblastine is substituted for vincristine, and procarbazine is used the same as in MOPP. The cycle is repeated monthly.

The survival and freedom from relapse for patients treated on the C4-6 studies are summarized in Fig. 2. The maximum follow-up is 6 years and the median follow-up is 3½ years. Thus far, the survival and freedom from relapse appear identical with the two different treatment approaches.

Table 7 compares the combined modality approach utilized on these studies for CS I-II disease with results of treatment in patients with PS I-II bulky disease treated on previous clinical trials at Stanford. In the current trial, patients have not undergone staging laparotomy and splenectomy whereas in previous trials laparotomy was employed routinely. Approximately one-third of patients in the current trial would be expected to have occult subdiaphragmatic disease. The radiotherapy in the previous studies was generally subtotal or total lymphoid irradiation, even when adjuvant chemotherapy was planned. In the current trial, the radiotherapy is to more limited fields, as described previously. With the elimination of staging laparotomy and limitation of the radiation fields, the duration and intensity of therapy for patients in the current trials is shorter than that of patients treated previously with combined modality therapy. All deaths in the current trial have been due to progressive Hodgkin's disease. Although still somewhat preliminary, it appears that the freedom from relapse with the current treatment programs may be slightly inferior to our historical experience. It is likely that this is a reflection of the admixture of an unknown proportion of patients with occult subdiaphragmatic disease. A comparative assessment of toxicity in these two treatment groups has not been completed.

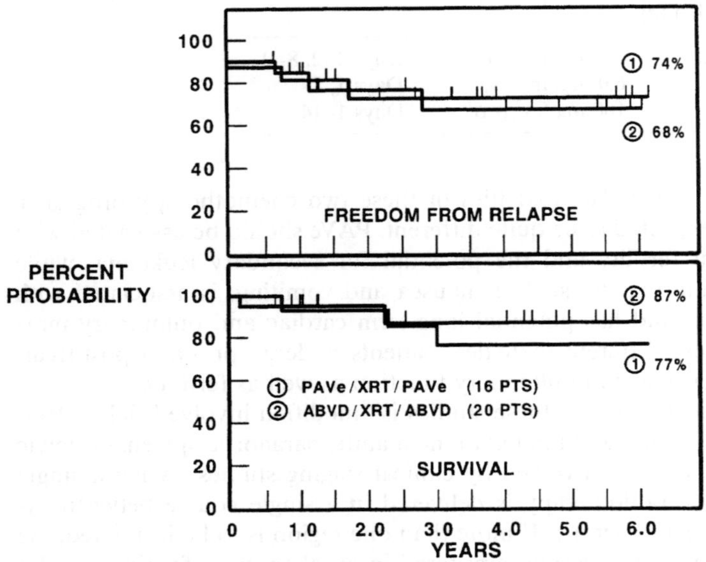

Fig. 2. Freedom from relapse and survival for the C4-6 studies for patients with "bulky" CS I-IIA, I-IIB, or IIIA Hodgkin's disease

Table 7. Treatment of patients with "bulky disease": Stanford trials

Trial(s)	Stage	*n*	Staging laparotomy	Radiotherapy fields	Chemotherapy	Survival (%)		Freedom from relapse (%)	
						3-year	5-year	3-year	5-year
H, S	PSI-IIA, B	14	Yes	STLI or TLI	None	93	84	45	45
H, S	PSI-IIA, B	27	Yes	STLI or TLI	MOPP or PAVe	93	84	81	81
C4, C5	CSI-IIA, B	25	No	Modified IF	PAVe or ABVD	86	80	70	70

C12-15 Studies

The C12-15 studies were designed for the management of patients with extensive Hodgkin's disease. This includes all patients with stage IIIB and all patients with stage IV (A or B) disease. In addition, patients who have undergone laparotomy and splenectomy and have stage IIISA disease with extensive involvement to the spleen (five or more microscopic nodules) are included in these studies.

The eligibility criteria and treatment programs for patients on these studies are summarized in Table 8. The intent of these studies is to compare combined modality therapy with chemotherapy alone. The combined modality program which was adopted was that which had been utilized previously in the S studies and shown to be quite effective in patients with stage IIIB and IV disease, a program of alternating chemotherapy and radiation with PAVe and total lymphoid irradiation (Hoppe et al. 1979). The chemotherapy alone option was chosen to be the alternating MOPP/ABVD program (Santoro et al. 1982).

Table 8. C12-15 studies

Eligibility	C12 PS $IIIA_S$+ extensive C13, C14 CS or PS IIIB C15 CS or PS IV A or B
Treatments	A. Alternating PAVe/TLI B. Alternating MOPP/ABVD

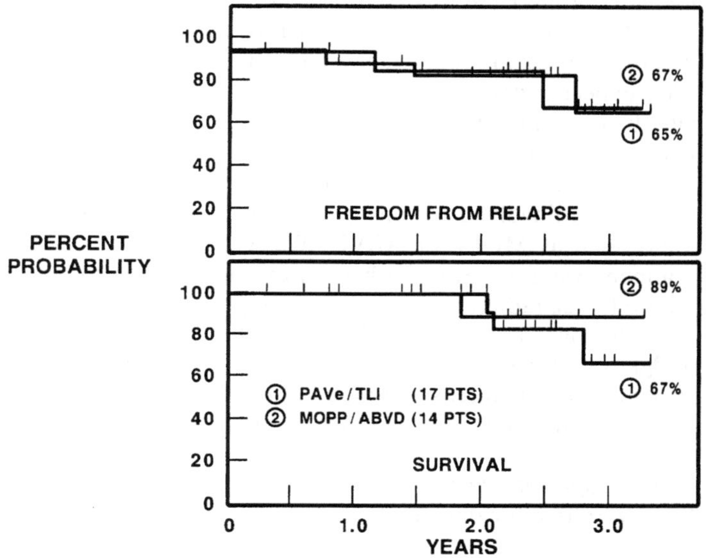

Fig. 3. Freedom from relapse and survival for the C12-15 studies for patients with PS IIIA with extensive splenic involvement or CS/PS IIIB, IVA, or IVB Hodgkin's disease

In the alternating PAVe/TLI program, patients receive a total of six cycles of PAVe chemotherapy and total lymphoid irradiation. The usual approach is to administer two cycles of PAVe chemotherapy followed by irradiation to the most extensively involved region, often the mantle. Two more cycles of PAVe are administered, followed by irradiation to the next region. A final two cycles of chemotherapy are then administered followed by irradiation to the final region, which is often the pelvis. The dose to all involved regions is 40-44 Gy. Treatment to the uninvolved pelvis or mediastinum is limited to 30 Gy.

For the alternative treatment program of MOPP/ABVD, patients receive MOPP and ABVD in an alternating fashion for a planned total of 12 monthly cycles. In addition to survival and freedom from relapse, patients are followed prospectively for comparison of toxicity. Patients undergo serial pre- and post-treatment assessment of pulmonary function, cardiac function, and fertility.

The C12-15 studies were initiated only in 1984. The maximum follow-up is 3.3 years and the median is 2.5 years. The results of both treatment approaches are quite similar (Fig. 3). The accrual of a large number of patients with a longer follow-up will be necessary before more detailed analyses can be completed.

Future Directions

Significant advances have been made in the management of Hodgkin's disease during the past 25 years. A major goal of the early Stanford trials, to develop curative programs for patients with all stages of Hodgkin's disease, has been achieved. Nevertheless, challenges remain. The price of cure for most patients is significant short-term morbidity and possible long-term complications. Both current staging (i.e., laparotomy) and treatment programs are associated with potential adverse effects. A major focus of future clinical trials will be a more tailored approach to therapy in order to minimize these effects.

We expect to develop criteria for more selective use of laparotomy, i.e., to eliminate its use among patients who are at particularly low or high risk for detection of subdiaphragmatic disease. We will explore further the use of VBM chemotherapy in combined modality treatment programs where it may be substituted for irradiation to uninvolved regions or used in place of more toxic chemotherapy programs such as MOPP or ABVD. We will attempt to refine combined modality treatment programs even further, for example, by the reduction in the extent of radiation fields, radiation doses, or number of cycles of chemotherapy administered.

Finally, one cannot overlook the fact that 20%–25% of patients will die of their Hodgkin's disease. For this cohort of patients, more innovative treatment approaches such as intensive induction therapy followed by autologous bone marrow transplantation will need to be tested.

Acknowledgments. The authors thank Marge Keskin for secretarial assistance and Anna Varghese for data analysis.

References

Bonadonna G, Zucali R, Delena M, Valagussa P (1977) Combined chemotherapy (MOPP or ABVD) - radiotherapy approach in advanced Hodgkin's disease. Cancer Treat Rep 61: 769–777

Hoppe RT, Portlock CS, Glatstein E, Rosenberg SA, Kaplan HS (1979) Alternating chemotherapy and irradiation in the treatment of advanced Hodgkin's disease. Cancer 43: 472–481

Hoppe RT, Coleman CN, Cox RS, Rosenberg SA, Kaplan HS (1982) The management of stage I–II Hodgkin's disease with irradiation alone or combined modality therapy: the Stanford experience. Blood 59: 455–465

Hoppe RT, Horning SJ, Rosenberg SA (1985) The concept, evolution, and preliminary results of the current Stanford clinical trials for Hodgkin's disease. Cancer Surv 4: 459–475

Horning SJ, Hoppe RT, Rosenberg SA (1988) VBM: an effective adjuvant in favorable Hodgkin's disease. J Clin Oncol 6: 1822–1831

Kaplan HS (1980) Hodgkin's disease, 2nd edn. Harvard University Press (Cambridge, MA)

Rosenberg SA, Kaplan HS (1985) The evolution and summary results of the Stanford randomized clinical trials of the management of Hodgkin's disease: 1962–1984. Int J Radiat Oncol Biol Phys 11: 5–22

Santoro A, Bonadonna G, Bonfante V, Valagussa P (1982) Alternating drug combinations in the treatment of advanced Hodgkin's disease. N Engl J Med 306: 770–775

Results and Prognostic Factors Following Optimal Treatment of Advanced Hodgkin's Disease

D. J. Straus, J. Gaynor, J. Myers, D. Merke, D. Chapman, J. Caravelli, L. Z. Nisce, B. J. Lee, and B. D. Clarkson

Memorial Sloan Kettering Cancer Center, 1275 York Avenue, New York, NY 10021, USA

Introduction

The introduction of the MOPP (mustine, oncovin, procarbazine, and prednisone) combination by DeVita and colleagues (Longo et al. 1986) represented a dramatic advance in the treatment of patients with advanced Hodgkin's disease. The complete remission (CR) rate at Memorial Hospital was at the lower end of those reported in the literature, perhaps reflecting the inclusion of advanced Hodgkin's disease patients with a worse prognosis than those seen in other centers.

In an attempt to improve upon our own results with MOPP for treatment of patients with advanced Hodgkin's disease, a new treatment program was started at Memorial Hospital in January 1975 employing two modifications of the standard MOPP regimen. The first was the addition of a drug combination potentially non-cross-resistant with MOPP, ABVD (Adriamycin, bleomycin, dacarbazine, vinblastine) (Santoro et al. 1987). The second was the use of adjunctive moderate-dose radiotherapy (RT) to areas initially involved with bulky disease which would be likely sites of relapse after chemotherapy alone.

Excellent results were achieved with the initial pilot study with MOPP/ABVD/RT as described below. A randomized trial of three alternating drug combinations [CCNU, melphalan, vindesine (CAD)/MOPP/ABV] and moderate-dose RT versus MOPP/ABVD/RT was initiated in 1979 (Straus et al. 1984). As described, the results of both arms of this trial were identical and the same as those in the initial pilot study with MOPP/ABVD/RT. The overall results with these trials are among the best reported, and for purposes of analysis of prognostic factors all of the results from these two trials have been combined.

Materials and Methods

Between January 1975 and December 1978, a total of 118 patients with advanced Hodgkin's disease were entered into the MOPP/ABVD/RT program. Sixty-seven patients were previously untreated, and 51 patients were treated in relapse. Clinical staging was performed according to the guidelines of the Ann Arbor Confer-

Recent Results in Cancer Research. Vol. 117

ence (Carbone et al. 1971), and pathologic staging by laparotomy was performed in 11 patients. Histologic classification was made according to the Rye modification of the Lukes and Butler scheme (Lukes et al. 1966). The MOPP program was the standard regimen of DeVita and colleagues (Longo et al. 1986). The second was ABVD: Adriamycin 25 mg/m^2 i.v., vinblastine 6 mg/m^2 i.v., and dacarbazine 250 mg/m^2 i.v. were all given on days 1 and 14, and bleomycin was administered in a dose of 2 mg subcutaneously on days 4-12 and 8-16. MOPP and ABVD were alternated during months 1-4, 6-9, and then every 2 months for an additional seven cycles of "maintenance" chemotherapy. Radiotherapy consisting of 2000-3000 rad tumor dose in 2-3 weeks was administered to areas of bulky disease (> 5 cm) in month 5. Relapsed patients who had received prior RT were not reirradiated. During the months between their bimonthly maintenance chemotherapy treatments in months 10-24, patients were randomized to receive levamisole 150 mg p.o. daily on days 1-6 and 15-20 or no treatment.

In January 1979 a new protocol was begun which randomized patients to MOPP/ABVD/RT or to a new MOPP/ABV/CAD/RT program which uses a three-drug combination in alternation and low-dose irradiation. Dacarbazine was dropped from ABVD because of the unpleasant side effects of severe nausea and vomiting. MOPP was continued, with the allowance of substitution of cyclophosphamide 650 mg/m^2 i.v. for nitrogen mustard in patients who experience intolerable nausea and vomiting with this drug. The third combination, CAD, used lomustine (CCNU) 100 mg/m^2 p.o. day 1, melphalan (Alkeran) 6 mg/m^2 p.o. days 1-4, and vindesine (DVA) 4 mg/m^2 i.v. day 1 and 8. The first two of these drugs have proven efficacy in Hodgkin's disease, and we have found the third to be effective. This combination was found to be active in patients who relapsed after MOPP or MOPP/ABVD. With the ten-drug regimen, MOPP/ABV/CAD, an attempt was been made to reduce the nausea and vomiting and thus make the treatment more acceptable and also to improve on results by adding a third potentially non-cross-resistant drug combination. Chemoterapy maintenance was dropped from both parts of the study, since there is no evidence that it is necessary. Previously untreated patients were randomized to nine cycles of either CAD/MOPP/ABV or MOPP/ABVD. An interruption was made between the sixth and seventh cycle of treatment for RT to involved nodal regions to doses of 2000-3000 rad. The details of this regimen have been reported previously (Straus et al. 1984).

In both trials, a complete remission (CR) was defined as disappearance of all measurable disease, and a partial remission (PR) as a 50% or greater reduction in the sum of the largest perpendicular diameters of all measurable disease. All responses had to be 1 month in duration.

Since the results of both trials were the same, they were combined for purposes of analysis of pretreatment prognostic characteristics of the patients. The outcome results selected was overall survival as determined by the method of Kaplan and Meier (1958). Multivariate analysis of survival was performed to to construct a proportional hazards model according to the method of Cox (1972).

Results

Between January 1975 and December 1978, 118 patients with advanced Hodgkin's disease were treated with MOPP/ABVD/RT. Ninety-nine of the 118 patients entered on the protocol were evaluable for response: 57 previously untreated patients, 16 patients who had relapsed after RT and/or small amounts of single-agent chemotherapy ("minimally pretreated"), and 18 heavily pretreated patients. Of the 27 patients not evaluable for response, 13 received less than 3 cycles of treatment because of severe nausea and vomiting, 6 had no measurable disease prior to treatment, and there were major protocol violations in the treatment of 8 patients. Among the evaluable patients there were 54 males and 37 females. Age ranged between 16 and 64 years. Fifty-seven patients had nodular sclerosis, 25 mixed cellularity and 9 not subclassified histologies. Fifteen patients were stage IIB, 13 stage IIIA, 28 stage IIIB, and 35 stage IV.

Among evaluable previously untreated patients, 50/57 (88%) achieved a CR and 7/57 (12%) a PR. Complete remission was attained by 11/16 (69%), PR by 3/16 (19%), and no response was seen in 2/16 (12%) of the minimally pretreated patients. Among the heavily pretreated patients, 9/18 (50%) achieved a CR, 4/18 (22%) a PR, and 5/18 (28%) failed to respond to treatment. Differences in response rates between patients in each stage for the previously untreated and minimally pretreated patients together were not statistically significant.

Overall, there were 18 relapses among the patients achieving a CR: 9 among previously untreated patients, 3 among minimally pretreated patients, and 6 among heavily pretreated patients, respectively. The relapse rate among previously untreated patients achieving a CR was 20% at 66 months. Twenty-three patients were randomized to receive and 25 patients not to receive levamisole. Eight patients relapsed from CR in the levamisole and four in the non-levamisole arms of the trial. The differences in remission duration and survival between patients who received and those who did not receive levamisole were not statistically significant.

The major toxicity was comparable to that of other programs for the treatment of advanced Hodgkin's disease (Straus et al. 1980). There have been 4 cases of acute leukemia among the total of 118 patients, 2 among prior-treated and two among previously untreated patients. In two instances this was preceded by a sideroblastic anemia. The ages of the patients were 41, 52, 54, and 64 years respectively, and all received RT below the diaphragm. There were six bleomycin-related pulmonary toxic events. Three patients had acute bronchospasm which resolved with discontinuation of bleomycin, one had an acute pneumonitis, and two had pulmonary fibrosis on chest X-ray. None of the six had permanent pulmonary symptoms, and all six completed treatment successfully with the omission of bleomycin. One patient, the oldest to receive low-dose mediastinal RT, developed symptomatic congestive heart failure during a relapse and died of progressive Hodgkin's disease. She was found to have cardiomyopathy at autopsy. No other patients developed cardiac symptoms, and resting and exercise radionuclide cardiac angiography and echocardiography in 19 asymptomatic patients studied at a median time of 2 years after completion of treatment showed cardiac function to be well preserved. There were two instances of graft-versus-host disease (GVHD) due to blood transfusion, one of which was proven by HLA typing (Dinsmore et al. 1980) and another suspected. A major problem for the patients receiving MOPP/ABVD/RT was severe nausea and vomiting, particularly related to dacarbazine. Thirteen patients (11%) dropped treatment before receiving 3 cycles, and 24 patients (28%) failed to complete maintenance treatment. Among the entire group of previously untreated patients, 67 evaluable and 10 unevaluable for response, there have been 15 deaths, 13 due to Hodgkin's disease and 2 due to secondary acute leukemia. There have been 11 deaths among the minimally prior treated and 20 deaths among the heavily prior treated groups, respectively.

Between January 1979 and January 1987 to the new protocol, 160 evaluable previously untreated patients were randomized; 80 received CAD/MOPP/ABV/RT and 80 MOPP/ABVD/RT. There were 89 males and 71 females. The median age was approximately 30 years. Two-thirds of the patients had nodular sclerosis histology. Thirty-six percent of the patients were CS IIB, 34% were CS IIIB, 28% were stage IVA and B, two patients were bulky CS IIAE, and one patient was CS IIIA. Among patients treated with CAD/MOPP/ABV/RT, 64 of 80 (80%) achieved a CR, 13 of 80 a PR (16%), and 3 (4%) progressed. Sixty-eight of 80 (85%) patients treated with MOPP/ABVD/RT achieved a CR, 10 of 80 (13%) a PR, and 2 (2%) progressed. There have been 8 relapses from CR, 7 from PR, and 12 deaths among the patients treated with CAD/MOPP/ABV/RT. Among the patients treated with MOPP/ABVD/RT, 9 have relapsed from a CR, 13 have relapsed from a PR, and 13 patients have died. The difference in CR percentages, CR durations, and survivals for patients on CAD/MOPP/ABV/RT and MOPP/ABVD/RT are not statistically significant. The median follow-up time for surviving patients on this protocol is 63 months.

The incidence of transient myelosuppression was greater for CAD/MOPP/ABV/RT than for MOPP/ABVD/RT. However, there were only three episodes of hemorrhage related to thrombocytopenia, three of sepsis related to neutropenia, and three of prolonged pancytopenia which were distributed among the patients on both arms of the trial. An important difference between the two regimens which greatly influenced their acceptability to patients was the incidence of severe nausea and vomiting. All patients on MOPP/ABVD/RT complained of this side effect, while it was seen in only 26% of patients treated with CAD/MOPP/ABV/RT.

Since the results of the initial pilot trial of MOPP/ABVD/RT and those of the subsequent randomized trial of CAD/MOPP/ABV/RT versus MOPP/ABVD/RT were the same, they were pooled together for purposes of analysis of prognostic factors. Overall 185 consecutive evaluable previously untreated patients with advanced Hodgkin's disease were entered into these two trials between January 1975 and October 1985. The follow-up time ranges between 33 and 150 months (median 74 months). Percentages of CR were 83.5% for CS IIB patients, 87% for CS IIIB patients, and 76% for stage IV patients ($P=$ NS). At 12 years, disease-free survivals were 84% for CS IIB, 81% for CS IIIB, and 51% for IVB ($P=0.04$). The 12-year survivals were 85% for CS IIB, 70% for CS IIIB, and 57% for IVB ($P=0.06$). Pretreatment patient characteristics found to be associated with a statistically significant decreased survival rate included age over 45 years, mixed cellularity as compared with nodular sclerosis histology, bone marrow involvement, mediastinal mass of greater than 0.45 of the thoracic diameter at the level of the carina, inguinal node involvement (a reflection of retroperitoneal disease), anemia, serum lactic acid dehydrogenase (LDH) greater than 400 U/liter, and serum albumin of less than 3.5 gm/dl. Multivariate analysis was performed to construct a model for overall survival according to the method of Cox (1972), and five initial presenting factors were slightly associated with a shorter survival: age greater than 45 years, serum LDH greater than 400 U/liter, inguinal node involvement, hematocrit below normal, and a mediastinal mass of greater than 0.45 thoracic diameter at the level of the carina. At 5 years all patients with none of these initial adverse prognostic factors are surviving. With one adverse factor, 94.4% are surviving. With two factors, 60.5% are surviving with further deaths past 5 years, and with three factors 20.5% are surviving at 5 years. These survival differences are highly statistically significant.

Discussion

These results represent a significant improvement over published results with MOPP alone for patients with advanced Hodgkin's disease (Longo et al. 1986). Preliminary results using MOPP on day 1 and ABV on day 8 in cyclical fashion are similar to those of MOPP/ABVD/RT at comparable follow-up time (Straus et al. 1980; Klimo and Connors 1985). Bonadonna et al. (1986) have reported similar results with MOPP/ABVD and have demonstrated superiority to MOPP for stage IV patients in a randomized trial. It seems that this approach has consistently achieved results among the best reported for advanced Hodgkin's disease. There are two remaining questions. The first is the need for adjuvant moderate-dose radiotherapy. The second is how to improve upon these results in the group of patients presenting with unfavorable prognostic factors. Autologous bone marrow transplantation has been a promising approach for relapsed patients (Jagannath et al. 1986) and might be employed earlier in poor prognosis patients. Future protocols will address these problems.

References

Bonadonna G, Valagussa P, Santoro A (1986) Alternating non-cross-resistant combination chemotherapy or MOPP in stage IV Hodgkin's disease. Ann Intern Med 104: 739-746

Carbone PP, Kaplan HS, Musshoff K, Smithers DW, Tubiana M (1971) Report of the committee on Hodgkin's disease staging. Cancer Res 31: 1860-1861

Cox DR (1972) Regression models and life tables (with discussion). J R Stat Soc Br 24: 187-200

Dinsmore RE, Straus DJ, Pollack MS, Woodruff JM, Garrett TJ, Young CW, Clarkson BD, Dupont B (1980) Fatal graft-vs-host disease following blood transfusion in Hodgkin's disease documented by HLA typing. Blood 55: 831-834

Jagannath S, Dicke K, Armitage J, Cabanillas FF, Horowitz LD, Vellekoop L, Zander A, Spizer G (1986) High-dose cyclophosphamide, carmustine, and etoposide and autologous bone marrow transplantation for relapsed Hodgkin's disease. Ann Intern Med 104: 163-168

Kaplan EL, Meier P (1958) Non-parametric estimation from incomplete observations. J Am Stat Assoc 53: 457-481

Klimo P, Connors J (1985) MOPP/ABV hybrid program: combination chemotherapy based on early introduction of seven effective drugs for advanced Hodgkin's disease. J Clin Oncol 3: 1174-1182

Longo DL, Young RC, Wesley M, Hubbard SM, Duffey PL, Jaffe ES, DeVita VT (1986) Twenty years of MOPP therapy for Hodgkin's disease. J Clin Oncol 4: 1295-1306

Lukes RJ, Craver LF, Hall TC, Rappaport H, Rubin P (1966) Report of the nomenclature committee. Cancer Res 26: 1311

Santoro A, Bonadonna G, Valagussa P, Zucali R, Vivani S, Villani F, Pagnoni AM, Bonfante V, Musumeci R, Crippa F, Tess JDT, Banfi A (1987) Long-term results of combined chemotherapy-radiotherapy approach in Hodgkin's disease. Superiority of ABVD plus radiotherapy versus MOPP plus radiotherapy. J Clin Oncol 5: 27-37

Straus DJ, Myers J, Passe S, Young CW, Nisce LZ, Lee BJ, Gee T, Clarkson BD (1980) The eight drug/radiation therapy program (MOPP/ABVD/RT) for advanced Hodgkin's disease. A follow-up report. Cancer 46: 233-240

Straus DJ, Myers J, Lee BJ, Nisce LZ, Koziner B, McCormick B, Kempin S, Mertelsmann R, Arlin Z, Gee T, Poussin-Rosillo H, Hansen H, Clarkson BD (1984) Treatment of advanced Hodgkin's disease with chemotherapy and irradiation. Controlled trial of two versus three alternating, potentially non-cross-resistant drug combinations. Am J Med 76: 270-278

Two Cycles of MOPP and Definitive Radiotherapy for Stage IIIA and IIIB Hodgkin's Disease

L. M. Fuller and F. B. Hagemeister

Departments of Clinical Radiotherapy and Hematology, University of Texas, M. D. Anderson Cancer Center, 1515 Holcombe Boulevard, Houston, TX 77030, USA

Combination chemotherapy is generally viewed as the treatment of choice for stage IIIB Hodgkin's disease (DeVita et al. 1980). However, trials that have employed adjuvant radiotherapy (XRT) have claimed better results (Horning et al. 1984; Prosnitz et al. 1982). Some of this improvement may be due to a lower incidence in nodal relapse, the primary pattern of recurrence in patients treated with chemotherapy only (Young et al. 1978; DeVita 1976). In the past, definitive XRT was advocated for stage IIIA (Rosenberg and Kaplan 1975). Recent studies have demonstrated that multiple-agent chemotherapy alone or in combination with XRT gives better disease-free survival rates than XRT alone for most patients with stage IIIA (Mauch et al. 1985; Kun et al. 1976; Lister et al. 1983; Bonadonna et al. 1979; Crowther et al. 1984). Most combined modality treatment programs have required a minimum of six cycles of chemotherapy, and the sequencing of radiotherapy has dependend on the individual programs. Despite the overall improvements in results that have been achieved with combined modality treatment, prognostic factors, other than constitutional symptoms and extent of abdominal disease, have not been well delineated (Carbone et al. 1971; Desser et al. 1977; Mazza et al. 1984; Stein et al. 1982; Hoppe et al. 1982).

In 1970 we initiated a combined modality program for both stage IIIA and IIIB Hodgkin's disease which utilized two cycles of MOPP only (mechlorethamine, vincristine, procarbazine, prednisone) followed by definitive radiotherapy (Rodgers et al. 1981). In this report we present 10-year results, with an analysis of prognostic factors and treatment tolerance for the first 197 patients who were entered in this study. This study indicates that this approach to treatment is appropriate for most patients with either stage IIIA or IIIB disease. However, it also defines subpopulations that require different treatment strategies.

Materials and Methods

Between 1970 and 1984, 197 patients were entered in this study. Of these, 119 patients had stage IIIA disease and 78 had stage IIIB. In approximately half of the patients staging was based on positive lymphangiogram findings. These patients

Recent Results in Cancer Research. Vol. 117

did not have staging laparotomies. Most patients with negative lymphangiogram studies had staging laparotomies. In almost all of these patients, the disease was limited to the celiac portal complex of nodes or the spleen. Our original plan for all patients was to initiate their treatment with two cycles of MOPP and to follow this with definitive radiotherapy administered sequentially to the mantle, abdomen, and pelvis (Fig. 1). In 1978, we modified our radiotherapy plan for two categories of patients. For patients with unfavorable mediastinal presentations, which include masses 7.5 cm in diameter or greater, hilar involvement, or direct extension to lung, we added low-dose whole-lung irradiation. Pelvic XRT was deleted for patients with abdominal disease that was limited to the celiac portal complex or the spleen. These changes were made to prevent pulmonary relapse and to preserve fertility. Radiotherapy doses ranged between 30 and 40 Gy to the mantle depending on the response to chemotherapy, delivered in 15-20 fractions over 3-4 weeks. Prophylactic lung irradiation was limited to 15 Gy delivered in 15 fractions or 3 weeks. Treatment to both the abdomen and pelvis was generally limited to 30 Gy delivered in 20 fractions over 4 weeks. The doses to the right lobe of the liver and the kidneys were limited to 15 and 18 Gy respectively by using appropriate shielding (Fig. 1).

For this analysis, patients were divided into three major groups based on the extent of their abdominal disease. In 97 patients whose disease was categorized as III_1, the involvement was limited to the celiac portal complex or the spleen. Patients with stage III_2 had paraaortic adenopathy, and patients with III_3 disease

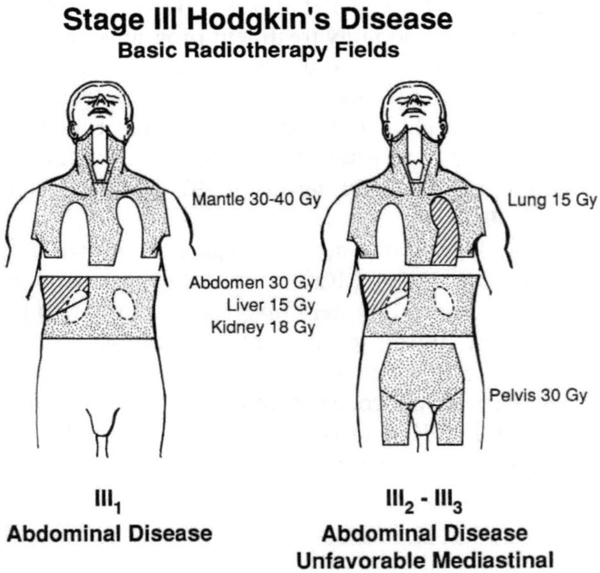

Fig. 1. *Left:* Current least amount (volume) of irradiation treatment that is given for stage III_1 Hodgkin's disease. *Right:* Current most treatment. This would be applicable for patients with stage III_2 or III_3 abdominal disease and unfavorable mediastinal disease. See text for tumor dose fractionation schema

had additional involvement of the pelvic nodes. Each of these three groups was further subdivided according to the presence or absence of unfavorable mediastinal disease. Kaplan Meyer curves were calculated for all patients to determine freedom from tumor progression (FTP) and to determine overall survival (which included all deaths), tumor-specific survival, and freedom from the combination of tumor mortality and acute toxicity (FTT). Gehan's modification of the generalized Wilcoxon method was used to test differences between various groups for significance.

Overall 10-Year Results

The overall 10-year survival rates including deaths from any cause was 56%. For patients with stage IIIA disease it was 58%. The corresponding figure for patients with stage IIIB disease was 51% (Table 1). The causes of death for patients are provided in Table 2. The tumor-specific survival rate for the entire group was 81%. It was 87% for patients with stage IIIA disease. The corresponding rate for patients with stage IIIB disease was significantly less, being 72% ($P=0.024$). Among the patients with stage IIIA disease there were four deaths that were attributable to acute toxicity which occurred during treatment. Two were due to hepatitis and two were related to myelosuppression and associated sepsis. When these were included with the deaths due to Hodgkin's disease, the survival rate for patients with

Table 1. Ten-year results: all patients

Patients	Survival				Freedom from progression	
	Absolute including all deaths	Tumor specific		Including acute toxicity deaths		
197	56%	81%		79%	76%	
119 IIIA	58%	87%	$> P=0.024$	84%	84%	$> P=0.0089$
78 IIIB	51%	72%		72%	64%	

Table 2. Causes of death in 58 patients

Hodgkin's disease	Acute toxicity during treatment	Acute myelogenous leukemia (AML)	Other second malignancies	Cardiac pathology	Other
28	4[a]	4[b]	6[c]	7[d]	7

[a] Two patients with hepatitis and two with myelosuppression and sepsis.
[b] Includes one patient who developed relapsing Hodgkin's disease in the bone marrow and AML concurrently.
[c] Includes three patients with non-Hodgkin's lymphomas.
[d] Includes six patients with coronary artery disease, one of whom did not receive radiotherapy to the mediastinum. The seventh patient died of pericarditis.

stage IIIA disease fell to 84%. Ten-year FTP rates were 76% for all patients, 84% for patients with stage IIIA disease, and 64% for patients with stage IIIB ($P = 0.0089$).

Prognostic Factors

Neither tumor-specific survival nor freedom from tumor progression were influenced significantly by sex or by histopathologic subtype. However, age had an adverse effect on tumor-specific survival. The 10-year tumor-specific survival rate for 146 patients under age 40 was 84% as compared with 73% for 51 patients who were 40 years of age or older ($P = 0.047$). All four deaths that were attributable to toxicity during treatment occurred among the older patients. The FTT for this older group was 66%. The difference between 84% for the younger group and 66% for the older group was significant at the $P = 0.026$ level (Table 3). FTP rates were 77% and 73% respectively.

Influence of Extent of Abdominal Disease

There was no significant difference between the tumor-specific survival rates for patients with III_1 abdominal disease and those with III_2 disease. Within each of these two groups, there was no significant difference in results between patients without constitutional symptoms and those with constitutional symptoms. Also there were no significant differences between the corresponding FTP (Table 4). The tumor-specific survival rate for all 102 patients with III_1 abdominal disease was 89%. For patients with III_1A disease it was 88% and for those with III_1B disease it was 91%. The corresponding figures for patients with III_2 abdominal disease were 86%, 89%, and 83%. Results for patients with III_3 abdominal disease were influenced adversely by the presence of constitutional symptoms. The tumor-specific survival rate for patients with constitutional symptoms was only 45% whereas the corresponding rate for patients with III_3A disease was 82%. FTPs for these two groups were 35% and 85% respectively ($P = 0.01$).

Influence of Extent of Mediastinal Disease

The effect of unfavorable mediastinal disease on results was compared for the three subgroups of patients with III_1, III_2, and III_3 abdominal disease (Table 5). There was no significant difference in either tumor-specific survival rates or FTP rates for patients with either III_1 or III_2 disease regardless of the status of the mediastinum or the presence or absence of constitutional symptoms. However, striking differences were observed among the results for patients with III_3 abdominal disease which were dependent on both the extent of mediastinal disease and on the presence or absence of constitutional symptoms.

Table 3. Effect of age on 10-year results

Age (Years)	Patients	Survival				Freedom from progression
		Tumor specific		Including acute toxicity deaths		
<40	146	84%	$P=0.047$	84%	$P=0.026$	77%
≥40	51	73%		66%		73%

Table 4. Ten-year results by abdominal substages

Stage	Patients			% Tumor-specific survival			% Freedom from progression		
	ALL	A	B	ALL	A	B	ALL	A	B
III ALL	197	119	78	81	87	72	76	84	64
					$P=0.024$			$P=0.0089$	
III_1	102	76	26	89	88	91	84	86	79
III_2	47	23	24	86	89	83	81	79	85
III_1] III_2	149	99	50	88	89	88	83	84	81
III_2] III_3	95	43	52	71	85	60	67	81	57
								$P=0.046$	
III_3	48	20	28	58	82	45	53	85	37
								$P=0.01$	

Table 5. Effect of unfavorable mediastinal disease on results

Abdominal disease	Mediastinal status	Tumor-specific survival (%)			Freedom from progression (%)		
		ALL	A	B	ALL	A	B
III_1	Unfavorable	81	80	83	83	82	85
	Favorable	95	94	100	85	88	73
III_2	Unfavorable	88	88 (7/8)	89	89	88 (7/8)	89
	Favorable	85	93	78 (8/10)	72	69	80 (8/10)
III_3	Unfavorable	27	67 (3/4)	17 (2/9)	29	75 (3/4)	11 (1/9)
	Favorable	76	88	68	63	88	50

"Favorable" includes patients with no mediastinal disease as well as those with relatively favorable prognostic features (see text).

Preservation of Fertility

Among the men who did not receive pelvic XRT, five had semen examination before and after all treatment. All five developed aspermia. However, at the time of this analysis, three had recovered normal counts. This occurred within approximately 1 year. One of these has become a father. Of 16 women who had normal menses, all 16 continued to have normal periods and 6 have had normal children.

Complications and Second Malignancies

Of the 197 patients who were treated on this protocol, 2 died as a result of myelosuppression and sepsis. Both of these patients, who were over age 40, developed a prolonged myelosuppression with each course of MOPP and radiotherapy. Early in this study, several patients developed hepatitis related to their treatment. With two exceptions, this was treated successfully with prednisone. Of the two patients who died, one was not treated for this complication having been seen in another center; the other, who was also managed in another center, was treated with inadequate doses of prednisone. In irradiating the abdomen at that time we were not reducing the dose to the right lobe of the liver, which therefore received 30 Gy in 20 fractions. Although this had been found to be a safe dose when we were using radiotherapy alone, it proved to be excessive when given after two cycles of MOPP chemotherapy. Since we adopted the policy of shielding the right lobe of the liver with 1 half-value layer of lead anteriorly and posteriorly there have been no further cases of hepatitis.

Acute myelogenous leukemia occurred in four patients (2%). All four of these cases were among the patients who were treated prior to 1978 before we modified our protocol to exclude the pelvis in patients with III_1 abdominal disease. Among eight second malignancies, three were non-Hodgkin's lymphoma. Of these eight patients, six are dead of the second malignancy. Of nine patients who developed cardiac symptomatology, one died of pericarditis believed to be secondary to mantle irradiation. Six of eight patients with coronary artery disease are dead. One of these patients who had a prior history of coronary artery disease did not receive irradiation to the mediastinum.

Discussion

Combined modality programs of combination chemotherapy and radiotherapy are now recognized to produce better disease-free survivals than definitive radiotherapy alone for patients with stage IIIA Hodgkin's disease. This has been particularly evident for those with III_2A and III_3A disease (Stein et al. 1982). Mauch et al. (1985) have suggested that patients with III_1A disease who have less than five splenic nodules can be treated with radiotherapy alone. However, our study suggests that patients with unfavorable mediastinal presentations regardless of abdominal substage require combination chemotherapy in addition to radiotherapy. Therefore, the majority of patients with stage IIIA disease require combined modality therapy. Although less well recognized, a combined modality approach can also be more effective treatment for patients with stage IIIB disease. This has been shown in this report and those of others for the majority of such patients. Most combined modality programs have included six or more cycles of combination chemotherapy prior to or after the administration of radiotherapy. In our study, two cycles of MOPP produced results similar to those in other programs, except for those patients with stage III_3B abdominal disease. These patients had inferior results following treatment with two cycles of MOPP and radiotherapy and other treatment strategies are needed for this category of patients.

Prevention of Pulmonary and Hepatic Relapse

In our original study, which began in 1970, low-dose whole-lung radiotherapy was not a part of the protocol. In patients with unfavorable mediastinal presentations, the incidence of pulmonary relapse was approximately half of that anticipated for similar patients treated with radiotherapy alone due to the effect of two cycles of MOPP. With the subsequent addition of low-dose whole-lung radiotherapy, there have been no pulmonary relapses in such patients. Fields for abdominal radiotherapy vary among institutions. One of the main areas of controversy involves the need for low-dose liver irradiation (Lee et al. 1984; Micaily and Brady 1984). We have always included the liver in our abdominal fields due to the high incidence of liver relapse observed in patients with clinical stage I and II disease who were treated with radiotherapy alone (Fuller et al. 1971). In the current series, there were no isolated initial hepatic relapses of Hodgkin's disease in these patients despite the fact that only two cycles of MOPP were administered.

Prevention of Treatment Complications

Sterility is one of the complications which follows treatment with six cycles of MOPP (Viviani et al. 1985). In our series of patients treated with only two cycles of MOPP and radiotherapy, there has been a gradual recovery of spermatic function and no menstrual difficulties in patients with stage III_1 disease who did not receive pelvic radiotherapy. These patients can have children without an increased risk of relapse of Hodgkin's disease.

The incidence of acute myelogenous leukemia in our patients is 2%. Patients who did not receive pelvic radiotherapy have not developed acute myelogenous leukemia. Long-term follow-up is necessary to ascertain the true incidence of acute myelogenous leukemia in this subgroup of patients. It is difficult to determine whether complications which include coronary artery disease and second malignancies other than acute myelogenous leukemia are due to radiotherapy and chemotherapy or are related to other factors present in the general population. Currently, we are in the process of reviewing all of our patients who have been treated with radiotherapy and chemotherapy to determine whether there is an increased risk of coronary artery disease and specific secondary malignancies in these patients relative to the general population.

References

Bonadonna G. Santoro A, Zucali R, Valagussa PM (1979) Improved 5-year survival in advanced Hodgkin's disease by combined modality approach. Cancer Clin Trials 2: 217-226

Carbone P, Kaplan H, Musshoff K, Smithers D, Tubiana M (1971) Report of the Committee on Hodgkin's Disease Staging. Cancer Res 31: 1860-1861

Crowther D, Wagstaff J, Deakin D et al. (1984) A randomized study comparing chemotherapy alone with chemotherapy followed by radiotherapy in patients with pathologically staged IIIA Hodgkin's disease. J Clin Oncol 2: 892-897

Desser RK, Golomb HM, Ultmann JE et al. (1977) Prognostic classification of Hodgkin's disease in pathologic stage III, based on anatomic considerations. Blood 49: 883-893

DeVita VT (1976) Hodgkin's disease: the salvage of radiation treatment failures. Int J Radiat Oncol Biol Phys 17: 252

DeVita VT, Simon RM, Hubbard SM et al. (1980) Curability of advanced Hodgkin's disease. Ann Intern Med 92: 587-595

Fuller LM, Gamble JF, Shullenberger CC, Butler JJ, Gehan EA (1971) Prognostic factors in localized Hodgkin's disease treated with regional radiation. Radiology 98: 641-654

Hoppe RT, Cox RS, Rosenberg SA, Kaplan HS (1982) Prognostic factors in pathologic stage III Hodgkin's disease. Cancer Treat Rep 4: 743-756

Horning SJ, Hoppe RT, Rosenberg SA (1984) The Stanford Hodgkin's disease trials 1968-1984: results of combined modality therapy in PS I, II and III. In: Salmon SE, Jones SE (eds) Adjuvant therapy of cancer IV. Grune and Stratton, New York, pp 633-642

Kun LI, DeVita VT, Young RC, Johnson RE (1976) Treatment of Hodgkin's disease using intensive chemotherapy followed by irradiation. Int J Radiat Oncol Biol Phys 1: 619-626

Lee CKK, Bloomfield CD, Levitt SH (1984) Liver irradiation in stage IIIA Hodgkin's disease with splenic involvement. Am J Clin Oncol 7: 149-157

Lister TA, Dorreen MS, Faux M, Jones AE, Wrigley PFM (1983) The treatment of stage IIIA Hodgkin's disease. J Clin Oncol 1: 745-749

Mauch, PM, Goffman T, Rosenthal DS, Canellos GP, Come SE, Hellman S (1985) Stage III Hodgkin's disease: improved survival with combined radiation therapy as compared with radiation therapy alone. J Clin Oncol 3: 1166-1173

Mazza P, Miniaci G, Lauria F et al. (1984) Prognostic significance of lymphography in stage III_s Hodgkin's disease (HD). Eur J Cancer Clin Oncol 20: 1393-1399

Micaily B, Brady L (1984) Hodgkin's disease: results from a program in radiation therapy. Am J Clin Oncol 7: 159-171

Prosnitz LR, Farber LR, Kapp DS, Bertino JR, Nordlund M, Lawrence R (1982) Combined modality therapy for advanced Hodgkin's disease: long term follow-up data. Cancer Treat Rep 66: 871-880

Rodgers RW, Fuller LM, Hagemeister FB et al. (1981) Reassessment of prognostic factors in stage IIIA and IIIB Hodgkin's disease treated with MOPP and radiotherapy. Cancer 47: 2196-2203

Rosenberg SA, Kaplan HS (1975) The management of stages I, II and III Hodgkin's disease with combined radiotherapy and chemotherapy. Cancer 35: 55-63

Stein RS, Golomb HM, Wiernik PH et al. (1982) Anatomic substages of stage IIIA Hodgkin's disease: follow-up of a collaborative study. Cancer Treat Rep 66: 733-742

Viviani S, Santoro A, Ragni G et al. (1985) Gonadal toxicity after combination chemotherapy for Hodgkin's disease. Comparative results of MOPP vs ABVD. Eur J Cancer Clin Oncol 21: 601-605

Young R, Canellos G, Chabner B, Hubbard S, DeVita V Jr (1978) Patterns of relapse in advanced Hodgkin's disease treated with combination chemotherapy. Cancer 42: 1001-1007

*Risk Factor Adapted Treatment of Hodgkin's Lymphoma in Childhood: Strategies and Results of Three Consecutive Multicenter Studies in the Federal Republic of Germany**

G. Schellong, I. Hörnig, E. W. Schwarze, and M. Wannenmacher

Abteilung für Hämatologie und Onkologie, Universitäts-Kinderklinik,
Albert-Schweitzer-Straße 33, 4400 Münster, FRG

Introduction

In the Federal Republic of Germany three consecutive multicenter studies on childhood Hodgkin's disease have been undertaken since 1978 (Brämswig et al. 1987; Breu et al. 1982; Schellong et al. 1985, 1986a, b, c). Up to 1987 more than 500 children under the age of 16 years have been treated in these studies at more than 60 centers, including some hospitals in Austria and The Netherlands.

The general objective of these studies was to minimize step-by-step radiotherapy and chemotherapy, as well as invasive staging procedures, in the context of a combined modality treatment concept. One particular aim was to reduce radiotherapy to the involved fields, using intermediate- and low-radiation doses, with the rationale that appropriate chemotherapy might suffice to eradicate occult microfoci in adjacent lymphatic areas. Concomitantly, the extent of chemotherapy and the exposure to alkylating agents were limited, depending on the stage of disease. Another purpose was to reappraise the need and indication for splenectomy and laparotomy.

Table 1 presents an overview of the staging methods and the therapeutic procedures in the three consecutive studies. The first study, HD-78, introduced a maximum program (Breu et al. 1982; Schellong et al. 1985), which was then reduced in the subsequent studies on the basis of the results obtained. At that time it was decided not to use lymphography as a routine measure because uniform evaluation in children is very difficult, if at all possible, in a multicenter study with so many hospitals participating. Consequently, all patients in the first study, with few exceptions, were submitted to laparotomy and splenectomy (Breu et al. 1982; Schellong et al. 1985). Detailed analyses of the staging findings obtained in the first study enabled splenectomy to be restricted to selected cases in the second study by the use of an intraoperative decisional strategy (Schellong et al. 1986a). All patients in the second study underwent laparotomy, but only 40% were splenectomized (Schellong et al. 1986b). In an analogous manner, statistical evaluations of the

* Supported by the Bundesminister für Forschung und Technologie of the Federal Republic of Germany.

Table 1. Three DAL studies in childhood Hodgkin's disease

	HD-78	HD-82	HD-85
Lymphography	-	-	-
Ultrasonography/CT	-	Most patients	All patients
Laparotomy	All patients	All patients	Selectively
Splenectomy	All patients	Selectively	Selectively
Chemotherapy	OPPA/COPP	OPPA/COPP	OPA/COMP
I/IIA	Two cycles	Two cycles	Two cycles
IIB/IIIA	Six cycles	Four cycles	Four cycles
IIIB/IV	Six cycles	Six cycles	Six cycles
Radiotherapy: field	Extended	Involved	Involved
dose	IF 36-40 Gy AF 36-40 Gy vs. AF 18-20 Gy	35, 30, 25 Gy	35, 30, 25 Gy

IF, involved fields; AF, adjacent fields.

ultrasonography and CT scan findings of the second study enabled us to develop a method allowing restriction of the indication for laparotomy in the third study (Schellong et al. 1986c).

All patients received a risk-adapted chemotherapy. In stages I and IIA only two cycles were given. In the first study all other patients were treated by six cycles (Breu et al. 1982; Schellong et al. 1985). Starting in the second study, three risk groups were formed, the middle group, which represented stages IIB and IIIA, receiving four cycles (Schellong et al. 1985, 1986b).

We avoided using nitrogen mustard because this drug is considered to have the strongest gonadotoxic activity and the highest cancerogenic effect in the MOPP combination. Nitrogen mustard was replaced by adriamycin (resulting in OPPA) or cyclophosphamide (resulting in COPP). In studies HD-78 and HD-82 the first two cycles consisted in OPPA (Table 2), the remaining two or four in COPP (Table 3). In study HD-85 it was attempted to eliminate procarbazine, after a gonadotoxic effect for this drug had also been found (Brämswig et al. 1987). In OPA procarbazine was eliminated without any replacement; in COMP it was replaced by methotrexate. Radiation treatment was still given as extended-field irradiation in the first study (Breu et al. 1982; Schellong et al. 1985). The involved fields received a dose of 36-40 Gy, the adjacent fields, in a randomized fashion, the same dose or 18-20 Gy. The results of this dose reduction in the adjacent regions were very favorable. Consequently, irradiation was then limited to the involved fields in the subsequent studies HD-82 and HD-85 (Schellong et al. 1985, 1986b). The radiation dose was also reduced, depending on the amount of chemotherapy given. After two, four or six cycles, respectively, 35, 30, or 25 Gy was used, i.e., the more chemotherapy the lower the radiation dose.

The following report is restricted to studies HD-82 and HD-85. However, one important result of the first study HD-78 should be stated: the high effectiveness of the OPPA combination (Table 2) as induction therapy for all stages of Hodgkin's disease in children could be documented (Breu et al. 1982; Schellong et al. 1985). The complete remission rate in 156 patients after two OPPA cycles was 71%.

Table 2. Application and dosage of drugs in one OPPA cycle

	Day	Dose
Adriamycin i.v.	1, 15	40 mg/m²/dose
Vincristine i.v.	1, 8, 15	1.5 mg/m²/dose (maximum single dose, 2.0 mg)
Procarbazine p.o.	1-15	100 mg/m²/day (maximum dose/day, 150 mg)
Prednisone p.o.	1-15	60 mg/m²/day

Table 3. Application and dosage of drugs in one COPP cycle

	Day	Dose
Cyclophosphamide i.v.	1, 8	500 mg/m²/dose
Vincristine i.v.	1, 8	1.5 mg/m²/dose (maximum single dose, 2.0 mg)
Procarbazine p.o.	1-14	100 mg/m²/day (maximum dose/day, 150 mg)
Prednisone p.o. (only second and fourth cycle)	1-14	40 mg/m²/day

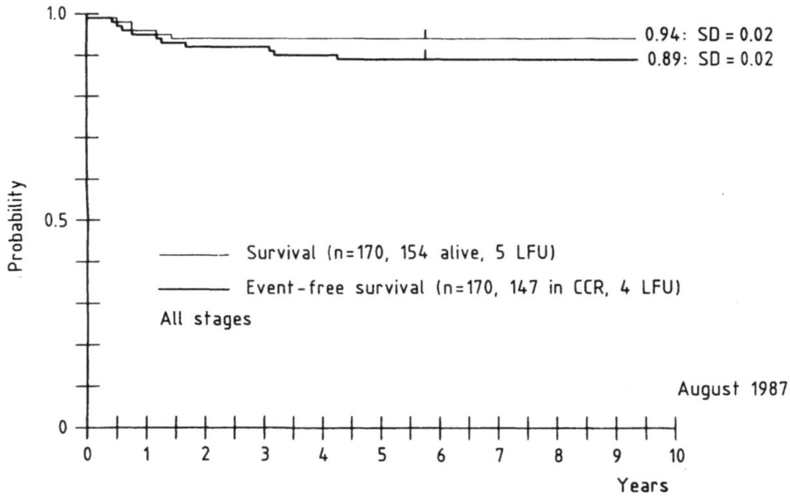

Fig. 1. Survival and event-free survival of 170 patients in study HD-78. The *vertical bars* represent the last patient of the study

One other fact of study HD-78 might be worth mentioning: not a single secondary malignancy has been observed so far in the 154 surviving patients of this study. The follow-up time of all patients in this study has reached the period of between 5 and 10 years which is critical for the appearance of acute nonlymphocytic leukemia (ANLL) (Fig. 1).

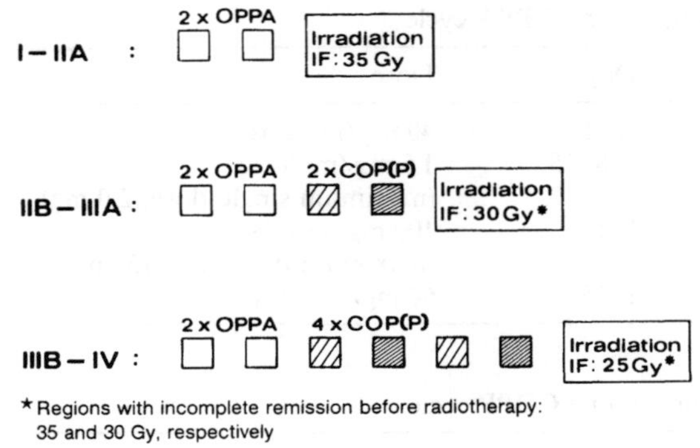

Fig. 2. Therapy protocol of study HD-82

DAL Study HD-82

Treatment Protocol (Fig. 2). Patients were classified into three groups according to pathological stages (I/IIA vs. IIB/IIIA vs. IIIB/IV), receiving two, four, or six cycles of chemotherapy, respectively (Schellong et al. 1985; 1986b). In the lowest risk group, chemotherapy consisted only of two OPPA cycles. Patients with more advanced stages also received two courses of OPPA (Table 2), but subsequently were also administered two or four cycles of COPP (Table 3). The following radiotherapy was given to the involved fields only. The dose was 35, 30, or 25 Gy, depending in the number of preceding courses of chemotherapy. In Patients of the medium- and high-risk group fields with residual tumor after completion of chemotherapy received a dose increased by 5 Gy, i.e., 35 or 30 Gy, respectively. Extralymphatic organs involved, e.g., lungs or liver, were also irradiated, receiving 12–15 Gy.

Diagnostic Procedures. Staging laparotomy was mandatory, whereas splenectomy was performed only selectively by use of an intraoperative decisional strategy (Fig. 3) developed on the basis of the statistical analyses in the first study. Children without changes in the splenic surface and without enlargement of the lymph nodes at the splenic hilus/pancreatic tail were not splenectomized. Lymphangiography was performed only in a few patients, whereas abdominal sonography and/or CT scan were applied in about two-thirds of them.

Results. The projected 5-year survival rate is 97% for the entire group of 203 protocol patients treated in this study; the event-free survival rate is 96% (Fig. 4). All 203 patients had achieved complete remission. In the further course, three children have died of intercurrent disease (one of varicella, two of sepsis), while five children have relapsed. The projected event-free survival rates for the three treatment groups after 5 years are 99%, 96%, and 90% (Fig. 5).

According to the strategy of selective splenectomy used (Fig. 3), the percentage of nonsplenectomized patients differs among the three risk groups: 83% (stages I/IIA), 53% (stages IIB/IIIA), and 28% (stages IIIB/IV) (Schellong et al. 1986b).

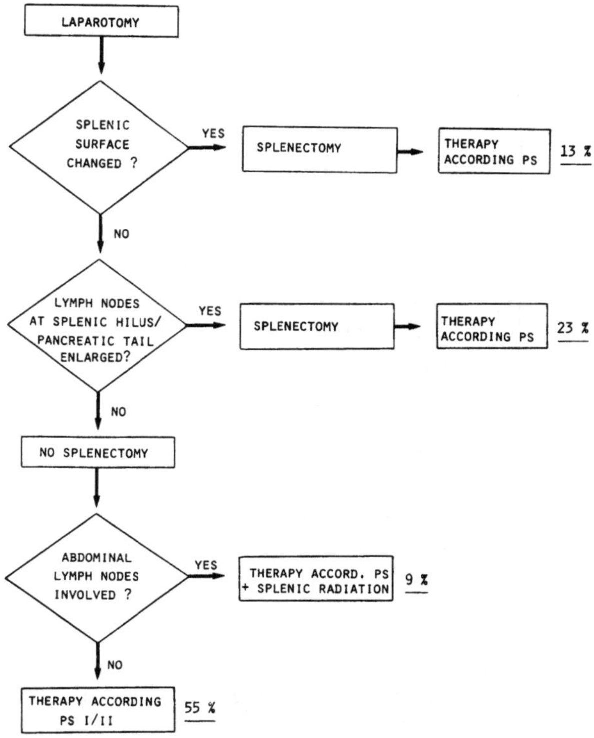

Fig. 3. Intraoperative decisional strategy for selective laparotomy derived from the statistical analysis in study HD-78

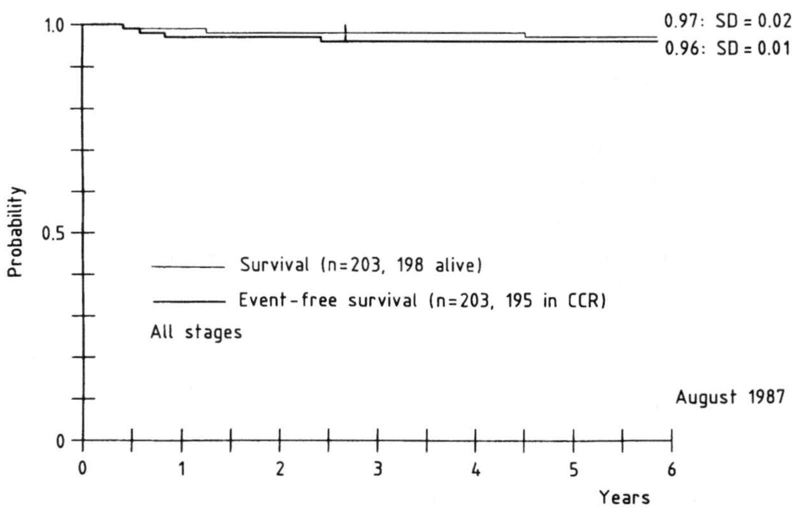

Fig. 4. Survival and event-free survival of 203 patients in study HD-82

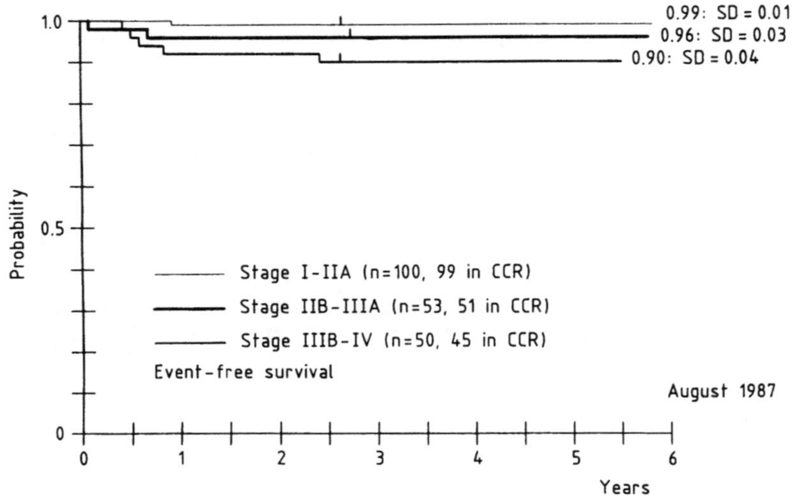

Fig. 5. Event-free survival in the three treatment groups of study HD-82. The *vertical bars* represent the last patient of the respective group

The following conclusions can be drawn from these results:

- Stage-dependent chemotherapy with two, four, or six cycles of OPPA/COPP is highly effective in eradicating occult microfoci, so that only involved-field irradiation is needed and splenic involvement recognizable during laparotomy may remain undetected, because smaller foci in the spleen are eradicated by two cycles of OPPA even without irradiation of this organ.
- There is a high probability of preventing local recurrence by combining radiation doses of 35, 30, or 25 Gy with the applied chemotherapy.
- The treatment results justify cautious attempts at further treatment reductions.

DAL Study HD-85

In the third study an attempt was made to eliminate the procarbazine from the chemotherapy. This decision rested on three considerations:

1. Procarbazine seems to cause long-term testicular damage in about 25%, 40%, or 60% of the boys treated, depending on the number of chemotherapy courses (Brämswig et al. 1987).
2. It was supposed that procarbazine has a cancerogenic effect and especially may induce ANLL (together with mustargen, in the MOPP combination).
3. The contribution of procarbazine to the effectiveness of chemotherapy combinations such as MOPP, COPP, OPPA, and others remained unclear.

Treatment Protocol. The overall treatment plan (Fig. 6) remained the same as in the preceding study HD-82, i.e., stratification in three risk groups, with two, four, or six cycles of chemotherapy, and involved-field irradiation with 35, 30, or 25 Gy, re-

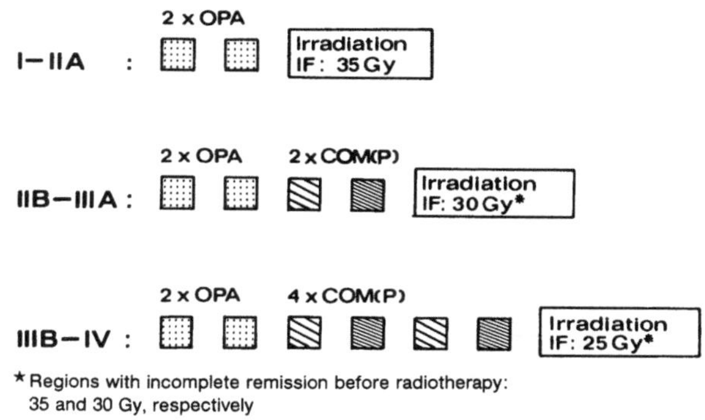

Fig. 6. Therapy protocol of study HD-85

spectively. Procarbazine was eliminated from OPPA, resulting in OPA, and replaced by methotrexate (40 mg/m^2 on days 1 and 8) in COPP, resulting in COMP (Schellong et al. 1988).

Diagnostic Procedures. Laparotomy was performed only selectively using two criteria derived from the statistical analyses of the data in the second study, namely pathological abdominal findings in SG/CT and/or enlargement of lymph nodes at the pulmonary hili (Schellong et al. 1986c). Children without these findings were not laparotomized. In patients with surgical staging, selective indication for splenectomy according to our decisional model (Fig. 3) was applied.

Results. Thirty-nine (40%) of the 98 protocol patients entered in this study between January 1985 and November 1986 were not laparotomized, and 67 (68%) were not splenectomized. The respective percentages in the three treatment groups are as follows: 66% and 94% in 53 patients with stages I/IIA, 14% and 67% in 21 patients with IIB/IIIA (including three patients with I_EA/II_EA), and 4% and 13% in 24 patients with IIIB/IV. The proportion of patients with progression and relapse was significantly higher than in the preceding study (Schellong et al. 1988). After 2 years 17 treatment failures out of a total of 98 patients had been reported to the study center. While we were still able to achieve remission in all of these cases by means of salvage therapy, so that no patient has died so far and the survival rate after 2 years is 100%, the projected rate of event-free survival is 74% (Fig. 7).

Differentiation of the three treatment groups shows that stages I and IIA so far do not show a deterioration of prognosis (Fig. 8). But the event-free survival curves for patients with medium and high risk show a rapid drop to 40%–60%. These two therapy branches were stopped in November 1986, because the results were considered not acceptable.

From this study it can be concluded that procarbazine is an important drug in the OPPA/COPP chemotherapy used in the preceding two studies. Methotrexate in the applied dosis is not suitable to replace procarbazine in this combination in the treatment of advanced stages. It is evident that a critical limit in the treatment reductions in our general therapy concept was crossed.

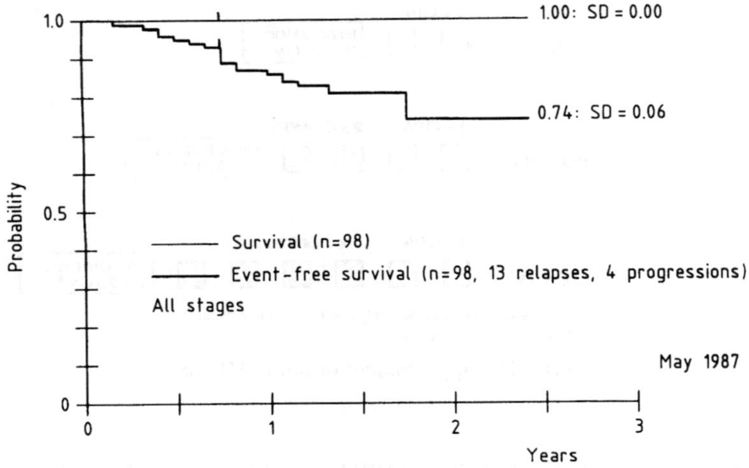

Fig. 7. Survival and event-free survival of 98 patients in study HD-85

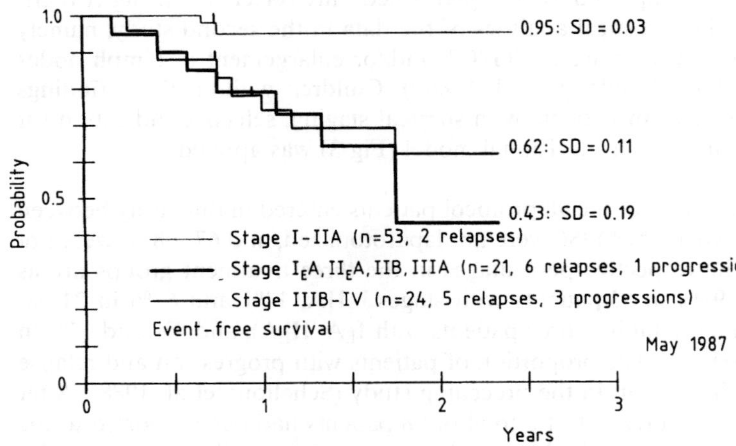

Fig. 8. Event-free survival in the three treatment groups of study HD-85. The *vertical bars* represent the last patient of the respective group

If we want to keep our basic chemotherapy concept which was so successfully applied in the first two studies, we have to look for a drug which, at specified doses, shows the same effectiveness as procarbazine but causes a lower rate of long-term sequelae.

References

Brämswig JH, Heermann E, Heimes U, Schlegel W, Nieschlag E, Schellong G (1987) Testicular function in boys treated for Hodgkin's disease: results of a longitudinal and cross-sectional study in 52 postpubertal boys (abstr). Acta Endocrinol (Copenh) 114: 20

Breu H, Schellong G, Grosch-Wörner I, Jobke A, Riehm H, Treuner J. Schwarze EW, Wannenmacher M (1982) Abgestufte Chemotherapie und reduzierte Strahlendosis beim Morbus Hodgkin im Kindesalter - Ein Bericht über 170 Patienten der kooperativen Therapiestudie HD-78. Klin Pädiatr 194: 233-241

Schellong G, Strauch ST, Waubke AK, Brandeis W, Pawlowsky J, Schwarze EW, Riehm H, Langermann HJ, Wannenmacher M (1985) Combined modality treatment with reduced chemotherapy and radiotherapy, and selective splenectomy, in children with Hodgkin's disease. In: Cavalli F, Bonnadonna G, Rozencweig M (eds) Malignant lymphomas and Hodgkin's disease. Experimental and therapeutic advances. Martinus Nijhoff, The Hague, pp 617-626

Schellong G, Waubke-Landwehr AK, Langermann HJ, Riehm HJ, Brämswig J, Ritter J (1986a) Prediction of splenic involvement in children with Hodgkin's disease. Significance of clinical and intraoperative findings. A retrospective statistical analysis of 154 patients in the German therapy study DAL-HD-78. Cancer 57: 2049-2056

Schellong G, Brämswig J, Ludwig R, Gerein V, Jobke A, Jürgens H, Kabisch H, Stollmann B, Weinel P, Gadner H, Schwarze EW, Wannenmacher M (1986b) Kombinierte Behandlungsstrategie bei über 200 Kindern mit Morbus Hodgkin: Abgestufte Chemotherapie, Involved-Field-Bestrahlung mit erniedrigten Dosen und selektiver Splenektomie. Ein Bericht der kooperativen Therapiestudie DAL-HD-82. Klin Pädiatr 198: 137-146

Schellong G, Lietzke S, Strauch ST, Kuhne B, Schneider B (1986c) Bedeutung sonographischer, computertomographischer und klinischer Befunde für die Erkennung eines Abdominalbefalls beim Morbus Hodgkin im Kindesalter - Eine retrospektive statistische Analyse bei 145 Patienten der Therapiestudie DAL-HD-82. Klin Pädiatr 198: 147-154

Schellong G, Hörnig I, Brämswig J, Bökkerink JPM, Steinhoff A, Ludwig R, Niethammer D, Reiter A, von Lengerke HJ, Heinecke H, Schwarze EW, Pötter R, Müller RP, Wannenmacher M (1988) Zur Bedeutung des Procarbazins in der Chemotherapie des Morbus Hodgkin - Ein Bericht der kooperativen Therapiestudie DAL-HD-85. Klin Pädiatr 200: 205-213

Hodgkin's Disease in Children: Adaptation of Treatment to Risk Factors

J. Lemerle[1], O. Oberlin[1], G. Schaison[2], G. Leverger[2], D. Olive[3], M. Bonnet[4], D. Dufillot[5], and L. de Lumley[6]

[1] Service de Pédiatrie, Institut Gustave Roussy, Rue Camille Desmoulins, 94800 Villejuif, France
[2] Hôpital Saint-Louis, Paris, France
[3] CHU, Nancy, France
[4] Hôpital Trousseau, Paris, France
[5] CHU, Bordeaux, France
[6] CHU, Limoges, France

Hodgkin's disease in children is very similar to the disease observed in adults, except for its incidence, which is approximately ten times less. Therefore it has long been considered that the considerable experience collected throughout the world in treating adults could be used in treating children, and for a long time the treatment strategies for children and adults were similar. But as time passed and the number of children surviving the disease long-term grew, it became obvious that sequelae of therapy, and especially of radiotherapy, were much more important in children and deserved special consideration. Pediatric oncologists then diverged from adult oncologists and developed treatment strategies adapted to the specific problem of children, for whom the expected sequelae of a given treatment are now to be included in the risk factors.

Risk Factors in Children: First Attempts To Take Them Into Account

Most of the presently known risk factors in Hodgkin's disease have been recognized when it was treated by radiotherapy alone, or associated with minimal single-drug chemotherapy. Stage, pathology, and general symptoms soon significantly emerged in children as well as in adults.

Therapy, as it improved, was adjusted to these factors and the information derived from the analysis of relapses and from staging laparotomies led to the concept of extended-field radiotherapy. Cure rates over 80% were obtained by radiotherapy alone, in children as well as in adults (Jenkin and Berry 1980).

Then multidrug chemotherapy, essentially the MOPP combination, was added to radiotherapy; recurrences became less frequent and cure rates approached 90% (Sullivan et al. 1982; Bayle-Weisgerber et al. 1984).

Complications and Sequelae of Combined Treatment in Children

Staging laparotomy, when performed in such a way that the maximum possible information is collected from most of the abdominal nodes, the liver, and the spleen, is a major surgical procedure which was associated with morbidity when it was

Recent Results in Cancer Research. Vol. 117

performed systematically. In a review of seven pediatric series (Jenkin and Berry 1980), small bowel obstruction was reported in 1%-10% of the cases, with no related death. Severe postsplenectomy infections were seen in 2%-10% of the cases. In the early 1970s (Chilicotte et al. 1976), 8 infections out of 18 were reported fatal in a population of 200 splenectomized children. More favorable reports were made later, especially in series reporting the experience of a single institution (Donaldson and Kaplan 1982; Donaldson and Link 1987).

Much more frequent, and in fact constant, is the growth impairment of soft tissues and bones which is caused by radiotherapy applied to young children (Donaldson and Kaplan 1982). These so-called cosmetic damages consist of a thin neck, short clavicles, and spinal and chest deformities, which severely impair the quality of life of these children when they have grown into young adults. It should be kept in mind that, like other growth defects related to radiotherapy, these can be fully assessed only at the end of growth and at least at the end of the pubertal period. The severity of these sequelae is directly related to the age at treatment, to the dose given, more generally to the radiotherapy technique, and also to the time elapsed between treatment and evaluation. Cardiac complications are not exceptional when 40 Gy is delivered to the mediastinum, essentially pericardial effusion which can occur several years after treatment and may result in acute heart compression. Lungs as well cannot be completely spared when mantle fields are used, especially in very young children.

The gonads, both of boys and girls, may be severely affected both by radiotherapy, mostly in girls receiving inverted Y radiotherapy, and by chemotherapy essentially in boys. Male sterility after MOPP (mustine, oncovin, procarbazine, and prednisone) is invariable after six courses (Aubier et al. 1986) and frequent after fewer courses. It is permanent in our experience, and not affected by the age at the time of treatment. Others have observed a similar effect of MOPP (Shalet 1980) and of other combinations where alkylating agents are given. A dose-effect relationship is clear in most reports (Da Cunha et al. 1984). Females are less affected by chemotherapy. In our experience all males have developed normal puberty after MOPP, even if totally azoospermic.

Second malignant neoplasms occurring in patients with cured Hodgkin's disease have become a major concern (Meadows et al. 1985; Tucker et al. 1984). In a group of 958 children treated for Hodgkin's disease and reviewed by the Late Effects Study Group, 16 developed solid tumors, which were all in an irradiated area, with a median interval of 10 years; 14 developed leukemias and all had received combined chemotherapy with procarbazine 1-10 years previously; 3 developed lymphomas. In this study, the calculated risk of a second tumor in patients with cured Hodgkin's disease is 18.5% at 20 years from treatment. The leukemia risk seems to be smaller, or even absent, when the ABVD (doxorubicin, bleomycin, vinblastine, dacarbazine) combination is used (Valagussa et al. 1986).

First Steps in Decreasing Therapy in Children

Staging laparotomy with splenectomy was first eliminated on the basis of data showing the effectiveness of chemotherapy in eradicating radiologically unapparent disease (Andrieu et al. 1981). Our data with no infradiaphragmatic radiotherapy given to 37 clinically staged (normal lymphangiogram) IA-IIA patients treated with MOPP followed by involved-field radiotherapy confirm these views since no infradiaphragmatic recurrence was observed in this group of patients (Oberlin et al. 1985). Most of the pediatric oncology groups now omit staging laparotomy.

Six to Three MOPP Courses. We reduced the number of MOPP courses, on consideration of data indicating that only a very small benefit could be expected of the last three courses when six cycles of MOPP were given (Fermé et al. 1985). Our own data (Oberlin et al. 1985) confirm that this reduction of chemotherapy has no adverse effect on outcome.

Reduction of Radiotherapy Fields and Doses. When chemotherapy is very active, radiotherapy reduction can be considered. But only in this case. This has to be clearly emphasized since, in most of the first attempts which were made to reduce the burden of radiotherapy given to children in view of reducing its sequelae, this advantage was balanced by the toxicity of the chemotherapy used, namely six courses of MOPP or MOPP equivalents.

In a study performed in 1975-1980, however, we gave three MOPP courses followed by 40 Gy in initially involved fields (IFs) in 20 stage I and IIA patients, and 6 Mopp + IF 40 Gy in 40 patients. Both groups had an equally good actuarial survival and disease-free survival, 93% and 86% respectively at 5 years (Oberlin et al. 1985).

Lower doses of radiotherapy have been used by others (Jenkin et al. 1982; Donaldson and Kaplan 1982; Donaldson and Link 1987) with good results. But these patients received six MOPP courses, and in Stanford they were surgically staged. In other centers, dose reductions were successfully tried, but again at the cost of aggressive chemotherapy containing high doses of alkylating agents.

The next problem then was clearly to reduce both radiotherapy in fields and doses and also chemotherapy in terms of aggressivity to the gonads and of potential carcinogenicity.

Current French National Study

In 1982 a study was started in France aiming at providing an answer to several pending questions related to the reduction of therapy in good cases and to the improvement of its efficacy in less favorable cases.

Staging Procedure

The staging procedure is uniform in all cases. The bipedal lymphangiogram is mandatory and has been performed in 95% of cases. No staging laparotomy is performed. A CT scan of the chest is optional, and is now performed in the majority of cases before and after treatment. Abdominal CT scan and echography are recommended in the rare cases with a negative lymphangiogram and clinically enlarged spleen. Bone biopsy is performed in cases with general symptoms and in stage IV.

The Ann Arbor clinical staging is used in this study. An erythrocyte sedimentation rate (ESR) over 70, even if clinical general symptoms are not found, is considered a reason to treat the patients like those with symptoms.

Aims and Design of the Study

As far as chemotherapy is concerned, the questions to be answered are the following:

1. In favorable cases (IA-IIA stages) is the ABVD combination alone as effective as alternating courses of MOPP and ABVD. The equivalence of both regimens has been suggested by clinical data but not yet demonstrated in a randomized way in children with low-stage disease treated with low-dose radiation. In addition we elected to use only four courses in these cases in order to minimize as much as possible the adverse effects of both regimens. In the case of equal efficacy, ABVD will be adopted for further studies since it contains neither mustard nor procarbazine, which are the most toxic agents in MOPP. The Adriamycin and bleomycin contained in ABVD are toxic respectively to heart and lungs but are given at low doses. In the case of isolated high cervical presentation, chemotherapy is not randomized and four ABVD cycles are given to all patients.
2. In less favorable cases, IB, IIB, III, and IV, the same chemotherapy is used in all cases, consisting of three courses of MOPP alternating with three courses of ABVD since this combination has been reported as being the most effective. An attempt is made to find out in a nonrandomized way whether it can achieve a high proportion of remissions in advanced stages and whether it can be combined with low-dose radiation to obtain a satisfactory cure rate.

For the radiotherapy, the French study aims at finding out whether, for patients achieving a good remission with the above-mentioned chemotherapy regimens, the dose of 20 Gy can eradicate the disease. "Good remission" is defined as either complete clinical and/or radiological disappearance of all tumor (CR) or as a tumor volume reduction greater than 70%. When a good remission is not attained the dose of 40 Gy is used. The extent of the fields is adapted to the initial stage of the disease. In stages IA - IIA, fields limited to the initially involved areas are used. In IB, IIB, III, IV, the paraaortic nodes and the spleen are also irradiated. "Adjacent areas" are irradiated in no case, since it is assumed that chemotherapy eradicates infraclinical disease. This protocol of treatment is considered as bearing a limited risk of producing important sequelae, according to our previous experience with similar doses of chemotherapy and radiotherapy given to patients with other malignancies but in the same age group. Children in the study, however, will be very carefully followed and assessed for late effects of therapy. The results of the study, apart from the randomized chemotherapy, will be evaluated in a historical way by comparison with our previous results and other published data, in terms of remission induction rate, survival, and disease-free survival.

Figure 1 summarizes the study protocol.

Clinical stage		Primary chemotherapy	Radiotherapy
		↗ 4 ABVD	20 Gy* in involved fields
IA - IIA	Random		
		↘ 2 MOPP+2 ABVD	
IB - IIB - III - IV	3 MOPP+3 ABVD		20 Gy* in involved fields +lumbosplenic fields
			*40 Gy if reduction < 70%

Fig. 1. French study protocol

Preliminary Results

From 1982 to May 1987, 174 patients from 29 centers in France have been included in the study; 157 have completed their treatment and may be evaluated. The mean age at inclusion was 10.8 years, ranging from 2 to 18 years. There were 6 patients with clinical stage (CS) IA limited to the upper cervical area; 84 CS IA and IIA; 18 CS IB - IIB; 29 CS IIIA and B; and 20 CS IV. At completion of primary chemotherapy, 147/157 patients were considered good responders (94%), 110 being in CR and 37 showing more than 70% regression. Ten cases were considered as poor responders (6%), 4 having regression smaller than 70% and 6 cases (all IVB) having had very early recurrences of the disease, before the radiotherapy started.

The radiotherapy was given according to the protocol. Among the 37 good responders in partial remission, 6 received localized boosts to 40 Gy (5/6 on the mediastinum). Of the 10 poor responders, 5 achieved complete remission with 40 Gy, 4 died with progressive disease, and 1 is still on therapy.

There were 5 relapses in the 147 good responders including 3 in an area treated with 20 Gy, which represents less than 1% of the initially involved nodal areas. Two of these patients have died.

Of the ten poor responders, all treated with 40 Gy, five have died (four of progressive disease and one of relapse) and five are alive, three in first CR and two after relapse.

The median follow-up of the population is 30 months, the actuarial overall survival is 95% at 3 years, and the disease-free survival is 88% (Fig. 2). But Fig. 3 showing the disease-free survival by stage indicates a clear-cut difference between CSI and III (93%-94%) and CSIV, 54%.

At this point no difference can be seen between the two randomized chemotherapy arms in stages IA - IIA.

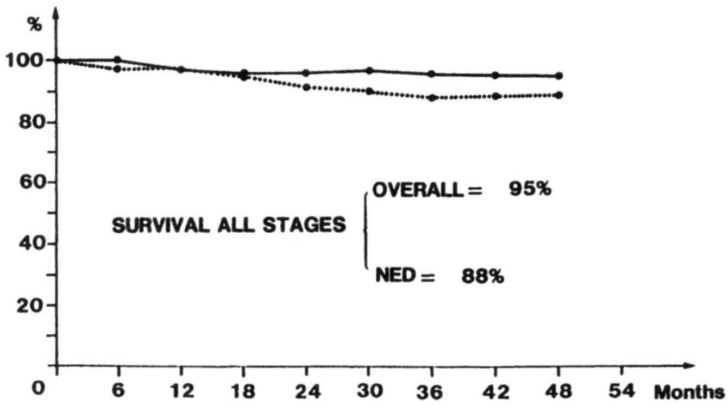

Fig. 2. Actuarial survival, all stages. Overall survival and disease-free *(NED)* survival

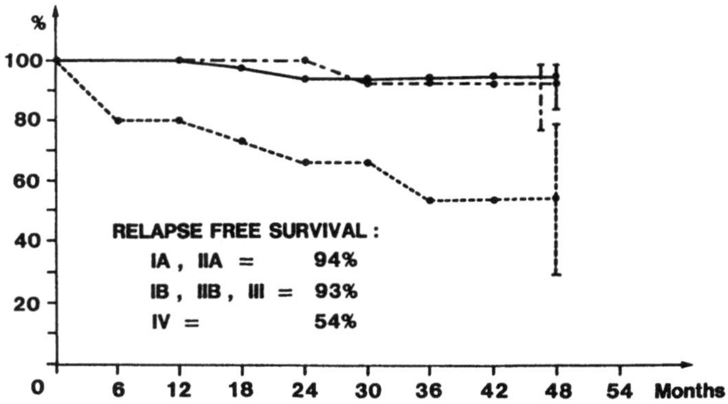

Fig. 3. Relapse-free survival by stage (actuarial)

Prognostic Factors

As expected in a study with a risk-adapted treatment, few prognostic factors could be identified despite the considerable decrease in therapy which was achieved in a comparatively large number of patients.

Age, histology, and size of the mediastinum predicted neither response to chemotherapy nor final outcome. Finally the two only prognostic indicators which emerge are the following. Stage IV is related both to poor response to chemotherapy since only 35% went into CR compared with 88% and 73% in other stages, and to poor disease-free survival. Within stage IV cases, however, it has not been shown that the site of extranodal involvement influenced the response to chemotherapy.

Response to primary chemotherapy also appeared in this study as a very strong prognostic indicator, in terms of survival and disease-free survival. Two out of 147 good responders have died so far, and 5 out of 10 poor responders.

Finally, the late effects of therapy which can be expected in our treated population are limited, since we have been able to avoid giving more than three MOPP courses to any of our patients, and only ten have received 40 Gy, and six localized boosts to 40 Gy.

Discussion

Tumor-related risk factors in Hodgkin's disease of children are likely to be the same as in adults. In addition, treatment-related risk factors are of paramount importance in children, and have to be taken into consideration when deciding on treatment plans. In children more strongly than in adults, the quality of life of cured patients has to be matched with the percentage of patients reported as "cured" or as long-term survivors.

With treatment combining chemotherapy and radiotherapy, the only prognostic indicators which can still be identified are clinical stage IV and poor response to primary chemotherapy. We therefore have been obliged to base our current studies on some of the previously known factors, stage, and general symptoms. Decreasing therapy to decrease sequelae in children with Hodgkin's disease is a priority, but has to be achieved very carefully not to jeopardize survival. Several studies are in progress in the world with the same aim (Behrendt et al. 1987; Vecchi et al. 1986; Fossati-Bellani et al. 1985; Schellong et al. 1985; Sackman Muriel et al. 1985). We report the preliminary results of a French National Study in which staging procedure, radiotherapy, and chemotherapy are balanced in a way which seems to be optimal when considering the available data on toxicity and efficacy of the proposed treatment. Twenty-gray IF and four-course ABVD in clinically staged favorable cases is probably the least aggressive and the most effective therapy which is proposed at present. A next step may be to find a less toxic chemotherapy for these cases. But we do not consider it is justified to give a toxic chemotherapy to cut further or even suppress completely radiotherapy.

In less favorable cases, our attempts completely to eliminate alkylating agents are to be continued, keeping the same radiotherapy scheme.

In stage IV and in poor responders, finding a more aggressive and more effective therapy is our goal, while avoiding late effects is not a priority in these patients with a life-threatening disease.

In view of the small numbers of cases, especially of stage IV patients, a European cooperative study is being prepared by the International Society of Pediatric Oncology (SIOP) in order to obtain answers to the pending questions on the treatment of childhood's Hodgkin's disease as rapidly as possible.

References

Andrieu JM, Asselain B, Bayle C et al. (1981) La séquence polychimiothérapie MOPP-iradiation sélective dans le traitement de la maladie de Hodgkin, stades cliniques IA - IIB. Bull Cancer (Paris) 68: 190-199

Aubier F, Flamant F, Caillaud JM, Lemerle J, Chaussain JL (1986) Male fertility after chemotherapy for solid tumors in children. Proceedings of the 18th International Society of Pediatric Oncology meeting, Belgrade

Bayle-Weisgerber C, Lemercier N, Teillet F, Asselain B, Gout M, Schweisguth O (1984) Hodgkin's disease in children, results of therapy in a mixed group of 178 clinically and pathologically staged patients over 13 years. Cancer 54: 215-222

Behrendt H, van Bunningen BNFM, van Leeuwen EF (1987) Treatment of Hodgkin's disease in children with or without radiotherapy. Cancer 59: 1870-1873

Chilicotte AR, Baehner AL, Hammond D (1976) Septicemia and meningitis in children splenectomized for Hodgkin's disease. N Engl J Med 295: 798-800

Da Cunha MF, Meistrich ML, Fuller LM (1984) Recovery of spermatogenesis after treatment for Hodgkin's disease: limiting dose of MOPP chemotherapy. J Clin Oncol 6: 571-577

Donaldson SS, Kaplan HS (1982) Complications of treatment of Hodgkin's disease in children. Cancer Treat Rep 66: 977-989

Donaldson SS, Link MP (1987) Combined modality treatment with low-dose radiation and MOPP chemotherapy for children with Hodgkin's disease. J Clin Oncol 5: 742-749

Fermé C, Teillet F, D'Agay MF, Boiron M (1985) Surgical restaging after 3 or 6 courses of MOPP chemotherapy in Hodgkin's disease. In: Cavalli F, Bonadonna G, Rozencweig M (eds) Malignant lymphomas and Hodgkin's disease: experimental and therapeutic advances. Nijhoff, Boston, pp 363-369

Fossati-Bellani F, Gasparini M, Kenda A et al. (1985) Limited field and low dose radiotherapy + ABVD chemotherapy for childhood Hodgkin's disease (abstr). Proceedings of the 17th International Society of Pediatric Oncology Meeting, Venice, pp 323-324

Jenkin RD, Berry MP (1980) Hodgkin's disease in children. Semin Oncol 7: 202-211

Jenkin RD, Chan H, Freedman M et al. (1982) Hodgkin's disease in children: treatment results with MOPP and low-dose, extended field irradiation. Cancer Treat Rep 66: 949-959

Meadows AT, Obringer AC, Landsberg P, Marrero O, Lemerle J for the Late Effects Study Group (1985) Risk of second malignant neoplasms in childhood Hodgkin's disease. Proc Am Assoc Cancer Res 26: 741

Oberlin O, Boilletot A, Leverger G, Sarrazin D, Schaison G, Lemerle J (1985) Clinical staging, primary chemotherapy and involved field radiotherapy in childhood Hodgkin's disease. Eur Paediatr Haematol Oncol 2: 65-70

Sackman Muriel F, Lobo Sanshuja F, Schwartzman E et al. (1985) Results in Hodgkin's disease in childhood: radio-chemotherapy versus chemotherapy alone (abstr). Proc Am Assoc Cancer Res 26: 184

Schellong G, Strauch ST, Waubke AK et al. (1985) Combined modality treatment with reduced chemotherapy and radiotherapy, and selective splenectomy in children with Hodgkin's disease. In: Cavalli F, Bonadonna G, Rozencweig M (eds) Malignant lymphomas and Hodgkin's disease. Experimental and therapeutic advances. Nijhoff, The Hague, pp 617-626

Shalet SM (1980) Effects of cancer chemotherapy on gonadal function. Cancer Treat Rev 7: 141-152

Sullivan MP, Fuller LM, Chen T et al. (1982) Intergroup Hodgkin's disease in children study of stages I and II: a preliminary report. Cancer Treat Rep 66: 937-947

Tucker MA, Meadows AT, Boice JD, Stovall M, Oberlin O, Hover RN, Fraumeni JF for the Late Effects Study Group (1984) Secondary leukemia after alkylating agents for childhood cancer. Proc Am Soc Clin Oncol 3: C333

Valagussa P, Santoro A, Fossati-Bellani F, Banfi A, Bonadonna G (1986) Second acute leukemia and other malignancies following treatment for Hodgkin's disease. J Clin Oncol 4: 830-837

Vecchi V, Brunelli R, de Bernardi B et al. (1986) Hodgkin's disease in childhood. First Italian multicentric study (AIEOP HD 83). Preliminary results of a chemoradiotherapeutic protocol (abstr). Proceedings of the 18th International Society of Pediatric Oncology meeting, Belgrade, pp 181-182

Role of Bone Marrow Transplantation in the Treatment of Hodgkin's Lymphoma

As becomes evident from the review on salvage therapy of Hodgkin's lymphoma, more efficient approaches are needed to improve the results in patients who do not achieve complete remission by the primary remission induction therapy, or relapse early after achieving a complete remission. In the following section we cover the most promising new approach: bone marrow transplantation. The results of bone marrow transplantation in Hodgkin's lmyphoma obtained by several groups are presented in the first part of this section. In the second part, we summarize problems and future directions as suggested by the participants of the workshop on the role of bone marrow transplantation in Hodgkin's lymphoma.

Conventional Salvage Therapy for Hodgkin's Lymphoma

M. Pfreundschuh, M. Schaadt, and V. Diehl

Medizinische Universitätsklinik, Joseph-Stelzmann-Straße 9, 5000 Köln 41, FRG

High remission rates are obtained in advanced Hodgkin's lymphoma utilizing combination chemotherapy such as MOPP (mustargen, vincristine, procarbazine, and prednisone) (DeVita et al. 1970; Longo et al. 1986) and ABVD (doxorubicin, bleomycin, vinblastine, and dacarbazine) (Bonadonna 1982), which is non-cross-resistant to MOPP, or a combination of the two (Bonadonna et al. 1986). About three-fourths of patients who do not present with very extensive disease or systemic symptoms generally have a long-term, disease-free survival; however, patients in advanced stages with certain risk factors, such as very extensive disease, B-symptoms, an excessively high erythrocyte sedimentation rate, or elevated alkaline phosphatase (Loeffler et al. 1988), have a cure rate of less than half when followed-up long enough (DeVita et al. 1980). Thus approximately one-third of all patients with Hodgkin's disease will eventually need salvage therapy.

Risk Factors in Patients Failing Primary Chemotherapy

While failure to obtain complete remission or relapse in Hodgkin's disease after combination chemotherapy is distressing, it is not always fatal and cure is still possible. This is especially true for a subgroup of patients whose residual or relapsing disease still has a relatively good prognosis *("low-risk failures")*, while there is some debate over the question whether patients not in CR or relapsing with disease with poor prognostic factors *("high-risk failures")* can be rendered free of disease for prolonged periods with conventional chemotherapy protocols that do not utilize bone marrow transplantation.

The following groups of patients are generally considered to have "low-risk failure": patients without systemic symptoms; those who do not obtain complete remission because of persistent disease limited to a single or few nodal sites; and those who relapse only in nodal sites after a prolonged unmaintained remission (> 12 months). On the other hand, treatment failures with opposite characteristics are considered to have a poor prognosis. In fact, some of the patients with solitary indolent nodal relapse may remain in stable disease for prolonged periods without retreatment (Bergsagel et al. 1987). In this respect a study should be recalled that

Recent Results in Cancer Research. Vol. 117

reported residual Hodgkin's disease in 16/18 autopsies of patients surviving beyond 10 years and dying from apparently unrelated causes (Strum et al. 1971).

Most patients with relapsing Hodgkin's lymphoma, however, show progressive disease requiring retreatment. For these patients, several options may be considered: radiotherapy, conventional chemotherapy, a combination of both, or bone marrow transplantation.

Salvage Therapy by Irradiation

Cure by wide-field radiotherapy is possible in patients who do not obtain complete remission with chemotherapy or relapse after combination chemotherapy if the "low-risk" cirteria are met, especially if the disease is limited to nodal sites which have not been irradiated previously. As salvage radiotherapy is usually well tolerated (Diehl et al. 1983; Mauch et al. 1987; Roach et al. 1987) and effective for this subgroup of patients, it is important to identify the respective patients. While usually considered to be rare, these patients may represent between 10% and 30% of all patients treated primarily with combination chemotherapy and up to 45% of all patients failing after such a therapy (Diehl et al. 1980; Fox et al. 1987).

Salvage Chemotherapy

There are data to suggest that patients with a long first remission may benefit from repeating their initial chemotherapy regimen. A 37% 5-year complete remission duration was reported for patients failing after MOPP chemotherapy when retreated with MOPP if their first remission had lasted >1 year (Fisher et al. 1979). However, it is doubtful whether a regimen that was unable to cure a patient when used as primary treatment can do so when reused for the treamtent of relapse, supposing the primary treatment has been given with strict adherence to the dosage and schedule of the protocol. It therefore seems to be preferable to use a drug combination for the treatment of relapse which has been proven to be non-cross-resistant with the combination used for the initial treatment.

Salvage Chemotherapy in MOPP Failures

Chemotherapy protocols used after MOPP failure are listed in Table 1. The greatest experience has been gained with the ABVD regimen. Although complete remission rates of up to 59% have been reported, long-term disease-free survival rates are often not quoted. In large series with long follow-up they rarely exceed 10%, and total survival ranges from 12 to 48 months.

Table 1. Salvage chemotherapy in MOPP failures

Protocol	Reference	No. evaluable	CR%	CR+PR%	Mean duration (months)		Survival (months)		% of patients with		
					CR	CR+PR	CR	All	Leukopenia grade III+IV	Thrombocytopenia grade III+IV	Vomiting grade III+IV
MOPP reinduction	Fisher et al. 1979	32	56	n.m.	21	n.m.	62	48	n.m.	n.m.	n.m.
MOPP/ABVD	Straus et al. 1981	21	18	50	72	9	n.m.	n.m.	4	n.m.	n.m.
ABVD	Sutcliff et al. 1979	41	7	63	n.m.	n.m.	25+	12	n.m.	n.m.	n.m.
ABVD	Krikorian et al. 1979	27	22	37	21	3	15+	20+	>3%	>3%	>3%
ABVD	Case et al. 1979	24	5	63	n.m.	6.5	n.m.	n.m.	n.m.	n.m.	n.m.
ABVD	Papa et al. 1982	18	55	55	36+	n.m.	60+	38+	0	n.m.	n.m.
ABVD	Santoro et al. 1982	54	59	72	17	7	60+	28	n.m.	n.m.	n.m.
ABDIC	Tannir et al. 1983	36	35	65	47	n.m.	70	24	40	55	n.m.
B-CAVE	Harker et al. 1984	48	71	27	24	n.m.	32	22	35	18	n.m.
SCAB	Levi et al. 1977	17	35	59	87	n.m.	16+	n.m.	25%	25%	n.m.
VABCD	Einhorn et al. 1983	18	44	88	21+	n.m.	28+	28+	>50	28	n.m.
CVB	Goldman et al. 1981	39	26	85	4.5+	n.m.	4.5+	n.m.	n.m.	n.m.	n.m.
EVA	Richards et al. 1988	19	32	37	n.m.	n.m.	n.m.	n.m.	58	>5	n.m.

ABVD, doxorubicin, bleomycin, vinblastine, dacarbazine; ABDIC, doxorubicin, bleomycin, dacarbazine, lomustine, prednisone; B-CAVE, bleomycin, lomustine, doxorubicin, vinblastine; SCAB, streptozotocin, lomustine, doxorubicin, bleomycin; VABCD, vinblastine, doxorubicin, bleomycin, lomustine, dacarbazine; CVB, lomustine, bleomycin, vinblastine; EVA, etoposide, vincristine, doxorubicin; n.m., not mentioned.

Table 2. Results of salvage therapy in patients refractory to MOPP or COPP and ABVD

Protocol	Reference	No. evaluable	CR%	CR+PR%	Mean duration (months)		Survival (months)		% of patients with		
					CR	CR+PR	CR	All	Leukopenia grade III+IV	Thrombocytopenia grade III+IV	Vomiting grade III+IV
CEP	Santoro et al. 1986	58	40	54	15	n.m.	24	17	10	16	84
CEP	Cervantes et al. 1986	15	26	40	12	5.5	n.m.	n.m.	3	3	27
TeCisPHex	Budmann et al. 1985	56	6	31	n.m.	4.8	n.m.	n.m.	12	14	18
CAD	Straus et al. 1983	15	13	46	n.m.	n.m.	n.m.	n.m.	13	7	7
CEM	Tseng et al. 1987	32	13	47	33+	5	n.m.	n.m.	19	19	n.m.
CEBLoM	Williams et al. 1985	5	20	20	n.m.	n.m.	n.m.	n.m.	50	21	79
DHAP	Velasquez et al. 1986	19	16	68	n.m.	n.m.	n.m.	n.m.	n.m.	n.m.	n.m.
MIME	Hagemeister et al. 1987	47	23	63	n.m.	n.m.	n.m.	12	75	60	n.m.
CEVD	Pfreundschuh et al. 1987	32	44	56	10+	8	26+	18	37	25	12

CEP, lomustine, etoposide, prednimustine; TeCisPHex, Teniposide, cisplatin, prednisone, hexamethylmelamine; CAD, Lomustine, melphalan, vindesine; CEM, lomustine, etoposide, methotrexate; CEBLoM, cisplatin, etoposide, bleomycin, lomustine, methotrexate; DHAP, dexamethasone, high-dose ara-C, cisplatin; MIME, mitoguazone, ifosfamide, methotrexate, etoposide; CEVD, lomustine, etoposide, vindesine, dexamethasone; n.m., not mentioned.

Salvage Chemotherapy After MOPP and ABVD Failure

The prognosis is even worse for patients who are resistant to both MOPP and ABVD (Table 2). Indeed, many oncologists do not believe that these patients can be cured by third-line chemotherapy. Complete remission rates are below 50%. The median duration of complete remissions rarely exceeds 1 year and the overall survival is less than 2 years. Results are similar, no matter whether patients fail after they have received MOPP and ABVD sequentially or as a monthly alternating combination chemotherapy (Santoro et al. 1986; Pfreundschuh et al. 1987). As resistance to MOPP (or COPP) and ABVD is defined differently by different groups, it is hard to compare the results of the different regimens listed in Table 2. It is evident, however, from studies that have included "poor-risk" failures (e.g., Pfreundschuh et al. 1987) that a small subgroup of such patients can obtain long-term disease-free survival by third-line therapy.

We do not believe that there are major differences between these chemotherapy regimens as far as complete remission rates, duration of complete remissions, and long-term disease-free survival are concerned. As significant toxicity can only be justified when the objective of treatment is cure, the choice of the optimal chemotherapy in this situation may depend mainly on the subjective (nausea and vomiting) and objective toxicity (mainly myelosuppression) of the protocols. In this respect, the CEVD (lomustine, etoposide, vindesine, dexamethasone) and CEP (lomustine, etoposide, prednimustine) protocols seem to be the most adequate for these often heavily pretreated patients. For patients who cannot tolerate drug combinations, vinblastine, lomustine, chlorambucil, procarbazine, or etoposide alone or in combination with steroids can offer substantial palliation (Mead et al. 1982).

Combined Modality Approach as Salvage Therapy

The indications for, and the value of adding chemotherapy to wide-field radiotherapy for patients with nodal relapses remain to be determined. In our opinion, this combined modality approach for "low-risk" relapse should not be given unless within a controlled prospective trial.

On the other hand, radiotherapy as an adjunct to chemotherapy-induced second or third remissions in "high-risk" failures seems to be a reasonable approach especially in patients who have not yet received extensive radiotherapy. This recommendation is also supported by our observation that a considerable proportion of patients who obtained complete remission by third-line CEVD relapsed with-nodal sites which had not been previously irradiated (Pfreundschuh et al. 1987). As wide-field irradiation in these extensively pretreated patients may cause massive myelosuppression, additive radiotherapy should be given to previously involved nodal sites or - if this is not possible - at least to the sites of the latest relapse.

Further Indications for Salvage Therapy and Future Directions

Besides from patients who fail to obtain complete remission by the first-line chemotherapy or relapse with progressive disease, there are other subgroups of patients for which additional therapy may be considered even if they achieve complete remission by the primary chemotherapy. These subgroups include patients known to have a high risk of relapse. Even though different risk factors emerge under different treatment strategies, it is generally accepted that patients whose disease responds slowly to therapy have a high risk of relapse (Carde et al. 1983). Moreover, patients with characteristics identifying them as having a less than a 50% chance for disease-free survival under a certain regimen (e.g., patients with an ESR >80 mm/h in the HD-3 trial of the German Hodgkin's Disease Study Group) should be entered into controlled trials which test the value of additional therapy for improving disease-free survival.

Another important task in the future will be to determine the indication for, and the value of intensified chemotherapy protocols with and without bone marrow transplantation. While the toxicity of bone marrow transplantation will be greatly reduced by the advent of recombinant stimulators of hematopoiesis such as GM-CSF and G-CSF (Brandt et al. 1988), the colony-stimulating factors at the same time will enable us to intensify conventional chemotherapy by time and dosis (Antmann et al. 1988). This should enhance the rates of complete remissions and disease-free survivors in patients with poor risk who are not candidates for bone marrow transplantation.

References

Antmann K, Griffin J, Elias A, Socinski M, Whitley M, Ryan L, Cannutra S, Frei E, Schnipper L (1988) Effect of rGM-CSF on chemotherapy induced myelosuppression in sarcoma patients. Proc ASCO 7: 619

Bergsagel DE (1987) Salvage treatment for Hodgkin's disease in relapse. J Clin Oncol 4: 525-526

Bonadonna G (1982) Chemotherapy strategies to improve the control of Hodgkin's disease. Cancer Res 42: 4309-4320

Bonadonna G, Valagussa P, Santoro A et al. (1986) Alternating non-cross-resistant combination chemotherapy or MOPP in stage IV Hodgkin's disease. Ann Intern Med 104: 739-746

Brandt SJ, Peters WP, Atwater SK, Kurtzberg J, Borowitz MJ, Jones RB, Shpall EJ, Bast RC, Gilbert CJ, Oette DH (1988) Effect of recombinant human granulocyte-macrophage colony-stimulating factor on hematopoietic reconstitution after high-dose chemotherapy and autologous bone marrow transplantation. N Engl J Med 318: 869-876

Budman DR, Anderson J, Obrecht JP, et al. (1985) Treatment of refractory Hodgkin's disease with teniposide, cisplatin, hexamethylmelamine and prednisolone: Cancer and Leukemia Group B trial 8171. Cancer Treat Rep 69: 719-720

Carde P, MacKintosh SR, Rosenberg SA (1986) A dose and time response analysis of the treatment of Hodgkin's disease with MOPP chemotherapy. J Clin Oncol 1: 146-153

Case DC, Young CW, Lee BJ (1977) Combination chemotherapy of MOPP-resistant Hodgkin's disease with adriamycin, bleomycin, dacarbazine and vinblastine (ABVD). Cancer 39: 1382-1386

Cervantes F, Reverter JC, Montserrat E et al. (1986) Treatment of advanced resistant Hodgkin's disease with lomustine, etoposide and prednimustine. Cancer Treat Rep 70: 665-667
DeVita VT, Serpick AA, Carbone PP (1970) Combination chemotherapy in the treatment of advanced Hodgkin's disease. Ann Intern Med 73: 891-895
DeVita VT, Simon RM, Hubbard SM et al. (1980) Curability of advanced Hodgkin's disease with chemotherapy. Long-term follow-up of MOPP treated patients at NCI. Ann Intern Med 92: 587-595
Diehl LF, Perry DJ, Terebelo H et al. (1983) Radiation as salvage therapy for patients with Hodgkin's disease relapsing after MOPP (mechlorethamine, vincristine, prednisone, and procarbazine) chemotherapy. Cancer Treat Rep 67: 827-829
Einhorn LH, Williams SD, Stevens EE, Bond WH, Cheoweth L (1983) Treatment of MOPP-refractory Hodgkin's disease with vinblastine, doxorubicine, bleomycin, CCNU, and dacarbazine. Cancer 51: 1348-1352
Fisher RI, DeVita VT, Hubbard SP et al. (1979) Prolonged disease-free survival in Hodgkin's disease with MOPP reinduction after first relapse. Ann Intern Med 90: 761-763
Fox K, Lippman SM, Cassady JR et al. (1987) Radiation therapy salvage of Hodgkin's disease following chemotherapy failure. J Clin Oncol 5: 38-45
Goldman JM, Dawson AA (1981) Chemotherapy for advanced resistant Hodgkin's disease. Lancet 2: 252
Hagemeister FB, Tannir N, McLaughlin P, et al. (1987) MIME Chemotherapy (methyl-GAG, ifosfamide, methotrexate, etoposide) as treatment for recurrent Hodgkin's disease. J Clin Oncol 5: 556-561
Harker WG, Kushlan P, Rosenberg SA (1984) Combination chemotherapy for advanced Hodgkin's disease after failure of MOPP: ABVD and B-CAVe. Ann Int Med 101: 440-446
Krikorian JG, Portlock CS, Rosenberg SA (1978) Treatment of advanced Hodgkin's disease with adriamycin, bleomycin, vinblastine and imidazole carboxamide (ABVD) after failure of MOPP therapy. Cancer 41: 2107-2111
Loeffler M, Pfreundschuh M, Hasenclever D et al. (1988) Prognostic risk factors in advanced Hodgkin's lymphoma. Report of the German Hodgkin Study Group. Blut 56: 273-281
Longo DL, Young RC, Wesley M et al. (1986) Twenty years of MOPP therapy for Hodgkin's disease. J Clin Oncol 4: 1295-1306
Mauch P, Tarbell N, Skarin A et al. (1987) Wide-field radiation therapy alone or with chemotherapy for Hodgkin's disease in relapse from combination chemotherapy. J Clin Oncol 5: 544-549
Mead GM, Harker WG, Hushlan P, Rosenberg SA (1982) Single agent palliative chemotherapy for end-stage Hodgkin's disease. Cancer 50: 829-835
Papa G, Mandelli F, Anselmo AP, de Luca AM, Maurizienrici R, Mauro F, Testi AM, Amadori S, Biagini C (1982) Treatment of MOPP-resistant Hodgkin's disease with adriamycin, bleomycin, vinblastine and dacarbazine (ABVD). Eur J Cancer Clin Oncol 18: 803-806
Pfreundschuh M, Schoppe WD, Fuchs R et al. (1987) Lomustine, etoposide, vindesine, and dexamethasone (CEVD) in Hodgkin's lymphoma refractory to cyclophosphamide, vincristine, procarbazine, and prednisone (COPP) and doxorubizin, bleomycin, vinblastine, and dacarbazine (ABVD): a multicenter trial of the German Hodgkin Study Group. Cancer Treat Rep 71: 1203-1207
Richard MA, Waxman JH, Man T et al. (1986) EVA treatment for recurrent or unresponsive Hodgkin's disease. Cancer Chemother Pharmacol 18: 51-53
Roach M III, Kapp DS, Rosenberg SA et al. (1987) Radiotherapy with curative intent: an option in selected patients relapsing after chemotherapy for advanced Hodgkin's disease. J Clin Oncol 5: 550-550
Santoro A, Bonfante V, Bonadonna G (1982) Salvage chemotherapy with ABVD in MOPP-resistant Hodgkin's disease. Ann Int Med 96: 139-142

Santoro A, Viviani S, Valagussa P et al. (1986) CCNU, etoposide, and prednimustine (CEP) in refractory Hodgkin's disease. Semin Oncol 13 (suppl 1): 23-26

Straus DJ, Passe S, Koziner B, Lee BJ, Young CW, Clarkson BD (1981) Combination chemotherapy salvage of heavily pretreated patients with Hodgkin's disease: an analysis of prognostic factors factors in two chemotherapy trials and the literature. Cancer Treat Rep 65: 207-211

Straus DJ, Myers J, Koziner B et al. (1983) Combination chemotherapy for the treatment of Hodgkin's disease in relapse. Results with lomustine (CCNU), melphalan (Alkeran) and vindesine (DVA) alone (CAD) and in alternation with MOPP and doxorubicin (Adriamycin), bleomycin and vinblastine. Cancer Chemother Pharmacol 11: 80-85

Strum SB, Rappaport H (1971) Persistence of Hodgkin's disease in long-term survivors. Am J Med 51: 222-240

Sutcliff SB, Wrigley PFM, Stansfeld AG, Malpas JS (1979) Adriamycin, bleomycin, vinblastine and imidazole carboxamide (ABVD) therapy for advanced Hodgkin's disease resistant to mustine, vinblastine, procarbazine and prednisolone (MVPP). Cancer Chemother Pharmacol 2: 209-213

Tannir N, Hagemeister F, Velasquez W, Cabainllas F (1983) Long-term follow-up with ABDIC salvage chemotherapy of MOPP-resistant Hodgkin's disease. J Clin Oncol 1: 432-439

Tseng C, Jacobs C, Coleman CN, Horning SJ, Lewis BJ, Rosenberg SA (1987) Third-line chemotherapy for resistant Hodgkin's disease with lomustine, etoposide and methotrexate. Cancer Treat Rep 71: 475-479

Velasquez W, Jagannath S, Hagemeister FB et al. (1986) Dexamethasone, high dose Ara-C and cisplatin (DHAP) as salvage treatment for relapsing Hodgkin's disease. Blood 68 (Suppl 1): 242 a (Abstr 846)

Williams SF, Gaynor E, Watson E et al. (1985) Management of advanced, resistant Hodgkin's disease (HD) and non-Hodgkin's lymphoma (NHL) with a non-cross resistant drug program: cisplatinum (P), VP-16 (E), bleomycin (B), lomustine (CCNU) or carmustine (BCNU) and methotrexate (MTX) with leucovorin (L). A phase II study. Proc ASCO 4: 204

High-Dose Combination Chemotherapy with Cyclophosphamide, Carmustine, Etoposide, and Autologous Bone Marrow Transplantation in 60 Patients with Relapsed Hodgkin's Disease: The M. D. Anderson Experience

J. A. Spinolo, S. Jagannath, K. A. Dicke, K. E. Smith, F. Cabanillas, F. B. Hagemeister, L. J. Horwitz, P. McLaughlin, F. Swan, and G. Spitzer

M. D. Anderson Cancer Center, University of Texas, Bone Marrow Transplantation/Box 65, 1515 Holcombe Boulevard, Houston, TX 77030, USA

Patients with Hodgkin's disease who do not respond to initial combination chemotherapy or who relapse after an initial response to chemotherapy have a poor prognosis. The complete remission rate with further combination chemotherapy is less than 50%, with short remission duration and low cure fraction (Hagemeister et al. 1987; Santoro et al. 1986; Tseng et al. 1987). In 1983, we started using high-dose chemotherapy with cyclophosphamide, carmustine, and etoposide (CVB) and autologous bone marrow transplantation (ABMT) for the treatment of these groups of patients. Other groups have also used high-dose chemotherapy and ABMT in similar patients (Ahmed et al. 1987; Carella et al. 1986; Goldstone et al. 1986; O'Reilly et al. 1987; Philip et al. 1986). This report will update our previously reported experience at the M. D. Anderson Cancer Center (Jagannath et al. 1986a, b).

Material and Methods

Eligibility Criteria

The eligibility criteria were as follows: informed consent as approved by M. D. Anderson Hospital Institutional review Board; chemotherapy-resistant disease or relapse after chemotherapy-induced complete remission; performance status 0–3 on the Zubrod scale; bone marrow cellularity $> 15\%$ without evidence of tumor in bilateral iliac crest biopsies at the time of bone marrow harvest; cardiac ejection fraction > 0.5; diffusing capacity for lung carbon dioxide ($DLCO_2$) $\geq 50\%$ of predicted; and normal renal and hepatic function.

Patient Population

Sixty patients with histological diagnosis of Hodgkin's disease were studied. The median age was 28.5 years (range 16–55 years). There were 38 males and 22 females. The median time from diagnosis to bone marrow transplantation was

Recent Results in Cancer Research. Vol. 117

34 months (range 6-122 months). At the time of relapse, 21 patients had disease confined to the lymph nodes, with the most common site being mediastinum. Four had extranodal disease only, and 31 had combined nodal and extranodal disease. Of the latter, 19 had a single extranodal site of involvement, 13 had two sites, 2 had three, and 1 had four sites. The commonest sites of extranodal disease were lungs (24 patients), pleura (10), bone (6), and liver (4). Performance status was 0 in 40 patients, 1-2 in 18 patients, and 3 in 2 patients. All patients but two had received a MOPP (nitrogen mustard, oncovin, procarbazine, and prednisone)-like regimen and ABVD (Adriamycin, bleomycin, vinblastine, and dacarbazine)-like combination before CBV-ABMT. The median number of previously failed combination chemotherapy regimens was two (range one to six); alternating schemata similar to MOPP/ABVD were counted as one regimen. Failure of a chemotherapy regimen was defined as lack of objective response, progression while on treatment or relapse after achievement of CR. Thirty-one patients had failed three or more regimens. Twenty-seven patients had not achieved a CR to their first-line chemotherapy regimen, and 15 patients had never obtained CR. Of the 52 patients who were receiving chemotherapy at the time of admission to the study, 19 had achieved an objective response to the current regimen and 33 were progressing on it. In the former 19 cases, the latest regimen was not counted as a failure.

Myeloablative Regimen

The schema of treatment is shown in Table 1. Cyclophosphamide (1.5 g/m^2) was given daily on days -6 to -3; BCNU (300 mg/m^2) was given on day -6, and etoposide (VP-16) was given every 12 h on days -6 to -4, to a total of six doses (Spitzer et al. 1980). The total dose of VP-16 was 600 mg/m^2 in 22 patients, 750 mg/m^2 in 16 patients, and 900 mg/m^2 in 22 patients. After a 2-day rest period, autologous bone marrow transplant was given (day 0). Patients received a median of 1.5×10^8 cells/kg body wt. (range 0.4-3.2×10^8/kg). All patients were hospitalized from day -7 to the time their absolute granulocyte count reached 0.5×10^9/liter.

Table 1. Chemotherapy schema for CBV-ABMT

Drugs	Day −6	Day −4	Day −4	Day −3	Day −2	Day −1	Day 0
Cyclophosphamide							
1.5 g/m^2	X	X	X	X			
BCNU							
300 mg/m^2	X						
VP-16							
200-300 mg/m^2	X	X	X				
ABMT							X

Evaluation of Response

Complete remission was defined as the disappearance of all clinical evidence of disease, with normalization of all radiographic studies and laboratory values. However, patients with no other symptoms or signs of disease, but with residual radiographical abnormalities in previously irradiated areas which did not change over a period of 6 months, were also considered to be in complete remission. Partial remission (PR) was defined as a reduction of 50% or more in measurable disease for at least 1 month. All other outcomes were considered failures.

Survival was calculated from the date of transplant (day 0) to the date of last follow-up or death. Event-free survival was calculated from day 0 to the date of documented progression of disease, last follow-up, or death.

The Kaplan-Meier method was used to obtain survival and event-free survival curves. The Brookmeyer-Crowley method was used to estimate confidence intervals.

Results

Sixty patients were treated. Of these, 56 had measurable disease at the time of CBV-ABMT, and 4 were intensified in their second (2) or third (2) complete remission (CCR patients). Twenty patients achieved CR, and 20 had PR, for an objective response rate of 71.4% in the group of measurable disease. CR was obtained in four of the patients who achieved PR, after involved-field radiation therapy was given to sites of residual nodal disease. The median follow-up is 13 months, and the median survival is 21 months (95% confidence interval, 14-40 months). The median event-free survival is 14 months (95% confidence interval, 9 to >70 months) (Fig. 1). One patient died 5 months after CBV-ABMT of cyto-

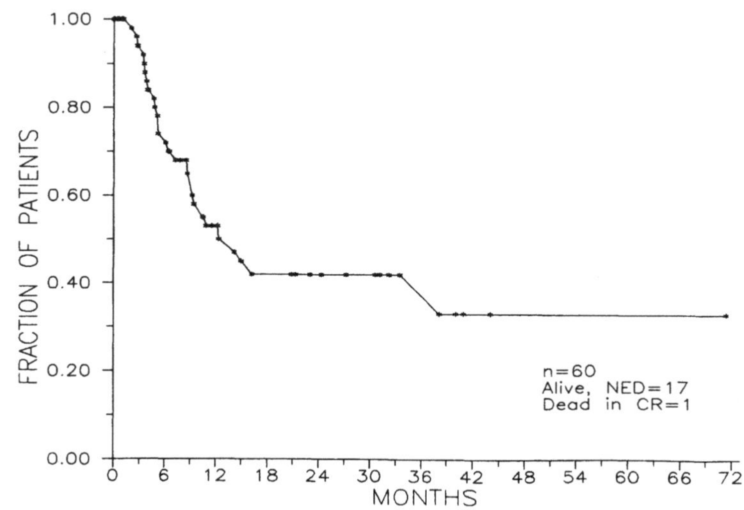

Fig. 1. CBV-ABMT for relapsed Hodgkin's disease: event-free survival

megalovirus and pneumocystitis pneumonia, while still in CR. Ten relapses have been observed among the 28 patients who were in CR after completion of CBV-ABMT (including the four patients who received radiation therapy), at a median time of 9 months (range 5-38 months). Results are summarized in Table 2. All relapses but one occurred within 18 months of transplantation. Seventeen patients are still alive and free of disease with a median follow-up of 27 months (range 7-70 months); only four of them have been followed for less than 18 months (Fig. 2). The median survival of the 28 complete responders has not been reached yet, and 87% are alive at 2 years. This contrasts with the median survival of the patients who only had PR or who failed to respond, which is 9 and 3 months, respectively (Fig. 3).

Table 2. CBV-ABMT in relapsed Hodgkin's disease: patients in CR after treatment

Response	No. of patients	Relapses (time to relapse in months)	Alive free of disease (follow-up in months)
CR	20	8 (5/6/7/9/9/12/14)	11 (10/12/23/24730/30/31/32/41/44/70)
CR(RT)	4	1 (38)	3 (21/21/37)
CCR	4	1 (9)	3 (7/9/27)
Total	28	10	17[a]

CR, complete response; CR(RT), CR achieved after involved-field radiation of residual nodal disease; CCR, continuous CR (patients treated in CR).
[a] One patient died of pneumonia while in CR.

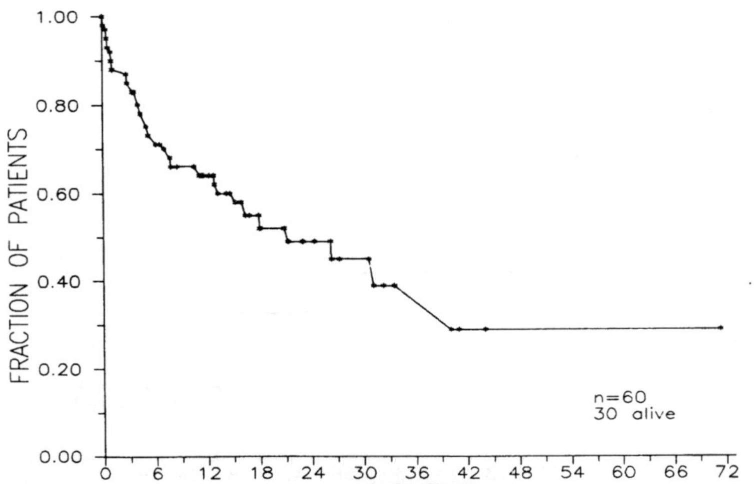

Fig. 2. CBV-ABMT for relapsed Hodgkin's disease: survival

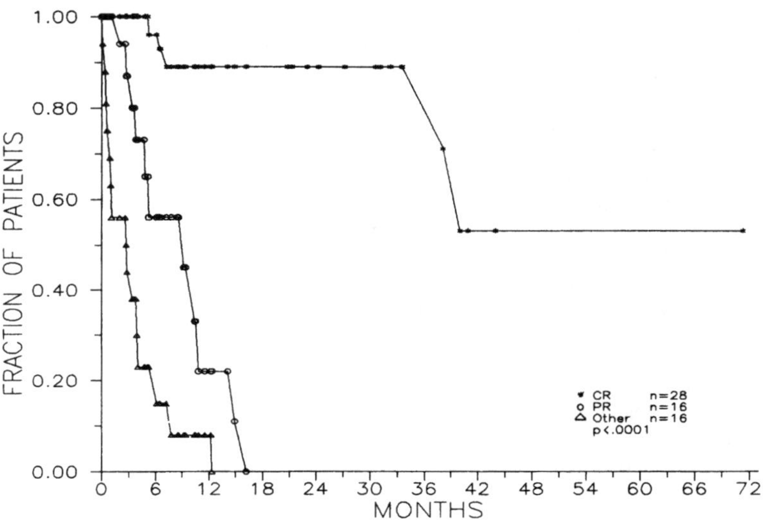

Fig. 3. CBV-ABMT for relapsed Hodgkin's disease: response and survival

Toxicity

There were five deaths in the first 4 weeks after BMT; all occurred before hematological recovery in patients who received the highest dose of VP-16 (900 mg/m^2). The acute toxicities observed during chemotherapy included nausea and vomiting (89%) and two cases of seizure of unclear etiology (3%). All patients developed severe marrow suppression after chemotherapy. Fifty-six (93%) of the patients had febrile episodes; 17 (28%) had pneumonia, and 23 (38%) had bacteremia. Hematuria was observed in 17% of the cases (10 patients), and congestive heart failure in one case. There was one instance of graft failure in a patient with extensive bone marrow disease, which appeared after bone marrow harvest.

References

Ahmed T, Gingrich SA, Ciavarella D et al. (1987) Myeloablative chemotherapy followed by autologous bone marrow transplantation (ABMT) for patients with refractory Hodgkin's disease. Proc ASCO 6: 150 (Abstr 589)

Carella AM, Congiu A, Santini G et al. (1986) High dose chemotherapy (HDC) and autologous bone marrow transplantation in advanced highly treated Hodgkin's disease (AHT-HD). Bone Marrow Transplant 1: 293-296

Goldstone AH, Anderson CC, Linch DC et al. (1986) High-dose therapy with autologous bone marrow transplantation for advanced relapsed adult Hodgkin's disease. Bone Marrow Transplant 1: 297

Hagemeister FB, Tannir N, McLaughlin P et al. (1987) MIME chemotherapy (methyl-GAG, ifosfamide, methotrexate, etoposide) as treatment for recurrent Hodgkin's disease. J Clin Oncol 5: 556-561

Jagannath S, Dicke KA, Armitage JO, Cabanillas F, Horwitz LJ, Vellekoop L, Zander AR, Spitzer G (1986a) High dose cyclophosphamide, carmustine and etoposide and autologous bone marrow transplantation for relapsed Hodgkin's disease. Ann Intern Med 104: 163-168

Jagannath S, Armitage JO, Spitzer G et al. (1986b) Prognostic factors predicting for response to high dose chemotherapy and autologous bone marrow transplantation (ABMT) for relapsed Hodgkin's disease. Blood 68 [Suppl 1]: 274a (Abstr 974)

O'Reilly S, Connors J, Voss N et al. (1987) High-dose cyclophosphamide, BCNU and etoposide (CBV) and autologous bone marrow transplantation in progressive Hodgkin's disease (HD). Proc ASCO 6: 196 (Abstr 774)

Philip T, Dumont J, Teillet F, Maraninchi D, et al. (1986) High dose chemotherapy and autologous bone marrow transplantation in refractory Hodgkin's disease. Br J Cancer 53: 737-742

Santoro A, Viviani S, Valagussa P et al. (1986) CCNU, Etoposide, and Prednimustine (CEP) in refractory Hodgkin's disease. Semin Oncol 13: 23-26

Spitzer G, Dicke KA, Litam J et al. (1980) High dose combination chemotherapy with autologous bone marrow transplantation in adult solid tumors. Cancer 45: 3075-3085

Tseng G, Jacobs C, Coleman CN et al. (1987) Third-line chemotherapy for resistant Hodgkin's disease with lomustine, etoposide and methotrexate. Cancer Treat Rep 71: 475-478

*Treatment of Resistant Hodgkin's Lymphoma with Bone Marrow Transplantation in Italy**

A. M. Carella and A. M. Marmont

Department of Hematology, Bone Marrow Transplantation Unit, Ospedale S. Martino, Genova, Italy

High-dose chemotherapy and autologous or allogeneic bone marrow transplantation are known to be useful in patients with hematological malignancies. The encouraging results of this approach in the treatment of acute leukemias prompted us to investigate its use in patients with Hodgkin's lymphoma not being cured by primary conventional chemotherapy.

Material and Methods

Eligibility Criteria

The eligibility criteria were as follows: informed consent; adequate bone marrow function in the absence of marrow involvement; Karnofsky performance status >40%, and resistance to MOPP and then later to radiation and ABVD or CEP (lomustine, etoposide, prednimustine; Santoro et al. 1986) or to the alternating regimen MOPP/ABVD + CEP and radiation, indicated by either progressive disease during therapy or relapsed disease after an initial complete remission.

Patient Population

There were 34 males and 16 females, with a median age of 27 years (range 12-45 years). Forty-one patients had advanced stage (16 III, 25 IV), and the majority had B symptoms (39 patients). Nine patients had only node involvement while 41 had lung involvement in association with bulky mediastinum (29 patients), liver (7 patients), bone (4 patients), and liver and bone (1 patient). Except for two, all patients were heavily pretreated.

All patients were truly refractory to conventional chemotherapy or in second or higher relapse after MOPP and ABVD therapy.

---

* This work was supported by the Associazione Italiana Ricera sul Cancro (A.I.R.C., 1989).

Recent Results in Cancer Research. Vol. 117

Myeloablative Regimen

Patients were treated with high-dose chemotherapy regimens consisting of cyclophosphamide 5 g/m^2, BCNU 600 mg/m^2, and etoposide 400 mg/m^2 (Carella et al. 1986; Carella et al. 1987; Jagannath et al. 1986). In the last 13 patients the total doses of etoposide and BCNU were escalated to 1000 mg/m^2 and 800 mg/m^2, respectively, and given in 4 days with bone marrow reinfusion on day 7, after a rest period of 2 days (CVB-2 protocol; Carella 1988). No CR patient received maintenance therapy, while patients achieving partial remission always received subsequent radiotherapy.

Evaluation of Response

Complete remission was defined as the disappearance of clinical and radiological evidence of Hodgkin's disease. Partial remission was defined as a reduction of 50% or more in measurable disease for at least 1 month.

Results

A total of 50 patients have been treated with high-dose chemotherapy and autologous bone marrow transplantation in Italy. The results are summarized in Table 1. Twenty-four (48%) patients achieved complete remission, with a median duration of 24 months. Sixteen patients (32) achieved a partial response, with a median duration of 9 months. Thus, the overall response rate was 80%. Twelve of 24 complete remissions subsequently relapsed 3–25 (median 7) months after ABMT. An assessment of the disease-free survival in the 12 patients still free of disease is still too early because 4 patients have a short follow-up. However, there are 8 patients with long-term disease-free survival at 14, 14, 18, 30, 36, 45, 59, and 79 months, respectively.

Table 1. Pretransplant characteristics and results of CVB protocol in 50 evaluable patients receiving transplants for advanced Hodgkin's lymphoma: The Italian experience

Patients	Response to first therapy	Median duration first CR		Sites of HL at ABMT		Response to CVB	Total survival	Freedom from progression
		<1 year	>1 year	Nodal	END			
50[a]	CR: 32 NR: 18	19	14	9	41	CR: 24 (48%) PR: 16 (32%) Total: 40/50 (80%)	24	12

CR, complete response; NR, no response; END, extranodal disease.
[a] Institutions contributing patient data: Genova 29 (Carella, Marmont), Bologna 8 (Mazza Tura), Roma 6 (Meloni, Cimino, Mandelli), Parma 3 (Mangoni, Rizzoli), Bolzano 2 (Coser), Verona 2 (Cetto).

Toxicity

The main toxicity after ABMT in the Italian experience was significant neutropenia and thrombocytopenia in all patients. Fever during neutropenia, which warranted intravenous antibiotics and antimycotic, antiviral, and high-dose immunoglobulin treatment, was observed in all patients. Six cases had severe mucositis and two had generalized herpes simplex virus infection, which was well controlled with intravenous acyclovir. Nausea, vomiting, and elevations of liver enzymes and/or alkaline phosphatase were observed in all patients. Three patients had lung toxicity attributable to carmustine, which resolved within a few months. Cardiotoxicity, as assessed by clinical evaluation, echocardiography, and calculation of pre-ejection period/left ventricular ejection time, was not observed.

References

Carella AM, Congiu A, Santini G et al. (1986) High dose chemotherapy (HDC) and autologous bone marrow transplantation in advanced highly treated Hodgkin's disease (AHT-HD). Bone Marrow Transplant 1: 293-296

Carella AM, Congiu A, Santini G et al. (1987) High-dose cyclophosphamide, etoposide and BCNU with autologous BMT (ABMT) rescue in advanced resistant Hodgkin's disease patients. Bone Marrow Transplant 2 [Suppl 1]: 210

Carella AM, Congiu A, Mazza P et al. (1988) High-dose chemotherapy with autologous BMT "rescue" in 50 advanced resistant Hodgkin's disease patients. An Italian Study Group Report. J Clin Oncol 6 [Suppl 9]: 1411-1416

Jagannath S, Dicke KA, Armitage JO, Cabanillas F, Horwitz LJ, Vellekoop L, Zander AR, Spitzer G (1986) High dose cyclophosphamide, carmustine and etoposide and autologous bone marrow transplantation for relapsed Hodgkin's disease. Ann Intern Med 102: 163-168

Santoro A, Bonfante V, Bonadonna G et al. (1982) Salvage chemotherapy with ABVD in MOPP-resistant Hodgkin's disease. Ann Intern Med 96: 139-143

Santoro A, Viviani S, Valagussa P et al. (1986) CCNU, etoposide and prednimustine (CEP) in refractory Hodgkin's disease. Semin Oncol 13 [Suppl 1]: 23-26

Preliminary Results of Autologous Bone Marrow Transplantation in the Management of Resistant Hodgkin's Disease: Experience of the Bloomsbury Transplant Group at University College, London

J. G. Gribben, A. H. Goldstone, and D. C. Linch

Department of Haematology, University College, Chenies Mews, London WC1, Great Britain

The preliminary results from our own (Goldstone et al. 1986; Gribben et al. 1987a;) and other centres (Appelbaum et al. 1985; Goldstone et al. 1987; Jagannath et al. 1987; Philip et al. 1986) prompted us to extend our ABMT trial in patients with Hodgkin's disease. For the purposes of planning our own randomized studies we somewhat subjectively suggested that a reasonable starting point is a projected 5-year survival of less than 35%. Since the early morbidity and mortality of intensive therapy is high in this patient group, we further proposed that no patient should receive an ABMT if the projected survival is greater than 65% at 2 years. We have used the British National Lymphoma Investigation (BNLI) experience of over 600 patients with advanced HD treated by MOPP-type regimens as a data base for designing future studies (Gribben et al. 1987b).

Within the Bloomsbury transplant group at University College, London, we have now treated 58 adult patients with advanced Hodgkin's disease by intensive therapy and autologous bone marrow transplantation. In the following we report the preliminary results obtained at our institution.

Material and Methods

Eligibility Criteria

The eligibility criteria were as follows: informed consent; primary resistant disease on first-line alternating chemotherapy or relapse after at least two regimens of chemotherapy; performance status 0–3 on WHO scale; bone marrow without evidence of tumour in bilateral iliac crest biopsies at the time of bone marrow harvest; and no major defects of cardiac, pulmonary, renal or hepatic function.

Recent Results in Cancer Research. Vol. 117

Patient Population

There were 41 males and 17 females, with a median age of 27 years. Four patients had primary resistant disease in first-line alternating chemotherapy, 54 had received two or more regimens of chemotherapy, with radiotherapy in addition in 29 patients.

Myeloablative Regimen

The initial three patients received total body irradiation-containing regimens but all subsequently received high-dose combination chemotherapy protocols. This consisted of the BEAM regimen (BCNU 300 mg/m^2 day 1, etoposide 100-200 mg/m^2 days 2-5, Ara-C 100-200 mg/m^2 days 2-5, melphalan 140 mg/m^2 day 6).

Evaluation of Response

The response to the high-dose chemotherapy was assessed clinically and radiologically by CT scan at 3 months post-ABMT and at 6-monthly intervals thereafter. WHO criteria were applied for the definition of complete and partial responses, as well as for progressive disease.

Results

Twenty-five patients (43%) have entered complete remission post-ABMT. Five patients have subsequently relapsed at 7-25 months post-ABMT and one patient died in CR at 8 months post-ABMT. Twenty-two patients (38%) have achieved a partial response (PR) from the autograft procedure. Three of these patients have received localized radiotherapy to sites of residual disease and have subsequently entered CR. A further two patients initially assessed as PR because of residual mediastinal masses have received no further treatment and have shown regression of the mass at 15 and 24 months post-ABMT. Of the remaining 17 PR patients, 4 have died of progressive disease and the remaining 13 patients remain in relapse of disease with a median follow-up of 12 months post-ABMT. Three patients showed no response to the high-dose chemotherapy, two of whom have subsequently died, while the third patient has received palliative radiotherapy and remains in relapse of disease. The actuarial survival of all patients is shown in Fig. 1. Those patients who achieve CR post-ABMT have an increased survival over those who do not (Fig. 2).

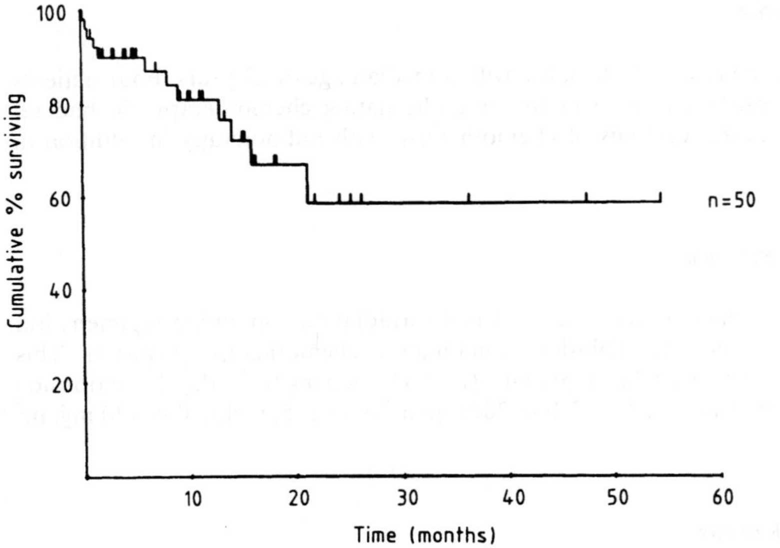

Fig. 1. Overall survival of 50 patients after autologous bone marrow transplantation for relapsed Hodgkin's disease

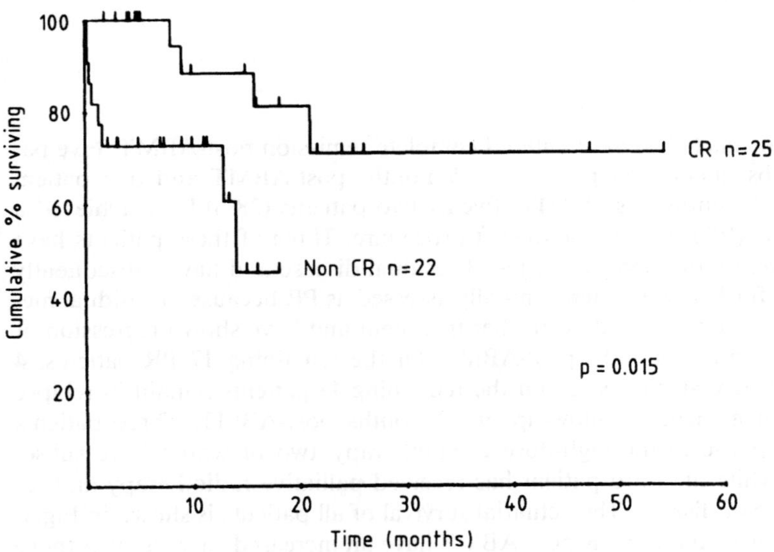

Fig. 2. Overall survival after autologous bone marrow transplantation: complete remission *(CR)* versus no complete remission

Toxicity

Six patients died during the aplastic phase, with sepsis being the cause of death in all cases. Four of these patients had previously been splenectomized and died of septicaemia with an organism not normally associated with life-threatening infections in the autografted patients. One patient aged 59 years died of acute cardiac failure at 8 months post-ABMT while in complete remission of disease and must also be considered as a treatment-related death. This gives a treatment-related mortality of 12%. This is more than twice the mortality in our patients autografted for acute leukaemia (Gribben et al. 1987a).

References

Appelbaum FR, Sullivan KM, Thomas ED, Buckner CD, Clift RA, Deeg HJ, Neiman PE, Sanders JE, Stewart P, Storb R (1985) Allogeneic marrow transplantation in the treatment of MOPP resistant Hodgkin's disease. J Clin Oncol 3: 1490-1494

Goldstone AH, Anderson CC, Linch DC et al. (1986) High-dose therapy with autologous bone marrow transplantation for advanced relapsed adult Hodgkin's disease. Bone Marrow Transplant 1: 297

Goldstone AH, Gribben JG, Dones L (1987) Fourth report of the ABMT experience of autologous bone marrow transplantation in lymphoma. Bone Marrow Transplant 2 [Suppl 1]: 200-203

Gribben J, Goldstone AH, Linch DL et al. (1987a) Autologous bone marrow transplantation in advanced relapsed adult Hodgkin's disease. Bone Marrow Transplant 2 [Suppl 1]: 219

Gribben JG, Vaughan Hudson B, Linch DC (1987b) The potential value of very intensive therapy with autologous bone marrow rescue in the treatment of malignant lymphomas. Hemat Oncol 5: 281-293

Gribben JG, Goldstone AH, Linch DC, Anderson CC, Richards JDM (1987c) Double autografting as potential curative therapy for acute leukaemia. Bone Marrow Transplantation 2 [Suppl 1]: 46-48

Jagannath S, Armitage JO, Dicke KA et al. (1987) Autologous bone marrow transplantation (ABMT) for relapsed Hodgkin's disease. Third International Conference on Malignant Lymphoma p 93 (Abstr 18)

Philip T, Dumont J, Teillet F, Maraninchi D et al. (1986) High dose chemotherapy and autologous bone marrow transplantation in refractory Hodgkin's disease. Br J Cancer 53: 737-742

Bone Marrow Transplantation in the Treatment of Hodgkin's Lymphoma: Problems, Remaining Challenges, and Future Prospects

J. O. Armitage, M. J. Barnett, A. M. Carella, K. A. Dicke, V. Diehl, J. G. Gribben, and M. Pfreundschuh

Department of Internal Medicine, University of Nebraska Medical Center, 42nd and Dewey Avenue, Omaha, NE 68105, USA

Table 1 summarizes the results of autologous bone marrow transplantation (ABMT) in more than 200 patients. The results seem to be rather consistent between the groups who have the largest published series of ABMT in Hodgkin's lymphoma. These results can be summarized as follows: about half of the patients not responding to primary induction chemotherapy or refractory to two or more chemotherapy regimens can obtain complete remission by high-dose chemotherapy and ABMT, and about half of the complete responders can be expected to remain in long-term disease-free survival. Even though toxicity and morbidity of this aggressive procedure are high, the death rate is about 10%. Thus, there is no doubt that high-dose chemotherapy with ABMT is an established approach in certain patients with Hodgkin's disease. However, as this approach is very toxic, several questions arise: what are the characteristics of patients with Hodgkin's disease who are likely to profit from this approach; which is the optimal high-dose myeloablative regimen; which is the optimal source of stem cells for hematopoietic reconstitution; is there any value in bone marrow purging; how can toxicity be reduced; and what are the future prospects of ABMT in Hodgkin's lymphoma?

Table 1. Autologous bone marrow transplantation in Hodgkin's disease

Reference	Number of patients	Treatment regimen	Toxic deaths	CR (%)	Relapse (%)	Free of disease (%)	Follow-up (months)
Carella et al. 1988	50	CVB	4%	48	50	24	2–79
Jagannath et al. 1986b	60	CVB	7%	47	36	30	7–70
Gribben et al. 1988	58	BEAM	10%	43	20	34	2–54
Total	168						

Recent Results in Cancer Research. Vol. 117

Definition of the Patient Population Most Likely To Benefit from ABMT

To date, most patients enrolled in ABMT protocols in Hodgkin's lymphoma have one or more of the following inclusion criteria: they do not obtain complete remission upon primary induction chemotherapy, they relapse within a short time (usually <12 months) after chemotherapy-induced CR, and they are in second or subsequent relapse after having received two or more chemotherapy regimens. The Seattle group (Sullivan et al. 1986) have suggested that all MOPP (mustargen, vincristine, procarbazine, predisone) failures should be considered as candidates for a transplant procedure. However, it is "uncatoun" whether the results obtained with ABVD (doxorubicin, bleomycin, vinblastine, dacarbazine) regimens after MOPP failure in a selected group of patients are significantly worse than those obtained in similar patients who meet the in- and exclusion criteria of ABMT. The analysis of Gribben and Goldstone in this volume shows that to date it is not possible to define a subgroup of patients achieving CR after MOPP with a sufficiently poor prognosis to merit ABMT at the time of first relapse on the basis of histology, stage, ESR, or time from CR to first relapse (Figs. 1, 3 in Gribben and Goldstone, this volume). This is different for patients who fail on an alternating regimen (for overview see Pfreundschuh et al., this volume) or fail second-line therapy. These patients have a median survival of less than 2 years, and less than 20% are alive after 5 years.

To build the recommendation for ABMT on a more objective basis, criteria for eligibility for ABMT should be based on survival data of subgroups of patients with defined risk factors. This, of course, would mean that the population eligible for ABMT would be different in every center using different conventional chemotherapy protocols with different results. It would also require a thorough search for, and analysis of, prognostic subgroups of patients within a given protocol.

Factors that have been suggested to have a negative prognostic impact in advanced-stage Hodgkin's disease are a high ESR or alkaline serum phosphatase (Löffler et al. 1988), histology, and hemoglobin levels (Vaughan Hudson et al. 1983). However, as has been pointed out earlier (Löffler et al., this volume), most of these analyses have been done retrospectively and on the basis of the results of single institutions rather than of comprehensive results obtained by the cooperation of several different groups. With these shortcomings of risk factor analyses in mind, it would, nevertheless, seem to be reasonable to recommend those patients for ABMT who have a limited probability of survival within a certain protocol. It may seem to be arbitrary where to set the cutoff point, but, at the present state of the art of ABMT in Hodgkin's lymphoma, it seems reasonable to recommend ABMT for all patients whose probability of survival after 5 years is <1/3. Similarly, ABMT does not seem to be advisable for patients whose probability of survival after 5 years is >2/3. These figures are similar to the ones suggested by the Bloomsbury Transplant Group (Gribben et al. 1987a).

On the other hand, there are also subgroups of *poor-prognosis patients* in whom ABMT adds only to toxicity without significant effect on survival: The results of many centers suggest that patients who never responded to conventional therapy and are completely refractory to chemotherapy are also bad candidates for ABMT and should be spared this aggressive procedure (Jagannath et al. 1986b).

For the time being, ABMT is most appropriate for patients with progression under conventional induction therapy, with failure to achieve CR, with relapse within 1 year after CR, or in second relapse. Such a strategy will concentrate the toxicity of ABMT on an appropriate group of patients and maximize the antineoplastic effects of the conditioning regimen by utilizing such regimens before a refractory disease state develops and before additional and ineffective conventional salvage treatment adds to the risk of eventual toxicity. The finding of an increased death rate in heavily pretreated patients with Hodgkin's lymphoma, in particular the sometimes severely delayed marrow recovery observed especially in Hodgkin's lymphoma patients, emphasizes the need for not postponing the ABMT for too long. Widening the indication for ABMT as primary therapy for patients with certain risk factors and utilizing ABMT as a consolidation treatment in first remission or as additive therapy at the time of maximal response do not seem to be justified for the time being and are merely experimental; however, the indication for prophylactic early marrow harvest should not be limited to the patients with the poor-prognosis criteria defined above (probability of survival after 5 years $<1/3$), but should be considered in patients with an intermediate prognosis (probability of survival after 5 years between 1/3 and 2/3), as these patients still have a considerable risk of relapsing after an effective conventional remission induction chemotherapy. If these patients do relapse, their autologous bone marrow which has been cryopreserved before exposure to long-term and multidrug therapy would be a better source of stem cells for ABMT in relapse than marrow which is harvested late in the course of the disease.

Optimal Myeloablative Regimen

Failure in patients with Hodgkin's lymphoma treated with ABMT is more likely to be due to recurrence of malignant cells not eradicated by an insufficient conditioning regimen than to the reinoculation of occult clonogenic malignant cells in the autologous marrow graft.

In many candidates for ABMT with Hodgkin's lymphoma usual chemoradiotherapy conditioning regimens are less than optimal because they would cause unacceptable toxicity. This holds especially true for total body irradiation (TBI) with doses of >10 Gy as patients who have received prior mediastinal radiotherapy or high doses of bleomycin have an excessive incidence of fatal interstitial pneumonitis (Phillips et al. 1986). Therefore regimens not employing TBI have become widely used for ABMT in Hodgkin's disease. Examples are shown in Table 1.

The most popular regimen is a combination of cyclophosphamide, BCNU, and etoposide, the CBV regimen developed by the M. D. Anderson group (Spitzer et al. 1980; Jagannath et al. 1986), of which several variants are currently in use (Ahmed et al. 1987; Carella et al. 1987; Kessinger et al. 1986; O'Reilly et al. 1987; Teillet et al. 1987). At the moment it is not clear whether further dose escalations of the CBV regimen, such as those used by the Italian group (Carella et al. 1988), the addition of regional radiotherapy, or preceding cytoreductive conventional chemotherapy really improve long-term results of ABMT or just add to toxicity.

Optimal Source of Stem Cells for Hematopoietic Reconstitution

Syngeneic grafts should be utilized whenever available. However, as the incidence of identical twinning is low, there is no published experience with this marrow source in Hodgkin's disease, and syngeneic grafts serve mainly as ideal models to study the toxicity and curative potential of conditioning regimes. Apart from syngeneic grafts, the optimal source of marrow (i.e., allogeneic versus autologous) has not been determined. *Autologous marrow* has the advantages of carrying no risk of graft-versus-host disease and an apparently lower incidence of interstitial pneumonia. On the other hand, *allogeneic marrow* has the possible advantage of a graft-versus-Hodgkin's disease effect and no risk of reinfusing autologous tumor cells. Moreover, it replaces an abnormal or damaged immune system with a normal one, at least in those patients who do not develop chronic graft-versus-host disease. For allogeneic transplantation all other sources of HLA-identical, or "one-antigen-mismatched" related donors are currently experimental (Gribben and Goldstone 1987). As only few candidates for bone marrow transplantation with Hodgkin's lymphoma have such a donor, the experience with allogeneic transplantation in Hodgkin's lymphoma is very limited. Thirteen cases have been reported and four of them are long-term survivors (Appelbaum et al. 1985; Maseret et al. 1986; Philip et al. 1986). However, 7/13 (63%) of patients died of transplant-related causes (3 of GVHD, 4 of infections), producing an overall survival rate of only 30% at 4-41 months after transplantation and confirming the experience of others that about one-third of patients with malignant lymphomas undergoing allogeneic transplantation die of transplantation-related causes. Therefore, we feel while allogeneic marrow transplantation has a role in the management of Hodgkin's disease it is a limited one. Allogeneic grafting should be reserved for patients with persistent primary involvement of bone marrow with Hodgkin's lymphoma, especially if they are less than 40 years of age.

The greatest concern about autologous grafting is the risk that the autologous bone marrow contains occult malignant cells and thus cells may be reinfused which are responsible for relapse. However, this concern must not be overestimated as <10% of Hodgkin's lymphoma patients eligible for transplantation have bone marrow involved by Hodgkin's or Reed-Sternberg cells. This rate may increase with more advanced disease; however, as pointed out earlier, patients with very far advanced disease are not optimal candidates for bone marrow transplantation anyway.

Finally, peripheral blood stem cells can be collected from some patients, and have been shown to reconstitute hematopoiesis after intensive conditioning (Kessinger et al. 1986). A rationale for this approach is the supposition that hematopoietic stem cells can be collected more easily from the peripheral blood than from the marrow, especially from those patients who have received extensive radiotherapy to pelvic fields. Blood stem cells may also be less contaminated by malignant cells when compared with marrow cells. This consideration may play a role in patients with a history of marrow involvement.

A few Hodgkin's disease patients have undergone peripheral blood stem cell harvest and cryopreservation with subsequent transplantation after intensive conditioning. In general, a very rapid and stable successful engraftment has ensued.

Results are preliminary, however, but the role of peripheral blood stem cell transplants in such patients appears to be greater than that of allogeneic transplantation.

Role of Purging

The selective removal of malignant clonogenic cells without excessive damage to normal hematopoietic stem cells is referred to as "purging." Intuition more than clinical evidence suggests that purging may have a role in autologous marrow transplantation for patients with lymphomas. The most widely applied purging techniques have been the use of monoclonal antibodies and complement lysis (MacIntyre 1986), and attempts are being made to purge the bone marrow of lymphoma cells in non-Hodgkin's lymphoma in several centers (Nadler et al. 1987; Yeager et al. 1984). However, techniques with demonstrated selective efficacy against malignant clonogenic cells in vitro are still unsatisfactory. In Hodgkin's disease immunological purging is additionally hampered by the uncertainty of the nature of the malignant cell and by the fact that monoclonal antibodies with specific cytotoxicity for Hodgkin's and Reed-Sternberg cells have only recently become available (Pfreundschuh et al. 1988). To date, we are not aware of any clinical studies of purging bone marrow from malignant cells in Hodgkin's lymphoma. Even if purging were effective in this disease, the beneficial effect of this approach would be difficult to prove: patients with a prior history of Hodgkin's lymphoma of the marrow are the most likely to benefit from purging, but a randomized trial for these patients would hardly be feasible because of their low number.

Toxicity

In general, the side effects of conditioning will depend on the selection criteria (especially regarding prior treatment), the specific conditioning regimen used, and the status of the marrow when reinfused. The fact that candidates for ABMT with Hodgkin's lymphoma have usually undergone more extensive chemoradiotherapy regimens when compared with ABMT patients with NHL is most likely the reason why the overall morbidity and mortality of ABMT in Hodgkin's disease seems to be greater than that seen in less heavily pretreated non-Hodgkin's lymphoma (Carella et al. 1986). In fact, in one center with considerable experience of autografting in malignant lymphomas, the only 2 of 98 patients who were treated with intensive chemoradiotherapy and cryopreserved autologous bone marrow transplantation who had severely delayed marrow recovery had Hodgkin's lymphoma (G. L. Phillips, unpublished observations). The treatment-related mortality rates of between 4% and 24% of ABMT in Hodgkin's lymphoma in many centers are more than twice those observed in patients autografted for acute leukemia (Gribben et al. 1987b). Thrombocytopenia can be treated adequately by platelet transfusion and hemorrhage is a rare cause of death in the absence of sepsis and diffuse intravascular coagulation.

However, the severe neutropenia challenges the skills of the therapist and his or her expertise of supportive care. Febrile episodes in the period of severe marrow suppression occur in the majority of patients, and in nearly half of them bacteremia can be demonstrated. Mucositis and pneumonia are other frequent infections. The former is often caused by herpes simplex infection and may be severe enough that patients are unable to eat for several days, requiring total parenteral nutrition in some patients. The incidence of interstitial pneumonitis depends to a large extent on the history and dose of mediastinal irradiation preceding ABMT, and was observed less frequently in protocols that do not employ TBI, immediate pretransplantation radiotherapy to the chest, or excessive doses (>1000 mg/m^2) of BCNU in the conditioning chemotherapy protocol. Nausea and vomiting are common, but can be managed without problems. Other rare nonhematological toxicities observed are cardiomyopathy, veno-occlusive disease, the hemolytic-uremic syndrome, and renal failure.

As the frequency and intensity of many of the side effects observed in ABMT in Hodgkin's lymphoma increase with the length of period of myelosuppression, and in particular the length of neutropenia, the advent of recombinant growth factors of hematopoiesis, such as G-CSF and GM-CSF, offers great hope to reduce significantly mortality and morbidity of ABMT. Indeed, administration of GM-CSF has recently been shown significantly to reduce chemotherapy-induced myelosuppression in patients not only after conventional chemotherapy (Antmann et al. 1988) but also after bone marrow transplantation (Brandt et al. 1988). With the integration of these factors into the armamentarium of the marrow transplanter it should be possible to reduce the risks of ABMT considerably in the very near future.

Future Aspects

As is often the case, a very important question to be answered in clinical trials is fairly mundane: specifically, there is an urgent need to determine the effect on event-free survival with a standardized conditioning regimen at the first sign of failure from primary chemotherapy. An attractive trial would be the comparison of ABMT with a salvage regimen with escalating doses of cytotoxic drugs. Such a dose escalation of conventional chemotherapy should be possible with the application of GM-CSF or G-CSF (Antmann et al. 1988). With the support of hematopoietic growth factors the maximal tolerated dose for chemotherapy without bone marrow transplantation would have to be redefined at a higher level. A randomized trial comparing such a "neoconventional" (i.e., growth factor-supported) chemotherapy with ABMT could answer many questions as to the relative efficacy, toxicity, and - last but not least - cost of the two approaches. It would also help to define more precisely subgroups of patients who benefit from either of the two procedures, thus enabling us better to meet the specific therapeutic needs of each individual patient with Hodgkin's lymphoma.

References

Ahmed T, Gingrich SA, Ciavarella D et al. (1987) Myeloablative chemotherapy followed by autologous bone marrow transplantation (ABMT) for patients with refractory Hodgkin's disease. Proc Am Soc Hematol 23: 239a

Antmann K, Griffin J, Elias A, Socinski M, Whitley M, Ryan L, Cannutra S, Frei E, Schnipper L (1988) Effect of rGM-CSF on chemotherapy induced myelosuppression in sarcoma patients. Proc ASCO 7: 619

Appelbaum FR, Sullivan KM, Thomas ED, Buckner CD, Clift RA, Deeg HJ, Neiman PE, Sanders JE, Stewart P, Storb R (1985) Allogeneic Marrow Transplantation in the Treatment of MOPP Resistant Hodgkin's Disease. J Clin Oncol 3: 1490-1494

Brandt SJ, Peters WP, Atwater SK, Kurtzberg J, Borowitz MJ, Jones RB, Shpall EJ, Bast RC, Gilbert CJ, Oette DH (1988) Effect of recombinant human granulocyte-macrophage colony-stimulating factor on hematopoietic reconstitution after high-dose chemotherapy and autologous bone marrow transplantation. N Engl J Med 318: 869-876

Carella AM, Santini G, Santoro A et al. (1985) Massive chemotherapy with non-frozen autologous bone marrow transplantation in 13 cases of refractory Hodgkin's disease. Eur J Cancer Clin Oncol 21: 607-613

Carella AM, Congiu A, Santini G et al. (1986) High dose chemotherapy (HDC) and autologous bone marrow transplantation in advanced highly treated Hodgkin's disease (AHT-HD). Bone Marrow Transplant 1: 293-296

Carella AM, Congiu A, Santini G et al. (1987) High-dose cyclophosphamide, etoposide and BCNU with autologous BMT (ABMT) rescue in advanced resistant Hodgkin's disease patients. Bone Marrow Transplant 2 [Suppl 1]: 210

Carella AM, Congiu A, Mazza P et al. (1988) High-dose chemotherapy with autologous BMT "rescue" in 50 advanced resistant Hodgkin's disease patients. An Italian Study Group Report. J Clin Oncol 6: 1411-1416

Gribben JG, Goldstone AH (1987) Bone marrow transplantation in NHL and Hodgkin's lymphoma in remission - autologous versus allogeneic. Bone Marrow Transpl 2 [Suppl 1]: 204

Gribben JG, Vaughan Hudson B, Linch DC (1987a) The potential value of very intensive therapy with autologous bone marrow rescue in the treatment of malignant lymphomas. Hemat Oncol 5: 281-293

Gribben JG, Goldstone AH, Linch DC, Anderson CC, Richards JDM (1987b) Double autografting as potential curative therapy for acute leukaemia. Bone Marrow Transplant 2 [Suppl 1]: 46-48

Jagannath S, Dicke KA, Armitage JO, Cabanillas F, Horwitz LJ, Vellekoop L, Zander AR, Spitzer G (1986) High dose cyclophosphamide, carmustine and etoposide and autologous bone marrow transplantation for relapsed Hodgkin's disease. Ann Intern Med 1024: 4163-4168

Jagannath S, Armitage JO, Spitzer G et al. (1986b) Prognostic factors predicting for response to high dose chemotherapy and autologous bone marrow transplantation (ABMT) for relapsed Hodgkin's disease. Blood 68 [Suppl 1]: 274a (Abstr 974)

Kessinger A, Armitage JO, Landmort JD et al. (1986) Cryopreserved autologous peripheral stem cell infusions following high-dose therapy result in complete restoration of hematologic function for patients with non-leukemic malignancies. Am Soc Hematol 68: 290a

Loeffler M, Pfreundschuh M, Hasenclever D et al. (1988) Prognostic risk factors in advanced Hodgkin's lymphoma. Blut 56: 273-281

MacIntyre EA (1986) The use of monoclonal antibodies for purging autologous bone marrow in the lymphoid malignancies. In: Goldstone AH (ed) Clinics in haematology. Saunders, London, pp 249-267 (Autologous Bone Marrow Transplantation, vol 15 (1))

Maseret B, Maraninchi D, Gastaut A et al. (1986) Treatment of malignant lymphoma with high-dose of chemo- or chemoradiotherapy and bone marrow transplantation. Eur J Cancer Clin Oncol 3: 461-471

Nadler LM, Takvorian T, Botwick L, Bast RC, Finberg R, Hellman S, Canellos GP, Schlossman SF (1984) Anti B1 monoclonal antibodies and complement treatment in autologous bone marrow transplantation for relapsed B cell non-Hodgkin's lymphoma. Lancet 25 (2): 427-431

O'Reilly S, Connors J, Voss N et al. (1987) High-dose cyclophosphamide, BCNU and etoposide (CVB) and autologous bone marrow transplantation in progressive Hodgkin's disease (HD). Proc ASCO 23: 196

Pfreundschuh M, Mommertz E, Meissner M, Feller AC, Hassa R, Krüger GR, Diehl V (1988) Hodgkin and Reed-Sternberg cell associated monoclonal antibodies HRS-1 and HRS-2 react with activated cells of lymphoid and monicytoid origin. Anticancer Res 8: 217-224

Philip T, Dumont J, Teillet F, Maraninchi D et al. (1986) High dose chemotherapy and autologous bone marrow transplantation in refractory Hodgkin's disease. Br J Cancer 53: 737-742

Phillips G, Barnett M, Buskard N et al. (1986) Augmented cyclophosphamide (C), BCNU (B) and etoposide (V) = CVB and autologous bone marrow transplantation (BMT). Blood 68: 277 a

Spitzer G, Dicke KA, Litam J et al. (1980) High dose combination chemotherapy with autologous bone marrow transplantation in adult solid tumors. Cancer 45: 3075-3085

Sullivan KM, Appelbaum FR, Horning SJ, Rosenberg SA, Thomas ED (1986) Selection of patients with Hodgkin's disease and non-Hodgkin's lymphoma for bone marrow transplantation. Int J Cell Cloning 4 [Suppl 1]: 94-106

Teillet F, Pulik M, Teillet-Thieband F et al. (1987) Autologous bone marrow transplantation (ABMT) in poor prognosis Hodgkin's disease. Bone Marrow Transplant 2 [Suppl 1]: 211

Vaughan Hudson B, Maclennan KA, Easterling MJ, Jellife AM, Haybittle JL, Vaughan Hudson G (1983) The prognostic significance of age in Hodgkin's disease: examination of 1500 patients. Clin Radiol 34: 503-506

Yeager AM, Kaizer H, Santos GW et al. (1986) Autologous bone marrow transplantation in patients with acute non-lymphocytic leukemia using ex vivo treatment with 4-hydroperoxycyclophosphamide. New Engl J Med 315: 141

Acute and Late Side Effects of Treatment

Quality of Life During and After Treatment of Hodgkin's Disease

B. Hœrni and H. Eghbali

Fondation Bergonié, 180, Rue de Saint-Genès, 33076 Bordeaux Cedex, France

During recent decades prognosis and survival in Hodgkin's disease have improved strikingly. For all patients, the rate of cure is approximately 75% and for the majority, with limited disease (stage I–IIIA), it is nearly 90% (Zittoun et al. 1985). Improving the quality of life for the majority of patients has also been a basic aim in recent years (Hœrni et al. 1983; Eghbali and Hœrni-Simon 1984).

Before we consider in detail how it is possible to improve quality of life, several factors must be underlined:

1. *The quality of life must be considered as early as possible* and not only after treatment. It must be taken into consideration as soon as the diagnosis is suspected or at least proved. Early discussion with the patient, early and accurate information, careful planning of staging procedures, and careful planning of treatment are very important to reduce the stress due to the disease and to improve practical organization of the treatment for reducing disturbances to the patient's daily life. Rehabilitation will be simpler if the treatment is less disturbing.
2. *Educating the patient is of major importance in achieving this objective.* Few studies have been devoted to this aspect. However, our daily experience as well as a formal cooperative investigation (Hœrni et al. 1986a, b) has convinced us of its importance. Information may largely reduce the fear of the disease, improve the patient's cooperation, and decrease needless inconvenience, provided the oncologist takes into account the many practical details in the patient's wishes. For this reason we devote one section of this paper to education.
3. *The quality of life is linked mainly to the quality of treatment.* Treatment must be well planned, perfectly executed from a technical point of view, and carefully followed. Many treatment complications have been reported. Most have arisen from use of somewhat outdated methods or have been from small centers which do not have the necessary experience and regular control, allowing the appearance from year to year of some imperfections. Some complications still exist and provide incentive for all oncologists to search constantly for improvement. One way is de-escalation of treatment when possible. Treatment of all patients in experienced centers is another method. However, improvement of quality of life depends less on decreasing complications than in searching positively for as

Recent Results in Cancer Research. Vol. 117

many means as possible for improving the patient's adaptation to his/her disease and treatment. Many reviews have been devoted to complications decreasing quality of life (see, for example, Bookman and Longo 1986) and we will thus emphasize ways of increasing it. Table 1 lists the main changes in quality of life, indicating their diversity (Fobair et al. 1986; Hœrni et al. 1986a, b; Wasserman et al. 1987).

4. We must recognize that *evaluation of quality of life is difficult*. Many studies have been devoted to its evaluation, showing that results depend largely on the method used (see Aaronson and Beckmann 1987). However, we have neither the specialized experience nor enough space to discuss this point, and such methodological discussions should not preclude concrete measures to act in favor of patients.

Finally, *Hodgkin's disease may be a good model for other cancers* on which less research has been carried out. It is observed in both sexes, sometimes in young patients who are not fully grown, and also in old patients. However it leads to no change in body image, involving mainly young adults in the upper social classes, by contrast with the majority of other cancers.

Table 1. Main changes in quality of life after Hodgkin's disease according to the patients

Declared disturbances
Personal
Loss of energy
Depression
Alopecia
Reduced sexual interest and activity
Other side effects of treatment
Fear of recurrence
Familial
Less marriage
Higher rate of divorce
Decreased fertility
Social
Unpleasant experiences with classmates
No job offered
Professional conflicts
Refusal of insurance
Termination of employment
Perceived benefits
"Spoiling" by family and friends
Improved outlook on life
Increased well-being
Increased involvement with other people
Increased religious feelings
Greater maturity
More patience and tolerance
Greater work ambition
Better educational achievement

The quality of life depends mainly on different factors: (1) Personal characteristics (sex, age, familial and work status) of patients constitute in some instances risk factors which lead oncologists to pay special attention to certain patients because of these. (2) Aspects of the disease have been improved thanks to early diagnosis, which enables less advanced disease to be recognized with fewer clinical consequences and more chance of cure with a light treatment. (3) The treatment may be alleviated by being adapted as precisely as possible to prognostic parameters, and in some cases the patient may have the choice between two equivalent protocols. (4) Tailoring information to the patient is necessary to obtain good cooperation and observance, and to keep the patient playing a positive role during and after treatment and thereby to decrease side effects and enhance rehabilitation. (5) Practical conditions of treatment may be arranged to decrease hospital time, to increase treatment on an outpatient basis, and to reduce as much as possible disturbances in the patients' daily life.

Personal Parameters

Oncologists must consider personal parameters in order to recognize certain risk factors decreasing the chance of adjustment in some persons to disease and treatment. On the other hand the weight of different side effects also depends on the patient, who is the ultimate referee to judge his/her quality of life.

Sex determines the type of side effects or their perception. For example loss of hair affects females worse than males, at least among adolescents (Wasserman et al. 1987). Sterility is more probable in males, due to some types of chemotherapy, whereas drugs do not affect ovocytes, which are mainly jeopardized by iliac radiotherapy: when such a treatment is planned, ovary displacement is mandatory in young females preferably outside the irradiated areas (near the iliac crest). In any case the contraceptive pill is useful: it is not given for contraception, which may be obtained by other means, such as an already present intrauterine device; the contraceptive pill could protect the inner ovary secretions, but it mainly maintains regular menses and may also enhance sexual activity. Usually female subjects express more concern for possible sterility than males (Wasserman et al. 1987). Males express more indifference but they have to be warned of possible sterility before any chemotherapy and offered the opportunity for a deposit of sperm for preservation. However, male sterility appears less frequent and less permanent after a chemotherapy combination different from MOPP such as ABVD (Santoro et al. 1981). In young boys there is not only an impairment of spermatogenesis but also an alteration of Leydig's cell function, both being mainly due to procarbazine (Brämswig et al. 1987), which may be thus omitted from the combination. In the series of Wasserman et al. (1987) there is a significantly higher rate of divorce in men after Hodgkin's disease. In the large investigation of Fobair et al. (1986) half of the divorces (34/69) are attributed by patients to Hodgkin's disease but the ways Hodgkin's disease causes divorce remain unclear: personal disturbances? sexuality? fecundity? uncertainty of future? Fear of recurrence is described more often by women than by men (Wasserman et al. 1987). Males are concerned about military service, depending on the country: some are glad to be relieved from these duties whereas in the United States military service may be detrimental for certain patients or jobs. Finally, in our experience we have observed that "moral fatigue" is significantly more marked in females than in males after treatment (Hœrni et al. 1986b).

Age is also an important parameter. Its role is particularly prominent in adolescents, who already have many adjustment problems (Hœrni 1985; Wasserman et al. 1987). Hodgkin's disease may be a supplementary factor of disturbances or, in other patients, may play a positive role by accelerating the maturation. The bad luck of the disease may stimulate patients in their studies, and young patients actually show better educational achievement (Wasserman et al. 1987). Patients younger than 30 years at the time of treatment are significantly more likely to report a return of energy to normal levels than older patients (Fobair et al. 1986). Age also determines the familial and social environment of patients.

Familial status is important in conditioning different kinds of support during and after the disease: parents, girlfriend or boyfriend, spouse more often than children. Adolescent females report more "spoiling" by parents and siblings than males (Wasserman et al. 1987). For single patients, marriage is less likely than for other people (Combes et al. 1977).

Job status also depends on age. By contrast with the majority of other cancers, patients with Hodgkin's disease are rarely retired. On the contrary they are young enough to be still students or not yet to have a stable working position. For these young people work difficulties demand the attention of physicians, particularly at such time when unemployement is widespread. For patients with a stable job, duration of its interruption is significantly linked to the type of work, longer if the work is poorly qualified or considered uninteresting (Combes et al. 1977). Regulations linking disease and work depend on the countries and on the type of companies.

Age and familial and professional status influence the patient's chances of applying for different types of life or health insurance or loans. Again there are a variety of regulations according to country.

Other personal parameters are more difficult to specify but may play a determining role in the capacity of the person to cope with the disease. As in other circumstances of daily life, there is large variation between persons in ability to adjust to health or other difficulties. In our investigation the variations in well-being indicated by patients on a linear scale from before to after the disease are slight: one-third of patients declared no change, one-third a decrease, and one-third an increase in their well-being (Hœrni et al. 1986b). No parameter could be correlated significantly to these changes or to their magnitude, which appear to be linked to very specific and personal choices or way of life.

Disease Characteristics

Prognosis and survival in Hodgkin's disease differ significantly from stage IA to stage IVB. Treatment necessary to cure such different stages is also different, varying from only 1 month of radiotherapy up to nearly 1 year of a combination of radiotherapy and chemotherapy or even in some selected cases of proposed bone marrow graft. Therefore the seriousness of disease decreases the quality of life both directly, by inducing for example fever or anemia, and indirectly through the treatment which is adapted in each case to the prognosis.

The primary way to improve the patient's quality of life is to recognize Hodgkin's disease as early as possible, at a limited stage without symptoms. In this period the patient's performance status may be quite normal. For example, sperm quality is unaltered if the disease is small and localized. Further cure of the patient will be obtained with a light and well-tolerated treatment. We may note here that the general prognosis of Hodgkin's disease has certainly been improved thanks to better treatment but also thanks to a larger proportion of limited disease at the time of diagnosis. This is shown by our experience where the proportion of stages I and II regularly increase so that stages III and IV are now quite rare. In the cooperative Pierre-et-Marie-Curie Group experience, two-thirds of the patients seen during the past decade were diagnosed as stages I-IIIA, which are highly curable. We have also observed that the mean delay between first symptom and diagnosis, which was 6 months in 1965, is now around 1 month (unpublished data).

Hodgkin's disease is now a well-known disease among physicians, who are aware of the critical importance of an early diagnosis and are thus prompted to ask for a biopsy in the case of a supraclavicular lymph node in many cases. It is also well known among the general population and patients and this greatly contributes to reducing the fear of the disease and ancillary disturbances. This knowledge and this decreased fear are also the factors prompting patients to see a physician when they discover an enlarged lymph node. Early diagnosis is the simplest and the best way of lengthening survival and improving its quality.

Treatment Type

The side effects of treatment are the worst memory of half the patients treated when they were adolescent and surveyed more than 5 years later (Wasserman et al. 1987). The length and weight of treatment and therefore its side effects are directly related to the gravity of the disease. This is truer now, when, after a period of therapeutic escalation (circa 1960-1975), there has been a de-escalation of treatment which is adapted to prognostic factors so as to offer each patient the best chance of cure without the needless toxicity of excessive treatment.

The length of treatment usually determines the length of interruption of work and then the difficulties in resuming work again. We see later how it is possible to give treatment without the patient stopping working if it is not too severe. On the whole, treatment intensity determines the occurrence of many complications such as herpes zoster (Guinee et al. 1985) and secondary acute leukemia (Henry-Amar 1988).

Radiotherapy side effects depend on the dose and the extension of irradiated areas. The standard dose is 40 Gy given over 4 weeks. Booster doses are no longer necessary since primary chemotherapy has been combined with radiotherapy for the treatment of bulky tumors. In children, a decrease to 20-25 Gy is intended to reduce growth disturbances. In some pediatric experiences radiation therapy can be omitted completely provided the involved lymph nodes are small enough and then curable by chemotherapy alone (Behrendt et al. 1987). Extended irradiation increases the variety of side effects evaluated by physicians as well as by patients (Hœrni et al. 1986b). Extended irradiation may also need 2 months rather than 1 month to be given, delaying the resumption of further chemotherapy when necessary; moreover in the latter situation, more irradiation of bone marrow may decrease the tolerance to cytotoxic drugs and then the doses which can be given. A small extension may have substantial consequences, for example, with dental complications when cervical fields are extended to the mandibula to include the tragion lymph nodes (Hœrni 1987). Trials have shown that irradiated fields may be reduced in the case of combination with chemotherapy (Zittoun et al. 1987). Reducing irradiation of uninvolved areas may also avoid some complications like pericarditis (Cosset et al. 1984) or late cardiac effects of mediastinal radiotherapy (Pohjola-Sintonen et al. 1987). Suppression of explorative laparotomy also dramatically decreases radiation injuries of the gastrointestinal tract when abdominal radiotherapy is required (Gallez-Marchal et al. 1984). This tendency might be reinforced if the high frequency of secondary solid tumors in irradiated areas is confirmed with a longer follow-up.

The side effects of chemotherapy are also dependent on its length and on the total number of doses given. Data suggesting that 3 months of MOPP would be as efficient as 6 months of the same regimen need to be confirmed (Ferme et al. 1984). MOPP no longer seems to be the reference regimen because of its toxicity. Associations like ABVD appear less toxic, at least if we consider secondary leukemias and azoospermia (Santoro et al. 1981). After the ABVD regimen, the substitution of doxorubicin with epirubicin, which appears less cardiotoxic (Torti et al. 1986), led to a much better combination (Zittoun et al. 1987). Recent data are in favor of a short and intense chemotherapy as in a regimen formerly used by the authors, which required hospitalization, but in which patients received only one 3-week course before and the same after radiotherapy (Chauvergne et al. 1973; Lagarde et al. 1988). We try to develop regimens as efficient as, and simpler or shorter than, the preceding ones (Hœrni et al. 1988).

Henceforth patients do not need the maximum treatment to attain the maximum chance of cure since this maximum security for the evolution of Hodgkin's disease has, on the other hand, more side effects. Actually in GPMC H76 trial dealing with stages I-IIIA the present figures of causes of death are equal for Hodgkin's disease and for other causes. The correct target is an optimum level where a maximum of efficacy is obtained for Hodgkin's disease and beyond which an increase of treatment increases the side effects and sequelae more than it decreases failures or relapses of the disease. This target requires an accurate evaluation of the prognostic factors (those linked to the disease as well as those linked to the patient), to deduce from them a precisely adapted treatment. In this search, to obtain early complete remission is a major aim for the future. However, its value needs to be more accurately assessed by further studies.

The patient's point of view has to be taken into consideration because side effects may have a different importance for each patient. For example, one patient may prefer to receive chemotherapy, which would be perhaps useless, rather than to undergo a laparotomy. In some situations it is unnecessary to give the free choice to the patient because the majority of them do not ask for it (Strull and

Charles 1984), but one may discuss with him/her to find out his/her preferences and take them into account for the final medical decision (Hœrni-Simon and Zittoun 1985; Zittoun 1985). In the GPMC H76 trial 1.5% of the patients were excluded early from the trial because of their probable poor cooperation and another 1% because they refused the planned treatment. It is certainly a complex task to collect all the necessary data, which are very heterogeneous, and to put them together for the therapeutic decision. A decision-analysis method may help (Zittoun 1985). We believe that the best final choice can be made only among specialized and trained oncological teams. Such refined thinking needs good daily expertise (Anonymous 1987). This is equally true of informing the patients and improving conditions in the treatment procedures.

Patient Education

This is more easily undertaken in Hodgkin's disease than in other cancers because the prognosis is relatively good and the information is thus not all bad news (Hœrni et al. 1986a). There are a number of reasons for fully informating a patient with Hodgkin's disease (Hœrni 1982).

Reasons for informing the patient are ethical and practical. Ethically, patients need to be informed about their disease. They need to know how it threatens their future life and their ability to organize their personal, familial, and professional life. Practically, the knowledge of the planned treatment, treatment data, type of treatment, and usual side effects usually improve the tolerance of patients.

It is not possible to give here all the useful information which may (must?) be given to patients and sometimes to his/her family with his/her agreement. We give only some examples.

After the diagnosis, it is usually possible and useful to give the name of the disease. Hodgkin's disease is known by many people and well known as a highly curable disease. To know the name of one's disease decreases the fear of the unknown. When one knows something, one is prevented from imagining something worse. Many patients are disturbed or even astonished that physicians do not know the origin of the disease since some contemporary rationalists think that medical science must explain all things. It is easy to tell them that it is preferable to know how to cure the disease without knowing its origin than vice versa. The physician also has to state definitely that it is not a hereditary disease neither for the parents nor for existing or yet to be born children. Furthermore it is not a contagious disease and the patient should not be isolated. The prognosis can be said to be good or very good thanks to a somewhat severe treatment due to the rather poor spontaneous evolution. The sequelae are rare and weak in the majority of cases. In summary this disease may be considered as the cause of some disturbances for a few months rather than a threat of death and thus the patient may consider it in a rather optimistic manner.

But it is not enough. Much more practical information should be given to reduce as much as possible the foreseeable disturbances. In the frequent cases of association of chemotherapy and radiotherapy, the conditions and the schedule of the treatment may be given to the patient to allow him/her to organize his/her life so that his/her usual life continues as well as possible. The reduction of usual activities depends more on the patient than on the disease or its treatment. If treatment is well planned, patients can organize around it other familial, leisure, or professional activities by taking into account the time of treatment as well as its foreseeable side effects. The disturbances due to the disease and to the treatment must be stated to be probable and not certain. There are sometimes some small changes and the physician must remain careful to avoid a change worrying or needlessly disturbing the patient. But even with a small amount of uncertainty, the majority of patients appreciate knowing what is planned and how, so that they can keep maximum control of their life. Usually Hodgkin's disease causes few disturbances by itself, its treatment can be planned precisely for a few weeks, and its efficacy is satisfactory so that the plans are not changed. This allows the patient to plan leisure, vacations, study, work, or anything else which constitutes a normal life, and reduces the demands of illness on the patient's time.

Finally there are several instructions to be given to patients for the application of their treatment: mouth and teeth care, sperm preservation (presented with skill so that the possibility of azoospermia does not induce impotency in males or the suspicion of the spouse in case of pregnancy), temporary contraception (hormonal contraception is mainly intended to maintain regular menses during chemotherapy), precautions to be taken to reduce side effects (for example not to eat too much around a chemotherapy injection or not to expose to the sun an irradiated area at least during the first summer following radiotherapy), appointments for consultations, investigations, and treatments. Too many patients think that it is impossible to continue a normal life when they are ill. The physician must often state precisely that many things are allowed beside the disease and the treatment. Many people have negative or erroneous ideas about disease and treatment or about diet or efforts so that their life is greatly disturbed by unnecessary precautions. Many recommendations of the physician will be positive to contradict false ideas: for example, a saltless diet is useless during short corticosteroid treatment; if this point is not made clear, many patients will omit salt from their food when given corticosteroids because they "know" that it is necessary. A final example will be given with sexual activity (Fobair et al. 1986), which for the usually young people with Hodgkin's disease is important. Sexual activity is often somewhat disturbed for psychological reasons due to the disease, or because of nauseas, fatigue, or mucitis due to treatment. However it is also often disturbed because the patient or his/her partner think that it must be suppressed or systematically reduced. In this field also, the patient must be at least offered information, eventually together with his/her partner, so that disturbances due to the disease are not unduly amplified for unjustifiable reasons.

We believe that more than the type of information, the manner of its delivery is important: this should be in a progressive, veracious, positive, and sympathetic way. Taking care over the exchanges between patient and physician is the best way to comfort the patient, to improve his/her cooperation, to reduce his/her fears, and to help him/her to cope with the disease and the treatment. It is also important to speak of the future: resuming studies or work, marriage, possible children – to help convince the patient that the present disease is only the source of temporary disturbances and that life will continue after wards, thus offering the patient some freedom to think about matters other than the present unpleasant ones. Fobair et al. (1986) have emphasized that it is useful to tell the patient that after the end of the treatment the total recovery may take approximately 1 year. The rapidity of the return to normal energy level depends on age, stage of the disease, intensity, and length of the treatment. Without this information, some patients may be anxious to recover very completely and quickly by the end of the treatment, may organize, as a rebound, a very active life again, and then will be worried when they encounter certain difficulties. By contrast such information will help them to recover progressively, reasonably, with patience but also with an active intention to resume a normal life and for many patients a better life than before the disease. Social workers may help them to overcome professional or insurance difficulties. They must be also encouraged not to misuse some of the advantages offered by the disease, for example not to claim systematically all the illness benefit to which they are entitled.

With such principles as routine, we have observed some interesting data in our original investigation within the GPMC (Hœrni et al. 1986a). As in other diseases, females appear to be better informed than males. A little more than half of the patients (52%) declare that they would have liked more information about their disease, the treatment, and above all its side effects. Patients with very good prognosis declare that they are better informed than those with a poorer prognosis. By taking into account this last point and even by considering patients with the same prognosis and the same treatment, there is a significant correlation between the information and the side effects as declared by the patient. The reason for this correlation is unclear: we were unable to elucidate further whether patients with more information experienced actually fewer or slighter side effects or whether they only considered the declared side effects to be lighter. Any interpretation of these facts should encourage the physician to impart more information. We also observed a significant link between information and plans in the professional field. Again we were unable to know whether better-informed patients were more likely to undertake projects in this field or whether patients with such projects were more likely to ask for information. Such and other data do not lead physicians to inform their patients as much as possible. Physicians have stated that information is usually insufficient (perhaps because the patients have an insatiable desire for information) and that information is usually harmless and useful, provided it is given in a manner adapted to each patient, i.e., at the right time, of the right type (answering the real question of the patient), with the right presentation. As for organic disturbances, the physician must satisfy the psychological needs of the patients. For this it is necessary to understand them, and to understand them it is necessary to listen to them. Since there are no general rules, the physician must adapt his/her attitude to each diseased person, who may vary with time.

Previous experience of patient education with a printed booklet, newsletters, and a variety of discussions has shown that such education represents an effective, useful, and inexpensive means of improving the psychological and social behavior of patients with Hodgkin's disease (Jacobs et al. 1983). By contrast, in the same study participation in a peer support therapy group did not result in a significant behavior change.

Again information management appears to be closely linked to the expertise of the team, including not only the physician(s) but also the nurses and all other persons providing care. The practical conditions of treatment depend on the attitudes and behavior of all these persons.

Treatment Conditions

Before going into detail, we must remember that technical quality of treatment is critical for the cure and for the quality of life of patients. We must also emphasize that the quality of all parts of the treatment is the primary aim of physicians. However, the same correct treatment may be given in very different practical conditions which deeply influence the patient's daily life during his/her disease. For example, if the patient can receive total or partial chemotherapy or radiotherapy near his/her home under the control of experienced physicians or nurses, it is not desirable to keep him or her hospitalized. We are particularly sensitive to this point in our cancer center in Bordeaux, which serves an area in southwest France almost twice as large as Switzerland. For each patient, depending on his/her geographical situation, we hold careful discussions with him/her and with his/her primary physician to determine whether and which part of the treatment must be taken the cancer center and which part may be given locally. If the displacement to the cancer center appears necessary, it will be worthwhile reducing the number of chemotherapy injections by increasing the dosage of drugs of each injection (Hœrni 1987).

For the majority of patients, all the treatment may be given on an outpatient basis. Thus treatment centers need to develop facilities like the day hospital. The organization of such facilities is critical. Patients are helped by the car park close to the hospital, by punctual appointments, and by reduction of waiting for consultations, blood cell results, radiotherapy or chemotherapy, administrative procedures, etc. For example some of our patients prefer to come to our center, even though they live 100 km away, because they lose less time than going locally for a chemotherapy course because of poor local organization of blood sampling, biological results, drug supplies, physician and nurse appointments, etc.

Such practical conditions depend on the willingness and cooperation of the patient. Usually young patients prefer to be free and spend only a short time in a treatment center, whereas some other patients prefer to spend more time in care. It is not always easy to find a good equilibrium between achieving perfect technical treatment and convenient conditions of its application. This equilibrium may change from the beginning to the end of the treatment following the patient's experience or the side effects of the treatment. Persons giving care should be very open minded to adapt if possible to the patient's wishes in a very flexible manner, but without losing the aim of quality of care. Information, discussion, and explanations are necessary throughout the treatment to take into account patients' personal considerations but also to expose them to what is medically necessary. Patients usually appreciate having their advice taken and considered for the whole organization of their treatment. They also appreciate knowing in advance what and when the next part of their treatment will be. Hospital staff are at their service and not the contrary.

We are aware that all these points appear quite trivial. However, we also know that treatment of Hodgkin's disease is not very simple and requires much attention from hospital staff, who may thus focus too exclusively on its technical conditions and forget somewhat the nonmedical life of the ill persons. Their quality of life depends on their life outside the disease and treatment as well as on the quality of treatment and on its results.

From the same point of view, oncologists must pay attention to the follow-up. Immediately after the completion of treatment, patients may be helped by consultations and the physician's advice about their rehabilitation. After 1 year, without an intervening relapse, the majority of patients are cured and no longer need medicine or physicians. Follow-up consultations should be avoided particularly for cooperative patients, who are advised that in the case of a health disturbance they can make an appointment with the physician. Follow-up is necessary for the patients and for the oncologist, mainly if patients are included in a controlled trial, but it may be undertaken without imposing too many or too frequent consultations. We are personally reluctant to have too many patients included in controlled studies because of the long-term side effects or complications. We agree that it is useful for oncologists to improve their knowledge of the late effects of their treatments. Repeated spermograms, endocrine dosages, respiratory function explorations, cardiac investigations, etc. have enabled oncologists to know better all the possible causes of decreasing quality of life of cured patients and then to treat new patients better. However, these investigations must have the patients' agreement and the number of the investigated patients must be as limited as possible to obtain accurate information as well as to avoid further disturbances to too many patients.

In summary the quality of life in Hodgkin's disease offers a good example of holistic medicine. Physicians and all other persons caring for the patient must consider the disease and the patient together with his/her survival and his/her quality of life, the technical and the practical treatment conditions, the present, and the future. This is true not only for Hodgkin's disease. But its high rate of cure leads to the idea that cure is not enough and that in any case we cannot accept that poor survival is worse than death. Enough cured patients are now surviving in satisfactory conditions to convince easily all persons caring for the patient that efforts to improve the quality of life are worthwhile.

References

Aaronson NK, Beckmann J (eds) (1987) The quality of life of cancer patients. Raven, New York

Anonymous (1987) Who should treat Hodgkin's disease and where? Lancet I: 605

Behrendt H, Van Bunningen BNFM, Van Leeuwen EF (1987) Treatment of Hodgkin's disease in children with or without radiotherapy. Cancer 59: 1870–1873

Bookman MA, Longo DL (1986) Concomitant illness in patients treated for Hodgkin's disease. Cancer Treat Rev 13: 77–111

Brämswig JH, Heiermann E, Heimes U, Schlegel W, Nieschlag E, Schellong G (1987) Testicular function in boys treated for Hodgkin's disease. Prospective evaluation in 68 boys of the DAL-HD-78 and DAL-HD-82 therapy study. Third International Conference on Malignant Lymphoma, Lugano (abstr p 25)

Chauvergne J, Hœrni B, Hœrni-Simon G, Durand M, Lagarde C (1973) Chimiothérapie de la maladie de Hodgkin associant procarbazine, vinblastine, cyclophosphamide et méthylprednisolone. Analyse d'une série de 124 cures. Z Krebsforsch 80: 179-188

Combes PF, Desclaux B, Malissard L, Pons A (1977) Poids du diagnostic de maladie de Hodgkin sur la qualité de la vie des sujets qui en sont guéris. Bull Cancer (Paris) 64: 395-408

Cosset JM, Henry-Amar M, Ozanne F, Le Bourgeois JP (1984) Les péricardites radiques. Etudes des cas observés dans une série de 160 maladies de Hodgkin irradiées en mantelet à l'Institut Gustave-Roussy, de 1976 à 1980. J Eur Radiothérapie 5: 297-308

Eghbali H, Hœrni-Simon G (1984) Qualité de la vie et maladie de Hodgkin. In: Schraub S, Brugère J, Hœrni B (eds) Qualité de la vie et cancers. Progrès en cancérologie, 4. Doin, Paris, pp 283-290

Ferme C, Teillet F, d'Agay MF, Gisselbrecht C, Marty M, Boiron M (1984) Combined modality in Hodgkin's disease. Comparison of six versus three courses of MOPP with clinical and surgical restaging. Cancer 54: 2324-2329

Fobair P, Hoppe RT, Bloom J, Cox R, Varghese A, Spiegel D (1986) Psychosocial problems among survivors of Hodgkin's disease. J Clin Oncol 4: 805-814

Gallez-Marchal D, Fayolle M, Henry-Amar M, Le Bourgeois JP, Rougier P, Cosset JM (1984) Radiation injuries of the gastrointestinal tract in Hodgkin's disease: the role of exploratory laparotomy and fractionation. A study of 19 cases observed in a series of 134 patients treated at the Institut Gustave-Roussy from 1972 to 1982. Radiother Oncol 2: 93-99

Guinee VF, Guido JJ, Pfalzgraf KA et al. (1985) The incidence of herpes zoster in patients with Hodgkin's disease. An analysis of prognostic factors. Cancer 56: 642-648

Henry-Amar M (1988) Quantitative risk of second cancer in patients in first complete remission from early stages of Hodgkin's disease. NCI Monogr 6: 65-72

Hœrni B (ed) (1982) L'information des personnes malades. SIMEP, Lyon-Villeurbanne

Hœrni B for Groupe Pierre-et-Marie-Curie (1985) Enquête psychosociale chez 28 adolescents traités pour une maladie de Hodgkin. Psychol Med 17: 1975-1976

Hœrni B, Hœrni-Simon G, Eghbali H, Richaud P (1983) Qualité de vie pendant et après association radio-chimiothérapique pour maladie de Hodgkin. Appréciation de 60 malades. Bull Cancer (Paris) 70: 284-288

Hœrni B, Zittoun R, Rojouan J et al. (1986a) Information des malades traités pour une maladie de Hodgkin. Liaisons avec la qualité de vie. Bull Cancer (Paris) 73: 305-310

Hœrni B, Zittoun R, Rojouan J et al. (1986b) Retentissement psychosocial du traitement de la maladie de Hodgkin. Appréciation par questionnaire auprès de 150 malades. Bull Cancer (Paris) 73: 620-626

Hœrni B, Orgerie MB, Eghbali H et al. (1988) Nouvelle association d'epirubicine, bléomycine, vinblastine et prednisone (EBVPII) avant radiothérapie dans les stades localisés de maladie de Hodgkin. Essai de phase II chez 50 malades. Bull Cancer (Paris) 75: 789-794

Hœrni-Simon G, Zittoun R (1985) Coût thérapeutique et qualité de vie pendant et après le traitement d'une maladie de Hodgkin. In: Hœrni-Simon G, Gisselbrecht C (eds) Maladie de Hodgkin. Progrès en hématologie, 7. Doin, Paris, pp 179-191

Jacobs C, Ross RD, Walker IM, Stockdale FE (1983) Behavior of cancer patients: a randomized study of the effects of education and peer support groups. Am J Clin Oncol 6: 347-353

Lagarde P, Eghbali H, Bonichon F et al. (1988) Brief chemotherapy associated with extended field radiotherapy in Hodgkin's disease. Long-term results in a series of 102 patients with clinical stages I-IIIA. Eur J Cancer Clin Oncol 24: 1191-1198

Pohjola-Sintonen S, Tötterman KJ, Salmo M, Siltanen P (1987) Late cardiac effects of mediastinal radiotherapy in patients with Hodgkin's disease. Cancer 60: 31-37

Santoro A, Bonfante V, Bonadonna G et al. (1981) Therapeutic and toxicologic effects of MOPP-ABVD combined with radiotherapy in Hodgkin's disease. In: Salmon SE, Jones SE (eds) Adjuvant therapy of cancer III. Grune and Stratton, New York, pp 85-91

Strull WM, Lo B, Charles G (1984) Do patients want to participate in medical decision making. JAMA 252: 2990-2994
Torti FM, Bristow MM, Lum BL et al. (1986) Cardiotoxicity of epirubicin and doxorubicin: assessment by endomyocardial biopsy. Cancer Res 46: 3722-3727
Wasserman AL, Thompson EI, Wilimas JA, Fairclough DL (1987) The psychological status of survivors of childhood/adolescent Hodgkin's disease. Am J Dis Child 141: 626-631
Zittoun R (1985) Indications thérapeutiques et processus décisionnel dans la maladie de Hodgkin. In: Hœrni-Simon G, Gisselbrecht C (eds) Maladie de Hodgkin. Progrès en Hématologie, 7. Doin, Paris, pp 193-215
Zittoun R, Audebert A, Hœrni B et al. (1985) Extended versus involved fields irradiation combined with MOPP chemotherapy in early clinical stages of Hodgkin's disease. J Clin Oncol 3: 207-214
Zittoun R, Eghbali H, Audebert A et al. (1987) Association d'épirubicine, bleomycine, vinblastine et prednisone (EBVP) avant radiothérapie dans les stades localisés de la maladie de Hodgkin. Essai de phase II. Bull Cancer (Paris) 74: 151-157

*Risk of Secondary Acute Leukemia and Preleukemia After Hodgkin's Disease: The Institut Gustave-Roussy Experience**

M. Henry-Amar, B. Pellae-Cosset, C. Bayle-Weisgerber, M. Hayat, J. M. Cosset, P. Carde, and M. Tubiana

Institut Gustave-Roussy, 94805 Villejuif Cedex, France

Introduction

Secondary acute nonlymphocytic leukemia (ANLL) and preleukemia have long been claimed to be one of the major long-term side effects of treatment of Hodgkin's disease (HD) with intensive treatment including alkylating agents and radiotherapy (Baccarani et al. 1980; Canellos et al. 1975; Castro et al. 1983; Coleman et al. 1982; Coltman and Dixon 1982; Tester et al. 1984) associated with an increased risk of ANLL (Bergsagel et al. 1982; Blayney et al. 1987; Boivin and Hutchison 1981; Collaborative Study 1984; Dorreen et al. 1986; Glicksman et al. 1982; Larsen and Brinker 1977; Pedersen-Bjergaard and Larsen 1982; Tolland et al. 1978; Tucker et al. 1987a). Among the various chemotherapy regimens, MOPP [mechloretamine (nitrogen mustard), Oncovin (vincristine), procarbazine, prednisone] has been most clearly demonstrated to be leukemogenic, either as first-line or as salvage treatment (Brusamolino et al. 1982; Coleman et al. 1977; Cornbleet et al. 1985; Cramer et al. 1985; Glicksman et al. 1982; Henry-Amar 1983; Jacquillat et al. 1983; Santoro et al. 1986; Toland et al. 1978; Valagussa et al. 1980; Valagussa et al. 1986). A significant increase in leukemic risk has been associated with many alkylating agents, including nitrogen mustard. A quantitative study of alkylators given in HD patients has recently been published, demonstrating for the first time a direct dose-response relation to leukemogenesis (Pedersen-Bjergaard et al. 1987; Van der Velden et al. 1988). Similar findings were observed in other tumor series (Boice et al. 1986; Cuzick et al. 1987; Greene et al. 1985; Greene et al. 1986; Haas et al. 1987; Pedersen-Bjergaard et al. 1985; Tucker et al. 1987b). The respective leukemogenicity for alkylating agents used in the current treatment of HD has recently been analyzed (Van der Velden et al. 1988).

The present study investigates the factors which may be associated with an increased risk of secondary ANLL, including age, sex, clinical stage, splenectomy, extent of radiotherapy, and cumulative dose of each alkylating agent given.

---

* Supported by grant number 85DIG from the Institut Gustave-Roussy and grant number BRI/87/05 from the International Agency for Research on Cancer (Lyon, France).

Recent Results in Cancer Research. Vol. 117

Patients and Methods

Patients and Treatments

From January 1960 to December 1984, 871 adult patients with previously untreated (96%) or previously treated (4%) HD with all treatment data available were treated at the Institut Gustave-Roussy. Patients with a follow-up time of less than 1 year are excluded from this series. The diagnosis of HD was assessed using the Rye classification. The stage was established clinically and radiologically according to the Ann Arbor classification. Patient characteristics and treatment categories are shown in Table 1. In 348 patients (40%), the staging procedure included laparotomy and splenectomy. Overall, radiotherapy (RT) was the sole treatment in 26% of the patients, while chemotherapy (CT) was given in combination with RT

Table 1. Patient characteristics and treatment categories

Age, mean in years (SD)	32.6 (13.3)
50 years or more	12%
Sex M/F	505/366
Histological types	
LP	7%
NS	56%
MC	31%
LD	3%
Unclassified	3%
Stage	
I	25%
II	47%
III	21%
IV	7%
Mediastinal involvement	51%
B symptoms	35%
ESR mean mm/1st h (SD)	43.7 (35.5)
Laparotomy-splenectomy	40%
Treatment categories (all treatments ever received)	
RT alone:	
limited	8%
extended	18%
CT alone:	
MOPP	4%
MOPP+ABVD	1%
Miscellaneous	1%
Combination RT-CT:	
MOPP	32%
MOPP+ABVD	5%
ABVD	2%
Miscellaneous	29%

LP, lymphocytic predominance; NS, nodular sclerosing; MC, mixed cellularity; LD, lymphocytic depleted.

Table 2. Distribution of chemotherapy agents given

	No. of patients (%)	Overall dose (mg) Mean (SD)
Alkylating agents		
BCNU	23 (3)	557 (333)
CCNU	69 (8)	624 (501)
Chlorambucil	45 (5)	303 (404)
Cyclophosphamide	97 (11)	7487 (5670)
Dacarbazine	83 (10)	3931 (6940)
Methyl-CCNU	5 (<1)	–
Nitrogen mustard	378 (43)	76 (40)
Procarbazine	532 (61)	11277 (7140)
Triethylene melanime	29 (3)	67 (133)
Natural products		
Adriamycin	124 (14)	308 (198)
Bleomycin	194 (22)	155 (106)
Vinblastine	420 (48)	230 (257)
Vincristine	404 (46)	19 (22)
Miscellaneous agents		
Etopsid	32 (4)	1807 (3136)
Teniposide	58 (7)	853 (773)
Vindesine	20 (2)	23 (35)
Patients who were administered at least one cycle of		
MOPP	352 (40)	
Bleo-MOPP	15 (2)	
ABVD	66 (8)	

in most of the patients (68%). MOPP or an equivalent combination such as bleo-MOPP was used in the treatment of 42% of the cohort. MOPP was used as first-line treatment in 33% of the patients, with (29%) or without (4%) associated RT. A total of 555 patients (64%) received alkylating agents (Table 2), 102 being treated with one alkylating agent, 295 with two, 97 with three, and 61 with four or more.

Biological Investigations

After treatment was completed, all patients were closely followed at least twice a year for relapse, myelodysplasia (MDP), or leukemia development by routine investigations including blood count and hemoglobin level. For patients who developed MDP, a bone marrow aspiration was performed in order to confirm the diagnosis. All abnormalities and the time of diagnosis were recorded. Leukemias were classified according to the FAB classification by one of us (CBW) (French-American-British Cooperative Group 1976). The cases of MDP and acute leukemias were combined in the present study.

Analysis of Data

The contribution to the risk of ANLL of initial characteristics (i.e., age at diagnosis, sex, histological type, stage) and type of treatment received (i.e., extent of radiotherapy and total dose of each drug recorded) were evaluated by the stepwise Cox proportional hazards model (Cox 1972) with the BMDP-2L program (Dixon et al. 1985). Initially, all factors were included in the model, and, for that analysis, the time at risk for secondary ANLL began at the initiation of therapy and ended at the date of ANLL, date of death, date of last known status, or 1 August 1987, whichever came first. In a second step, all factors that did not have a significant prognostic influence ($P > 0.20$) or for which the following equation $|\beta.|.SD(Z_i)$ was <1 (Peduzzi et al. 1980) were removed from the analysis. The BMDP-2L program was then rerun with time-dependent procedure where the time at risk began at the initiation of RT in patients treated by RT alone, and at first administration of specific CT in patients treated by combination RT-CT. In such an analysis a patient who was first treated by RT and 2 years later by CT for relapse is included in the group of patients treated by RT alone for 2 years until CT was administered, and then is included in the group of patients treated by a combination of RT and CT.

The total dose of each drug ever received by the patients was computed instead of total dose per square metre since height and weight were not available for all patients. Logarithms were used in the analysis.

The cumulative proportion of ANLL from initiation of specific therapy (see above) was estimated by the Kaplan-Meier method (Kaplan and Meier 1958). In addition, a comparison of cancer incidence between the population of patients with HD and the general population was used. In this approach, the exposure to the risk was based on the accumulation of person-years of observation. The ratio of the number of ANLLs observed among the patients with HD *(O)* to the expected number *(E)* was computed from general population cancer incidence data. Expected numbers were computed with the use of age-, sex-, and calendar year-specific cancer incidence rates published by the Bas-Rhin Tumor Registry in France (Schaffer and Lavillaureix 1981). The confidence limits (CL) of O/E were obtained assuming a Poisson distribution. This assumption was also used for significance testing. Relative risks *(RR)* greater than 1 were considered of interest and a one-sided test was used.

Results

ANLL Observed

Overall, ten leukemias and nine MDPs were observed 13–147 months after initial treatment for HD (Table 3). Of these, 3 occurred in patients treated by extended-field RT (patient Nos 5, 15, 19), while the other 16 occurred in patients who had been treated by combination of RT and MOPP, either as initial treatment (12 patients) or salvage treatment (4 patients).

Table 3. Characteristics of the patients who developed secondary ANLL

	Age (years)	Sex	Histological type	Stage	Splenectomy	Treatment	Delay from HD to ANLL (months)	FAB classification	Status (follow-up)
1.	34	M	MC	IVB	No	MOPP × 7.5 + paraaortic RT	32	M1	Dead
2.	53	M	NS	IIA	No	Mantle RT + MOPP × 7	81	M2	Alive (7 months)
3.	33	M	NS	IIIA	No	MOPP × 3 + ABVD × 3 + STNI	30	M2	Alive (4 months)
4.	28	M	Not done	IIIA	Yes	MOPP × 6 Extended RT for relapse + PCZ/Teniposide	67	M3	Dead
5.	38	M	NS	IA	Yes	Mantle + Paraaortic RT	13	M3	Dead
6.	29	F	NS	IB	Yes	Limited RT + MOPP × 8 for relapse	82	M4	Alive (3 months)
7.	30	M	MC	IIIA	Yes	TNI MOPP × 6 for relapse	60	M4	Alive (7 months)
8.	17	F	NS	IIIA	Yes	Mantle RT + MOPP × 5 Inverted Y RT for relapse	22	M5	Alive (5 months)
9.	51	M	MC	IIB	Yes	MOPP × 6 + TNI	33	M6	Alive (24 months)
10.	28	M	MC	IVA	Yes	MOPP × 12 + TNI + ABVD × 6	71	Myelogenous unclassified	Alive (4 months)

11.	36	M	NS	IIA	No	STNI MOPP × 6 for relapse	81	MDP	Dead
12.	48	M	LP	IIA	Yes	STNI MOPP × 4 for relapse + inverted Y RT	118	MDP	Dead
13.	28	M	MC	IIB	No	MOPP × 6 + TNI	41	MDP	Alive (14 months)
14.	23	F	NS	IVA	Yes	Polychemotherapy including MOPP × 5 and ABVD × 6	32	MDP	Alive (1 months)
15.	54	M	MC	IIB	No	TNI	48	MDP	Dead
16.	28	M	MC	IIB	No	MOPP × 6 + Mantle RT + MOPP × 9 + ABVD × 6 for relapse	147	MDP	Dead
17.	26	M	MC	IIA	Yes	MOPP × 6 + Mantle RT	114	MDP	Dead
18.	47	M	MC	IVB	No	MOPP × 6 + TNI	33	MDP	Dead
19.	64	M	MC	IA	No	STNI	24	MDP	Alive (7 months)

MDP, myelodysplasia (preleukemia); TNI, total nodal irradiation; STNI, subtotal nodal irradiation.
When a patient is alive, the follow-up time from ANLL diagnosis is specified in parentheses.

Risk of ANLL in Relation to Initial Characteristics and Treatment

After stepwise multivariate analysis, only age at diagnosis of HD, stage, splenectomy, and dose of nitrogen mustard were considered as having a significant prognostic value on the risk of ANLL occurrence. These factors plus sex and the extent of RT were therefore included in a time-dependent analysis. Age was divided into four categories, i.e., <30, 30–39, 40–49, and ≥50 years; stage was grouped into two categories, i.e., I–II vs. III–IV; and the extent of RT was classified as limited to one side of the diaphragm or extended, i.e., subtotal nodal irradiation (STNI) or total nodal irradiation (TNI). The prognostic value of nitrogen mustard was estimated as follows:

1. Qualitatively: nitrogen mustard ever given, no = 0, yes = 1
2. Quantitatively with the logarithm of the total dose given
3. Qualitatively subgrouping the dose into three groups: 1–59 mg, 60–119 mg, and ≥120 mg, i.e., equivalent to one to three cycles of MOPP, four to six cycles, and ≥seven cycles.

The three successive analyses gave similar results, showing an excess risk associated with the use of nitrogen mustard ($P<0.001$), with evidence of a dose-response relationship (Table 4). By contrast, no significant excess of risk was observed with age (although there was a trend), sex, stage, splenectomy, or the extent of RT. To estimate the effect of initial therapy on the risk of secondary ANLL we analyzed the data from patients who were successfully treated and did not relapse. In such an analysis, the time at risk is computed from start of specific therapy (RT, CT, or

Table 4. Cox model (time-dependent): coefficient of risk (β) and relative risks (RR) for a model allowing all the variables simultaneously ($P<0.001$)

Variables	β/SE	RR	*P* value
Age (years) <30	0	1.0	
30–39	0.03/0.58	1.03	>0.20
40–49	0.30/0.80	1.35	>0.20
≥50	0.98/0.63	2.66	0.12
Sex M	0	1.0	
F	−0.86/0.65	0.42	0.19
Stage I–II	0	1.0	
III–IV	0.05/0.50	1.05	>0.20
Splenectomy No	0	1.0	
Yes	0.28/0.49	1.32	>0.20
Radiotherapy Limited	0	1.0	
Extended	0.60/0.48	1.83	0.20
Total dose of nitrogen mustard			
0	0	1.0	
1–59 mg	1.10/1.16	3.00	>0.20
60–119 mg	2.46/0.70	11.67	0.004
≥120 mg	3.52/0.75	33.82	<0.001

first administration of nitrogen mustard) to the date of relapse, date of occurrence of ANLL, or date of last-known vital status. Twelve ANLLs were observed, three after extended RT only, the other nine after combination RT and CT including nitrogen mustard. The most important ANLL-related factor was the use of nitrogen mustard ($P<0.0001$), with a similar dose-response relationship to that reported in Table 4; the second was age at diagnosis above 50 years (RR = 12, $P<0.005$); and the third was extended-field RT (RR = 4.71, $P=0.01$).

Six treatment categories were therefore defined: limited RT, no CT; extended RT, no CT; combination RT and CT not including nitrogen mustard; and combination RT and CT including nitrogen mustard in the three subgroups defined above. Patients in these six categories did not differ by age at diagnosis, the proportion of patients with mediastinal involvement or who underwent staging laparotomy and splenectomy. On the other hand, they differed for sex, B symptoms, mean pretreatment erythrocyte sedimentation rate (ESR), clinical stage, and histological type, patients treated by combination RT and CT with or without nitrogen mustard being more often males, with B symptoms, elevated ESR, clinical stage III or IV, and mixed cellularity or lymphocytic depletion histological types.

General population comparisons were computed on the basis of these results (Table 5). For these comparisons, a patient may be included in one or more treatment categories depending on the treatment he or she received in succession. In patients treated by limited RT only, no secondary ANLL occurred, while in those treated by extended RT only the associated RR of ANLL was equal to 68 (95% CL 14-199; $P<0.001$). In patients treated with 1-59 mg nitrogen mustard, the RR of ANLL was equal to 45 (95% CL 1.15-253; $P<0.001$) similar to that of patients treated with extended RT alone. The RR significantly increased ($P<0.001$) with

Table 5. General population comparisons: risk of secondary ANLL

Mode of therapy	No. of patients at risk	Person-years at risk	Number of ANLLs observed	RR	95% CL	*P* value
Limited RT, no chemotherapy	445	753	0	0	0-115	NS
Extended RT, no chemotherapy	426	1185	3	68	14-199	<0.001
Radiotherapy + chemotherapy without nitrogen mustard	645	2192	0	0	0-35	NS
Radiotherapy + chemotherapy with nitrogen mustard						
All patients	369	1648	16	219	125-356	<0.001
Total dose nitrogen mustard						
1-59 mg	120	457	1	45	1.15-253	<0.001
60-119 mg	201	927	8	211	91-415	<0.001
⩾120 mg	48	264	7	636	256-1311	<0.001
All patients	871	5778	19	75	45-118	<0.001

Table 6. Cumulative proportions of secondary ANLL from start of specific treatment at 5, 7, and 10 years by treatment category

Mode of therapy	No. of patients at risk	Percentage of ANLLs (SE) at year		
		5	7	10
All patients	871	1.5 (0.5)	3.1 (0.8)	4.1 (1.1)
Limited RT, no CT	445	0	0	0
Extended RT, no CT	426	2.4 (1.4)	2.4 (1.4)	2.4 (1.4)
RT and CT, no nitrogen mustard	645	0	0	0
RT + nitrogen mustard				
All patients	369	3.5 (0.9)	7.9 (2.0)	12.4 (3.5)
Total dose nitrogen mustard				
1-59 mg	120	0	0	4.8 (3.1)
60-119 mg	201	4.7 (1.7)	4.7 (1.7)	10.6 (6.0)
⩾120 mg	48	5.8 (3.3)	23.5 (7.7)	38.8 (11.5)

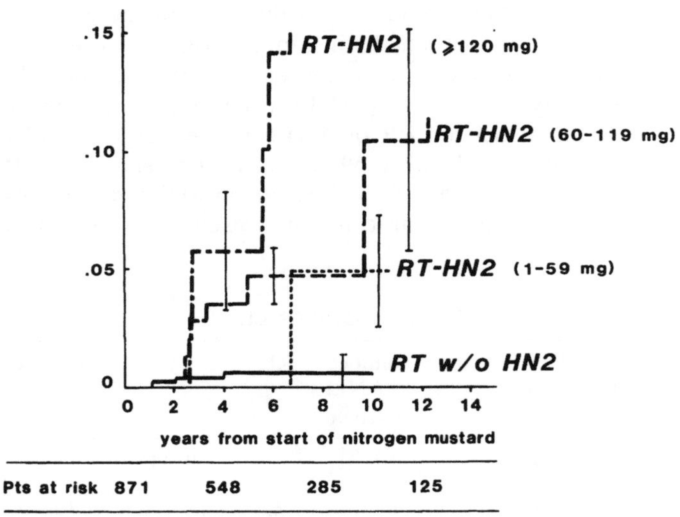

Fig. 1. Cumulative proportions of secondary ANLL (*error bars*, 95% confidence interval) by treatment category from start of specific therapy (see text). *HN2*, nitrogen mustard

the dose of nitrogen mustard given: RR = 211 (95% CL 91-415) in patients treated with 60-119 mg, and RR = 636 (95% CL 256-1311) in those treated with ⩾120 mg.

In the overall population, the 10-year cumulative proportion of ANLL was 4.1% (Table 6). It was 0.6% in patients treated with RT alone (2.4% in patients treated with extended RT), 0% in patients treated with a combination of RT and CT not including nitrogen mustard, and 12.4% in patients who were treated with RT and nitrogen mustard (Fig. 1). Among the latter, the 10-year cumulative proportion of ANLL significantly increased with the dose of nitrogen mustard given.

Among the six treatment categories the overall 10-year survival was 92% (SD 3%) in the group of patients treated with limited RT only; 84% (4%) in that treated with extended RT only; 61% (3%) in that treated with RT and CT without nitrogen mustard; 43% (6%), 66% (7%), and 68% (12%) in patients who were treated with RT and CT including a total dose of nitrogen mustard equal to 1-59 mg, 60-119 mg, and ⩾120 mg respectively.

Discussion

Our results are consistent with those of most series, which report a significant increased risk of ANLL associated with intensive treatment for HD (Bergsagel et al. 1982; Blayney et al. 1987; Boivin et al. 1984; Collaborative Study 1984; Dorreen et al. 1986; Glicksman et al. 1982; Larsen and Brinker 1977; Pedersen-Bjergaard and Larsen 1982; Toland et al. 1978; Tucker et al. 1987a; Van der Velden 1988). In particular, a 10-year cumulative proportion of 12.4% in patients treated with nitrogen mustard is similar to that reported by Pedersen-Bjergaard et al. (1987) after alkylating agent therapy. Moreover, their results and ours are superimposed when the total dose of alkylating agents received by the patients is subdivided into three subgroups: low dose, intermediate dose, and high dose. The consistency of the findings in two independent cohorts of HD patients demonstrates that:

1. Therapy with certain alkylating agents leads to a high risk of secondary ANLL
2. In the MOPP combination, nitrogen mustard appears to be the most leukemogenic alkylating agent with a dose-response relationship.

In a case-control study, Van der Velden et al. (1988) reported a 10-fold increase in the risk of ANLL associated with nitrogen mustard, but also a significant increase in risk with combined vincristine and procarbazine. Since these three components have usually been administered in the MOPP combination, it seems difficult to conclude which of these drugs has the highest leukemogenicity.

The leukemogenicity of other alkylating agents, such as chlorambucil or cyclophosphamide, might have been underestimated in this series since only few patients have been treated with these drugs. Among the 16 patients who developed ANLL following nitrogen mustard chemotherapy, only two were also treated with one of these two drugs. The other alkylating agents used in the treatment of these 16 patients were CCNU in three and dacarbazine in three.

Nitrogen mustard was combined with procarbazine in 68% of the patients treated with alkylating agents, and in all of the 16 patients with secondary ANLL following MOPP chemotherapy. Procarbazine was given as the sole chemotherapy agent or associated with other alkylating agents but not nitrogen mustard in 28% of the patients. Among these, no secondary ANLL was observed although the mean total dose of procarbazine given was not significantly different from that of patients treated by both nitrogen mustard and procarbazine. This fact is in accordance with what has been previously reported (Carde et al. 1985; Henry-Amar 1985; Coleman 1986).

The lack of evidence of leukemogenicity for the cytotoxic drugs other than nitrogen mustard confirms the lower leukemogenicity of the ABVD combination

(Adriamycin, bleomycin, vinblastine, dacarbazine) compared with MOPP as it was previously reported (Amadori et al. 1983; Valagussa et al. 1988).

No independent influence of RT was found in this series, nor was it found in the Danish study (Pedersen-Bjergaard et al. 1987), when patients were treated with combination RT and CT. Moreover, no influence of age was observed although it was reported by other investigators (Coltman and Dixon 1982; Henry-Amar 1983, 1985; Van Leeuwen et al. 1987).

Recently, Van Leeuwen et al. (1987) have reported that splenectomy might predispose to secondary ANLL. This observation was confirmed by Van der Velden et al. (1988). This finding was not observed in the present study, although significant numbers of patients underwent staging laparotomy and splenectomy.

Compared with other studies from population-based registries (Curtis et al. 1985; Kaldor et al. 1987; Storm et al. 1985), the RR of ANLL estimated from hospital-based registries is much greater, roughly five fold in the present study, a discrepancy that cannot be explained only by combining acute leukemias and MDP in hospital-based registry studies. In the present study, there were 10 cases of acute leukemias, providing a RR = 40. Another explanation may be that in the hospital-based series a nonnegligible proportion of patients might be lost for follow-up while in the population-based series the follow-up of the patients is much more exhaustive, based on a national registry of death certificates. Also, hospital series might be composed of more intensively treated patients.

All in all, beside the risk of developing a treatment-related ANLL, one must keep in mind that the use of cytotoxic agents has dramatically improved survival rates in all stages of HD. This might balance the observed risks of ANLL, which can be considered as a price to pay for cure (Anonymous 1985; Coleman 1986). Moreover, the cumulative proportions of secondary ANLL after 10 years should be balanced against the fact that only a few patients remained at risk in the six treatment categories. Therefore, the occurrence of an ANLL at that time can induce a high percentage increase of the risk.

Conclusions

This study confirms previous findings relative to the leukemogenicity of alkylating agents used in the treatment of Hodgkin's disease. Additionally, it was possible to identify nitrogen mustard as the major leukemogenic compound in the MOPP regimen. The finding of a dose-response relationship between nitrogen mustard and secondary ANLL risk should be helpful clinically in determining the mode of therapy used. Although the benefit of the use of cytotoxic drugs in the treatment of Hodgkin's disease patients cannot be denied, the use of nitrogen mustard at total doses higher than 60 mg is questionable. Similar analysis should be performed for series that have used alkylating agents other than nitrogen mustard (such as ChVPP or C-MOPP) and, if similar results are found, new combinations with similar efficiency and less hematological toxicity should be used.

Summary

From 1960 to 1984, 871 patients were treated for Hodgkin's disease at the Institut Gustave-Roussy. Twenty-six percent of the cohort were treated with radiotherapy (RT) alone, 6% with chemotherapy (CT) alone, and 68% with a combination of RT and CT, either at first line or for salvage treatment. MOPP chemotherapy was given to 42% of the patients. Overall, 19 secondary acute leukemias or preleukemias were observed, 3 of them after extended RT alone, the other 16 after a combination of RT and MOPP. Among the alkylating agents used, only nitrogen mustard (mechloretamine) was shown in a multivariate analysis to be significantly associated ($P < 0.001$) with an increased risk of secondary leukemia. A dose response was observed, with the risk relative to general population incidence rates being 45 in patients having been treated with 1-59 mg (total dose) of nitrogen mustard, 211 in those treated with 60-119 mg, and 636 in those treated with ≥120 mg. No other factors were found to be associated with leukemia risk. The 10-year cumulative incidence of leukemia was zero in patients treated with limited RT alone, 2.4% in those treated with extended RT alone, 0% in those treated with a combination of RT and CT without nitrogen mustard, and 12.4% in those treated with RT+nitrogen mustard. Whether other alkylating agents give a similar result remains to be determined; these data suggest that the use of nitrogen mustard at a higher total dose than 60 mg is questionable in the treatment of Hodgkin's disease.

Acknowledgments. We are grateful to Mrs. A. Auquier for statistical advice, Dr. S. Friedman for linguistic corrections and suggestions, and Mrs. S. Iacobelli for technical assistance in the preparation of this paper.

References

Amadori S, Papa G, Anselmo AP et al. (1983) Acute promyelocytic leukemia following ABVD (doxorubicin, bleomycin, vinblastine and dacarbazine) and radiotherapy for Hodgkin's disease. Cancer Treat Rep 67: 603-604

Anonymous (1985) Second malignancies in lymphoma patients. Lancet ii: 1163-1164 (editorial)

Baccarani M, Bosi A, Papa G (1980) Second malignancy in patients treated for Hodgkin's disease. Cancer 46: 1735-1740

Bergsagel DE, Alison RE, Bean HA et al. (1982) Results of treating Hodgkin's disease without a policy of laparotomy staging. Cancer Treat Rep 66: 717-731

Blayney DN, Longo DL, Young RC et al. (1987) Decreasing risk of leukemia with prolonged follow-up after chemotherapy and radiotherapy for Hodgkin's disease. N Engl J Med 316: 710-714

Boice JD, Green MH, Killen JY et al. (1986) Leukemia after adjuvant chemotherapy with semustine (methyl-CCNU). Evidence of a dose-response effect. N Engl J Med 314: 119-120

Boivin JF, Hutchison GB (1981) Leukemia and other cancers after radiotherapy and chemotherapy for Hodgkin's disease. JNCI 67: 751-760

Boivin JF, Hutchison GB, Lyden M et al. (1984) Second primary cancers following treatment of Hodgkin's disease. JNCI 72: 233-241

Brusamolino E, Lazzarino M, Salvaneschi L et al. (1982) Risk of leukemia in patients treated for Hodgkin's disease. Eur J Cancer Clin Oncol 18: 237-242

Canellos GP, De Vita VT, Arseneau JC et al. (1975) Second malignancies complicating Hodgkin's disease in remission. Lancet i: 947-949

Carde P, Henry-Amar M, Tubiana M et al. (1985) No increased incidence of second leukemias in patients treated with vinblastine alone or associated with procarbazine following radiotherapy in two successive controlled trials in clinical stages I-II Hodgkin's disease. Proc Am Assoc Clin Oncol 4: 212

Castro C, Chrumka K, Coldman AJ, Connors JM (1983) Second malignant neoplasm in patients with Hodgkin's disease. Proc Am Assoc Clin Oncol 2: 209

Coleman CN (1986) Editorial: second malignancy after treatment of Hodgkin's disease: an evolving picture. J Clin Oncol 4: 821-824

Coleman CN, Williams CJ, Flint A et al. (1977) Hematologic neoplasia in patients treated for Hodgkin's disease. N Engl J Med 297: 1249-1252

Coleman CN, Kaplan HS, Cox R et al. (1982) Leukaemias, non-Hodgkin's lymphomas and solid tumours in patients treated for Hodgkin's disease. Cancer Surv 1: 733-744

Collaborative study (1984) Radiotherapy of stage I and II Hodgkin's disease. Cancer 54: 1928-1942

Coltman CA, Dixon DO (1982) Second malignancies complicating Hodgkin's disease: a Southwest Oncology Group 10-year follow-up. Cancer Treat Rep 66: 1023-1033

Cornbleet MA, Vitolo U, Ultman JE et al. (1985) Pathologic stages IA and IIA Hodgkin's disease: results and treatment with radiotherapy alone (1968-1980). J Clin Oncol 3: 758-768

Cox DR (1972) Regression models and life tables (with discussion). J R Stat Soc (B) 34: 187-220

Cramer P, Andrieu JM, Julien C (1985) Secondary acute non lymphoblastic leukemias after MOPP chemotherapy plus radiotherapy for Hodgkin's disease. Role of infradiaphragmatic radiotherapy. Proc Am Assoc Clin Oncol 4: 213

Curtis RE, Boice JD, Kleinerman RA et al. (1985) Summary: multiple primary cancers in Connecticut, 1935-82. NCI Monogr 68: 219-242

Cuzick J, Erskine S, Edelman D, Galton DAG (1987) A comparison of the incidence of the myelodysplastic syndrome and acute myeloid leukemia following melphalan and cyclophosphamide treatment for myelomatosis. Br J Cancer 55: 523-529

Dixon WJ, Brown MB, Engelman L et al. (eds) (1985) BMDP statistical software. University of California Press, Los Angeles

Dorreen MS, Gregory WM, Wrigley PFM et al. (1986) Second primary malignant neoplasms in patients treated for Hodgkin's disease at St Bartholomew's hospital. Hematol Oncol 4: 149-161

French-American-British (FAB) Cooperative Group: Bennett JM, Catovsky D, Daniel MT, Flandrin G, Galton DAG, Gralnick HR, Sultan C (1976) Proposals for the classification of the acute leukemias. Br J Haematol 33: 451-458

Glicksman AS, Pajak TF, Gottlieb A et al. (1982) Second malignant neoplasms in patients successfully treated for Hodgkin's disease: a Cancer and Leukemia Group B study. Cancer Treat Rep 66: 1035-1044

Greene MH, Boice JD, Strike TA (1985) Carmustine as a cause of nonlymphocytic leukemia. N Engl J Med 313: 579

Greene MH, Harris EL, Gershenson DM et al. (1986) Melphalan may be a more potent leukemogen than cyclophosphamide. Ann Int Med 105: 360-367

Haas JF, Kittelmann B, Mehnert WH et al. (1987) Risk of leukemia in ovarian tumour and breast cancer patients following treatment with cyclophosphamide. Br J Cancer 55: 213-218

Henry-Amar M (1983) Second cancers after radiotherapy and chemotherapy for early stages of Hodgkin's disease. JNCI 71: 911-916

Henry-Amar M (1985) Second cancers after treatment in two successive cohorts of patients with early stages of Hodgkin's disease. In: Cavalli F, Bonadonna G, Rozencweig M (eds)

Malignant lymphomas and Hodgkin's disease: experimental and therapeutic advances. Martinus Nijhoff, Boston, pp 417-428

Jacquillat C, Auclerc G, Weil M et al. (1983) Leucémies aigües et tumeurs solides dans l'évolution de la maladie de Hodgkin. Bull Cancer (Paris) 70: 61-66

Kaldor JM, Day NE, Band P et al. (1987) Second malignancies following testicular cancer, ovarian cancer and Hodgkin's disease: an international collaborative study among cancer registries. Int J Cancer 39: 571-585

Kaplan EL, Meier P (1958) Non parametric estimation from incomplete observations. Am Stat Assoc J 53: 457-481

Larsen J, Brinker H (1977) The incidence and characteristics of acute myeloid leukemia arising in Hodgkin's disease. Scand J Haematol 18: 197-206

Pedersen-Bjergaard J, Larsen SO (1982) Incidence of acute nonlymphocytic leukemia, preleukemia and acute myeloproliferative syndrome up to 10 years after treatment of Hodgkin's disease. N Engl J Med 307: 965-971

Pedersen-Bjergaard J, Ersbøll J, Sørensen HM et al. (1985) Risk of acute nonlymphatic leukemia and preleukemia in patients treated with cyclophosphamide for non-Hodgkin's lymphomas. Ann Int Med 103: 195-200

Pedersen-Bjergaard J, Specht L, Larsen SO et al. (1987) Risk of therapy-related leukemia and preleukemia after Hodgkin's disease. Relation to age, cumulative dose of alkylating agents, and time from chemotherapy. Lancet ii: 83-88

Peduzzi PN, Hardy RJ, Holford TR (1980) A stepwise variable selection procedure for nonlinear regression models. Biometrics 36: 511-516

Santoro A, Viviani S, Villarreal CJR et al. (1986) Salvage chemotherapy in Hodgkin's disease irradiation failures: superiority of doxorubicin-containing regimens over MOPP. Cancer Treat Rep 70: 343-348

Schaffer P, Lavillaureix J (1981) Le cancer dans le Bas-Rhin: incidence des nouveaux cas de 1975 à 1977. Economica, Paris

Storm HH, Jensen OM, Ewertz M et al. (1985) Summary: multiple primary cancers in Denmark, 1943-80. NCI Monogr 68: 411-430

Tester WJ, Kinsella TJ, Waller B et al. (1984) Second malignant neoplasms complicating Hodgkin's disease: the National Cancer Institute experience. J Clin Oncol 2: 762-769

Toland DM, Coltman CA, Moon TE (1978) Second malignancies complicating Hodgkin's disease: the Southwest Oncology Group experience. Cancer Clin Trials 1: 27-33

Tucker MA, Coleman CN, Cox RS et al. (1987a) Risk of second cancers following Hodgkin's disease after 15 years. Proc Am Assoc Clin Oncol 6: 192

Tucker MA, Meadows AT, Boice JD et al. (1987b) Leukemia after therapy with alkylating agents for childhood cancer. JNCI 78: 459-464

Valagussa P, Santoro A, Kenda R et al. (1980) Second malignancies in Hodgkin's disease: a complication of certain forms of treatment. Br Med J [Clin Res] 280: 216-219

Valagussa P, Santoro A, Fossati Bellani F et al. (1986) Second acute leukemia and other malignancies following treatment for Hodgkin's disease. J Clin Oncol 4: 830-837

Valagussa P, Santoro A, Fossati Bellani F et al. (1988) Hodgkin's disease and second malignancies. Proc Am Assoc Clin Oncol 7: 227

Van der Velden JW, Van Putten WLJ, Guinee VF et al. (1988) Subsequent development of acute nonlymphatic leukemia in patients treated for Hodgkin's disease. Int J Cancer 42: 252-255

Van Leeuwen FE, Somers R, Hart AAM (1987) Splenectomy in Hodgkin's disease and second leukemias. Lancet ii: 210-211

MIX
Papier aus verantwortungsvollen Quellen
Paper from responsible sources
FSC® C105338

If you have any concerns about our products,
you can contact us on
ProductSafety@springernature.com

In case Publisher is established outside the EU,
the EU authorized representative is:
Springer Nature Customer Service Center GmbH
Europaplatz 3, 69115 Heidelberg, Germany

Printed by Libri Plureos GmbH
in Hamburg, Germany